AF566768

Drug Stereochemistry

CLINICAL PHARMACOLOGY

Series Editor

Murray Weiner

Division of Clinical Pharmacology and Toxicology
University of Cincinnati College of Medicine
Cincinnati, Ohio

1. Nicotinic Acid: Nutrient · Cofactor · Drug, *Murray Weiner and Jan van Eys*
2. Drugs in Central Nervous System Disorders, *edited by David C. Horwell*
3. Clinical Application of Interferons and Their Inducers: Second Edition, *edited by Dale A. Stringfellow*
4. Beta-Lactam Antibiotics for Clinical Use, *edited by Sherry F. Queener, J. Alan Webber, and Stephen W. Queener*
5. Receptor Binding in Drug Research, *edited by Robert A. O'Brien*
6. Drug Therapy in Hypertension, *edited by Jan I. M. Drayer, David T. Lowenthal, and Michael A. Weber*
7. Adverse Reactions to Muscle Relaxants, *edited by Isaas Azar*
8. Clinical Pharmacology in the Elderly, *edited by Cameron G. Swift*
9. Anti-Inflammatory Compounds, *edited by W. R. Nigel Williamson*
10. Nonsteroidal Anti-Inflammatory Drugs: Mechanisms and Clinical Use, *edited by Alan J. Lewis and Daniel E. Furst*
11. Drug Stereochemistry: Analytical Methods and Pharmacology, *edited by Irving W. Wainer and Dennis E. Drayer*
12. Drug Interaction Compendium, *Murray Weiner*
13. Receptor Pharmacology and Function, *edited by Michael Williams, Richard A. Glennon, and Pieter B. M. W. M. Timmermans*
14. Adverse Reactions to Drug Formulation Agents: A Handbook of Excipients, *Murray Weiner and I. Leonard Bernstein*
15. Cardiovascular Function of Peripheral Dopamine Receptors, *edited by J. Paul Hieble*
16. Rational Therapeutics: A Clinical Pharmacologic Guide for the Health Professional, *edited by Roger L. Williams, D. Craig Brater, and Joyce Mordenti*
17. Ulcer Disease: Investigation and Basis for Therapy, *edited by Edward A. Swabb and Sandor Szabo*
18. Drug Stereochemistry: Analytical Methods and Pharmacology. Second Edition, Revised and Expanded, *edited by Irving W. Wainer*

ADDITIONAL VOLUMES IN PRODUCTION

DRUG STEREOCHEMISTRY

Analytical Methods and Pharmacology

Second Edition, Revised and Expanded

Edited by

Irving W. Wainer

McGill University and Montreal General Hospital
Montreal, Quebec, Canada

Marcel Dekker, Inc. New York • Basel • Hong Kong

Library of Congress Cataloging-in-Publication Data

Drug stereochemistry : analytical methods and pharmacology / edited by
Irving W. Wainer. -- 2nd ed., rev. and expanded.
p. cm. -- (Clinical pharmacology ; v. 18)
Includes bibliographical references and index.
ISBN 0-8247-8819-2 (alk. paper)
1. Chiral drugs. 2. Pharmaceutical chemistry.
3. Stereochemistry. I. Wainer, Irving W. II. Series.
[DNLM: 1. Chemistry, Pharmaceutical--methods. 2. Molecular,
Conformation. 3. Stereoisomers. W1 CL764H v. 18 1993 / QV 744 D943
1993]
RS429.D78 1993
615'.19--dc20
DNLM/DLC
for Library of Congress 92-48364
CIP

This book is printed on acid-free paper.

MARCEL DEKKER, INC.
270 Madison Avenue, New York, New York 10016

Current printing (last digit):
10 9 8 7 6 5 4 3 2 1

PRINTED IN THE UNITED STATES OF AMERICA

To my wife, Pamela Zulli—my best friend,
companion, and constant source of inspiration.

Series Introduction

In the beginning, medicinal therapy consisted of concoctions coadministered with incantations. Little thought was given to dissecting out the contributions of the pharmacology of the concoction from the spiritual consequences of the incantations. Eventually, potions and extracts were recognized to have predictable activity that could be observed and described objectively. Then came the capacity to estimate potency, leading to concentration and purification. And finally came chemical identification and synthesis and the currently fashionable demand that a modern medicament must be a pure single chemical entity. With synthesis replacing nature, the chemical aspect of pharmaceutical science had gone as far as it could go—or had it?

The longing to know exactly what is in a medicament is axiomatically rational beyond challenge. It is the first step in the intelligent preparation and use of a therapeutic agent. The task is much simplified if the medicinal is a single chemical entity. In some scientific medicinal circles, acceptability of a medicine is inversely proportional to the square of the number of its active components.

But what is meant by a simple active component? Most modern pharmaceuticals are weak acids or bases that exist in ionized and unionized form as a strict consequence of the nature of their solvent, and particularly the pH of aqueous solutions. These distinct forms are radically different in terms of several biologically critical characteristics, including physiological disposition and receptor binding. While such agents are generally formulated as a single substance, usually as a particular salt or as the free acid or base, on dissolution in the body there is rapid conversion into a specific equilibrium of charged and uncharged molecules. Nevertheless, such agents are quite properly considered a single component biologically, even though a mixture of chemical species exists.

But what about stereoisomers? Not too long ago the question was considered largely theoretical, since many synthetic drugs were racemates for which there were no practical means of resolution. Efforts to prepare and isolate pure enantiomorphs were generally perceived as interesting chemical exercises of little practical pharmaceutical importance because it was presumed that both enantiomorphs were equally active, or one of the pair was totally inert, or the enantiomorphs would spontaneously racemize in solution. Now we know better. The *d* versus *l* forms of amphetamine are perhaps one of our best and oldest examples of important pharmacological differences between enantiomorphs of the same agent.

Some natural products, such as heparin, have not yet succumbed to the onslaught of synthesis or genetic engineering and continue to be produced and used as a spectrum of closely related chemical entities, rather than a single species of precisely known chemical structure. Such exceptions to the single-component objective are becoming increasingly rare, and unresolved racemates as drugs are undergoing a similar fate.

In view of the current attitude toward mixtures or drug combinations, it is as difficult to justify the use of a racemate as it is to use a mixture of other analogs. With genetic engineering to help the synthetic chemist, even complex structures with a high degree of stereospecificity are being developed as the rule rather than the exception. It is therefore particularly timely to have available an in-depth review of the problems and opportunities resulting from our current knowledge of the stereochemical aspects of pharmaceuticals.

If one accepts the logic of seeking to deal with pure "single entity" drugs, should one also accept the proposal that drugs which undergo metabolic conversion should be avoided since, in effect, they behave as mixtures? Such a decision would indeed disrupt our present armamentarium of useful drugs. The prohibition of racemates would also be disruptive, since there are still instances where therapy with a cheaper, easy-to-make racemate is fully as effective as with a more expensive specific enantiomorph.

As long as we accept drugs that undergo metabolic conversion in vivo, we will continue to be obliged to accept some racemates and even mixtures of chemical analogs when it is hard to come by an equivalent single active component. Nevertheless, a single active moiety remains the preferred goal. The experts need to understand and evaluate how the recent advances in stereochemistry can both solve and create problems for the pharmaceutical manufacturer, research worker, and physician. In this revised and expanded second edition of *Drug Stereochemistry*, Dr. Wainer has put together a volume that deserves careful review and will be a valuable resource to pharmaceutically oriented chemists, biologists, and clinicians who can no longer ignore the question of stereoisomerism in relation to drugs.

Murray Weiner

Preface to the Second Edition

The first edition of this book contained the observation that the stereoisomeric composition of drug substances was rapidly becoming a key issue in the development, approval, and clinical use of pharmaceuticals. During the past four years, this assertion has become a reality, and stereochemistry is now a major theme in all branches of pharmaceutical science. The interest in this topic is reflected in the content of articles appearing in natural and biological science journals and two new international journals, *Chirality* and *Tetrahedron Asymmetry*. Moreover, a variety of conferences and courses have been devoted to this subject. In addition, the European, North American, and Japanese drug regulatory agencies have expressed an interest in a closer examination of the stereoisomeric composition of chiral drugs and the therapeutic and toxicological consequences of this composition. The pharmaceutical industry has quickly responded to this situation by considering stereochemistry in its initial drug evaluation strategies. In fact, at the present time, a number of companies have made the decision to market only single-isomer drugs.

In light of these developments, we felt that it was necessary to review the content of the first edition and to update or supplement the information presented in this work. The new topics examined in this edition include: (1) enzymatic synthesis and resolution of enantiomerically pure compounds (Chapter 8); (2) toxicological consequences and implications of stereoselective biotransformations (Chapter 9); (3) stereoselective transport across epithelia (Chapter 10); and (4) assessment of bioavailability and bioequivalence of stereoisomeric drugs (Chapter 11). The chapter on stereoselective protein binding (Chapter 12) has been completely rewritten and new contributions are presented on the regulatory, industrial, and clinical aspects of stereoisomeric drugs (Chapters 13–16). In addition, the chapters discussing stereoselective chromatographic separations (Chapters 4–6) have been revised and expanded.

As with all volumes of this type, this project could not have been completed without the contributors, and I would like to thank them for their superb efforts. I am indebted to the following people for their help: Ann Samson, David Lloyd, Camille Granvil, Karen Fried, Hiltrud Fieger, Tanja Alebic-Kolbah, Anne-Françoise Aubry, Gabriella Masolini, Xiao Ming Zhang, and Pierre Wong. The assistance of the publisher's staff is also appreciated, particularly the efforts of Sandra Beberman, Assistant Vice President, Editorial, and Ted Allen, the production editor.

Irving W. Wainer

PREFACE TO THE FIRST EDITION

The stereoisomeric composition of drug substances is rapidly becoming a key issue in the development, approval, and clinical use of pharmaceuticals. This volume is designed to cover the current debate on this topic from the academic, regulatory, and industrial points of view. We have endeavored to focus attention not only on what is already known about stereoisomeric drugs, but also on how this issue will be approached in the future. To accomplish this task, we have investigated three aspects of this question: (1) the preparation of stereochemically pure drugs, (2) the pharmacological differences between drug stereoisomers, and (3) perspectives on the use of stereochemically pure drugs.

The introductory section of this book covers the early history of stereochemistry and defines a number of stereochemical terms. These contributions are designed to provide the reader with a background for the concepts discussed in the rest of the work.

The second section addresses the issue that if a single isomer of a stereoisomeric drug is going to be used, it is necessary to prepare the substance and to be able to prove its purity. This section of the book includes chapters dealing with chromatographic and nonchromatographic methods for the resolution of drug enantiomers and for the determination of the purity of stereoisomeric drugs. The section also includes a discussion of the rapidly growing field of stereospecific synthesis.

The drive to produce stereoisomerically pure drugs is based on the recognition that there are pharmacological differences between drug stereoisomers. These differences are addressed in the third section of this book, which discusses the pharmacokinetic, plasma protein binding, efficacy, toxicity, and biotransformation of stereoisomeric drugs.

The debate on whether to produce and use only stereochemically pure

1

The Early History of Stereochemistry

From the Discovery of Molecular Asymmetry and the First Resolution of a Racemate by Pasteur to the Asymmetrical Chiral Carbon of van't Hoff and Le Bel

Dennis E. Drayer* *Cornell University Medical College, New York, New York*

The first half of the nineteenth century was the great age of geometrical optics. Several French scientists studied diffraction, interference, and polarization of light. In particular, linear polarization of light and rotation of the plane of polarization very quickly attracted attention because of the possible relationship between these phenomena and the structure of matter. Optical activity, the ability of a substance to rotate the plane of polarization of light, was discovered in 1815 at the College de France by the physicist Jean-Baptiste Biot. In 1848 at the Ecole Normale in Paris, Louis Pasteur (Fig. 1) made a set of observations that led him a few years later to make this proposal, which is the foundation of stereochemistry: Optical activity of organic solutions is determined by molecular asymmetry, which produces nonsuperimposable mirror-image structures. A logical extension of this idea occurred in 1874 when a theory of organic structure in three dimensions was advanced independently and almost simultaneously by Jacobus Henricus van't Hoff (Fig. 2) in Holland, and Joseph Achille Le Bel (Fig. 3) in France. By this time it was known from the work of Kekule in 1858 that carbon is tetravalent (links up with four other groups or

*Present affiliation: American Cyanamid Company, Pearl River, New York.

FIGURE 1 Pasteur. [From Vallery-Radot (17).]

FIGURE 2 van't Hoff. [From *Zeitschrift Für Physikalische Chemie, 31* (1899).]

atoms). van't Hoff and Le Bel proposed that the four valances of the carbon atom were not planar, but directed into three-dimensional space. van't Hoff specifically proposed that the spatial arrangement was tetrahedral. A compound containing a carbon substituted with four different groups, which van't Hoff defined as an asymmetric carbon (*asymmetrisch koolstofatoom*), would therefore be capable of existing in two distinctly different nonsuperimposable forms. The asymmetric carbon atom, they proposed, was the cause of molecular asymmetry and therefore optical activity.

FIGURE 3 Le Bel. [From Snelders (18).]

The purpose of this chapter is to describe the observations and reasoning that led Pasteur, van't Hoff, and Le Bel to make these epochal discoveries. In several instances the protagonists will speak for themselves. More detailed accounts of their work are presented in Weyer (1), Partington (2), and Riddell and Robinson (3). Also, the three methods discovered by Pasteur to resolve for the first time an optically inactive racemate into its optically active components (enantiomers) will be discussed. To truly appreciate the contributions of these three chemists, one should remember that during their time even the existence of atoms and molecules was questioned openly by many scientists, and to ascribe shape to what seemed like metaphysical concepts was too much for many of their contemporaries to accept.

Ordinary tartaric acid has been known since the eighteenth century and is a by-product of alcoholic fermentation obtained in great quantities from the tartar deposited in the barrels. This acid has been especially important in medicine and dyeing. Paratartaric acid (also called racemic acid), discovered in certain industrial processes in the Alsace region of France, came to the attention of chemists only in the 1820s, when Gay-Lussac established that it possessed the same chemical composition as ordinary tartaric acid. Because of their importance for the emerging concept of isomerism, the two acids thereafter attracted considerable notice. On January 20 and February 3, 1860, Pasteur gave lectures before the Council of the Société Chimique of Paris describing the principal results of his research (done from 1848–1850) on tartaric acid and paratartaric acid, from which evolved his proposals on the molecular asymmetry of organic products. The excerpts below are taken, with permission, from an English translation made by the Alembic Club (4). An English translation is also found in Pasteur (5). Additional insight is found in Mauskopf (6). The headings and interspersed comments below are mine. To better understand what follows, ordinary tartaric acid is now called dextro-tartaric acid and paratartaric acid is the racemate, *d,l*-tartaric acid.

I. HEMIHEDRAL CRYSTAL STRUCTURE

Pasteur begins his first lecture by discussing the precedents that led up to his research and then defines hemihedral crystals. These are cubical crystals with four little facets inclined at the same angle to the adjacent surfaces and arranged alternately so the same edge of the cube does not contain two facets (Fig. 4). Under these conditions, no point or plane of symmetry exists in the cube.

II. MOLECULAR ASYMMETRY AND OPTICAL ACTIVITY

Pasteur now describes the research that led to his conclusion about the causal relationship between molecular asymmetry and optical activity.

> When I began to devote myself to special work, I sought to strengthen myself in the knowledge of crystals, foreseeing the help that I should draw from this in my chemical researches. It seemed to me to be the simplest

crystals hemihedral to the right deviated the plane of polarisation to the right, and that those hemihedral to the left deviated it to the left (here Fig. 5); and when I took an equal weight of each of the two kinds of crystals, the mixed solution was indifferent towards the light in consequence of the neutralisation of the two equal and opposite individual deviations.

Thus, I start with paratartaric acid; I obtain in the usual way the double paratartrate of soda and ammonia; and the solution of this deposit, after some days, crystals all possessing exactly the same angles and the same aspect. To such a degree in this case that Mitscherlich, the celebrated crystallographer, in spite of the most minute and severe study possible, was not able to recognise the smallest difference. And yet the molecular arrangement in one set is entirely different from that in the other. The rotatory power proves this, as does also the mode of asymmetry of the crystals. The two kinds of crystals are isomorphous, and isomorphous with the corresponding tartrate. But the isomorphism presents itself with a hitherto unobserved peculiarity; it is the isomorphism of an asymmetric crystal with its mirror image. This comparison expresses the fact very exactly. Indeed, if, in a crystal of each kind, imagine the hemihedral facets produced till they meet, I obtain two symmetrical tetrahedra, inverse, and which cannot be superposed, in spite of the perfect identity of all their respective parts. From this I was justified in concluding that, by crystallisation of the double paratartrate of soda and ammonia, I had separated two symmetrically isomorphous atomic groups, which are intimately united in paratartaric acid. Nothing is easier to show than that these two species of crystals represent two distinct salts from which two different acids can be extracted.

The announcement of the above facts naturally placed me in communication with Biot, who was not without doubts regarding their accuracy. Being charged with giving an account of them to the Academy, he made me come to him and repeat before his eyes the decisive experiment. He handed over to me some paratartaric acid which he had himself previously studied with particular care, and which he had found to be perfectly indifferent to polarised light. I prepared the double salt in his presence, with soda and ammonia which he had likewise desired to provide. The liquid was set aside for slow evaporation in one of his rooms. When it had furnished about 30 to 40 grams of crystals, he asked me to call at the College de France in order to collect them and isolate them before him, by recognition of their crystallographic character, the right and the left crystals, requesting me to state once more whether I really affirmed that the crystals which I should place at his right would deviate to the right, and the others to the left. This done, he told me that he would undertake the rest. He prepared the solutions with carefully measured quantities, and when ready to examine them in the polarising apparatus, he once more invited me to come into his room. He first placed in the apparatus the more interesting solution, that which ought to deviate to the left. Without even making a measurement, he saw by the appearance of the tints of the two images, ordinary and extraordinary, in the analyser, that there was a strong deviation to the left. Then, very visibly affected, the illustrious old man took me by the arm and said:

"My dear child, I have loved science so much throughout my life that this makes my heart throb."

Indeed there is more here than personal reminiscences. In Biot's case the emotion of the scientific man was mingled with the personal pleasure of seeing his conjectures realized. For more than thirty years Biot had striven in vain to induce chemists to share his conviction that the study of rotatory polarisation offered one of the surest means of gaining a knowledge of the molecular constitution of substances.

Let us return to the two acids furnished by the two sorts of crystals deposited in so unexpected a manner in the crystallisation of the double paratartrate of soda and ammonia. I have already remarked that nothing could be more interesting than the investigation of these acids.

One of them, that which comes from crystals of the double salt hemihedral to the right, deviates to the right, and is identical with ordinary tartaric acid. The other deviates to the left, like the salt which furnishes it. The deviation of the plane of polarisation produced by these two acids is rigorously the same in absolute value. The right acid follows special laws in its deviation, which no other active substance had exhibited. The left acid exhibits them, in the opposite sense, in the most faithful manner, leaving no suspicion of the slightest difference.

The paratartaric acid is really the combination, equivalent for equivalent, of these two acids, is proved by the fact that, if somewhat concentrated solutions of equal weights of each of them are mixed, as I shall do before you, their combination takes place with disengagement of heat, and the liquid solidifies immediately on account of the abundant crystallisation of paratar-

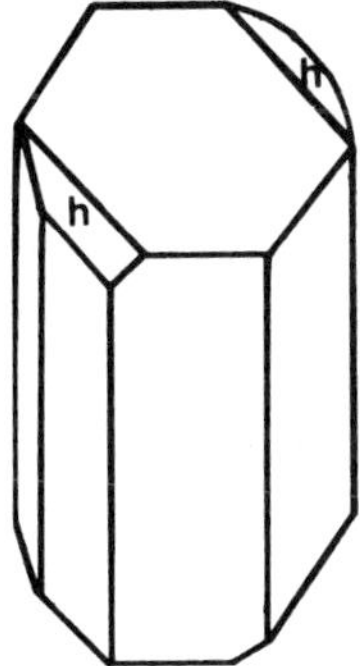

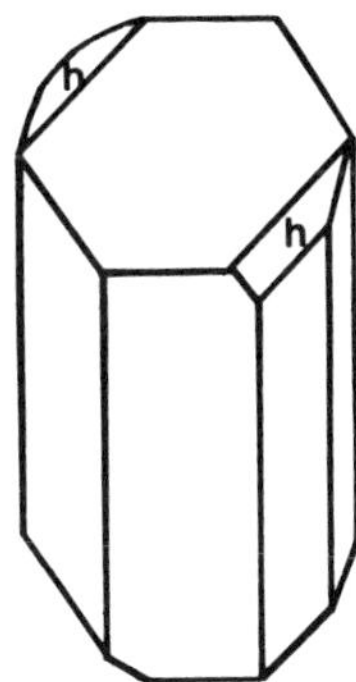

FIGURE 5 Paratarate of soda and ammonia formed by an equal mixture of hemihedral crystals of levo-tartrate (on left) and dextro-tartrate (on right). The anterior hemihedral facet "h" is on the left side of the observer in the levo-tartrate and on his or her right in the dextro-tartrate. [From Descour (17).]

taric acid, identical with the natural product. (This beautiful experiment called forth applause from the audience.)

Pasteur ends the first lecture with the following summary:

1). When the elementary atoms of organic products are grouped asymmetrically, the crystalline form of the substance manifests this molecular asymmetry in nonsuperposable hemihedry.

The cause of this hemihedry is thus recognised.

2). The existence of this same molecular asymmetry betrays itself, in addition, by the optical rotative property.

The cause of rotatory polarisation is likewise determined.

3). When the non-superposable molecular asymmetry is realised in opposite senses, as happens in the right and left tartaric acids and all their derivatives, the chemical properties of these identical and inverse substances are rigorously the same.

In the second lecture, Pasteur gives a further discussion of his fundamental idea that optical activity of organic solutions is related to molecular geometry. This insight was far ahead of the organic structural theory of the time.

We saw in the last lecture that quartz possesses the two characteristics of asymmetry—hemihedry in form, observed by Hauy, and the rotative phenomenon discovered by Arago! Nevertheless, molecular asymmetry is entirely absent in quartz. To understand this, let us take a further step in the knowledge of the phenomena with which we are dealing. We shall find in it, besides, the explanation of the analogies and differences already pointed out between quartz and natural organic products.

Permit me to illustrate roughly, although with essential accuracy, the structure of quartz and of the natural organic products. Imagine a spiral stair whose steps are cubes, or any other objects with superposable images. Destroy the stair and the asymmetry will have vanished. The asymmetry of the stair was simply the result of the mode of arrangement of the component steps. Such is quartz. The crystal of quartz is the stair complete. It is hemihedral. It acts on polarised light in virtue of this. But let the crystal be dissolved, fused, or have its physical structure destroyed in any way whatever; its asymmetry is suppressed and with it all action on polarised light, as it would be, for example, with a solution of alum, a liquid formed of molecules of cubic structure distributed without order.

Imagine, on the other hand, the same spiral stair to be constructed with irregular tetrahedra for steps. Destroy the stair and the asymmetry will still exist, since it is a question of a collection of tetrahedra. They may occupy any positions whatsoever, yet each of them will nonetheless have an asymmetry of its own. Such are the organic substances in which all the molecules have an asymmetry of their own, betraying itself in the form of the crystal. When the crystal is destroyed by solution, there results a liquid active towards polarised light, because it is formed of molecules, without arrangement, it is true, but each having an asymmetry in the same sense, if not of the same intensity in all directions.

III. RESOLUTION OF RACEMATES

Pasteur devised three methods to resolve paratartaric acid: the first was manual, the second was chemical, and the third could be considered biological or physiological. Because paratartaric acid (also called racemic acid) was the first inactive compound to be resolved into optical isomers (enantiomers), an equimolar mixture of two enantiomers is now called a racemate.

A. Manual Separation

As indicated in the first lecture, Pasteur, using a hand lens and pair of tweezers, laboriously separated a quantity of the sodium ammonium salt of paratartaric acid into two piles, one of left-handed crystals and the other of right-handed crystals, and in this way accomplished the first resolution of a racemate. After purifying the free tartaric acids from the separate salt solutions, he found one acid to be identical to the previously characterized ordinary tartaric acid (which was dextrorotatory) and the other acid to be the previously unknown levorotatory isomer. Pasteur was extremely fortunate in this area of his research. The tartrate used by him is one of the very few substances that undergo a spontaneous separation into enantiomeric (hemihedral) crystals, thereby allowing resolution by hand. That is, most enantiomers do not form enantiomeric crystals. Moreover, this separation takes place only below 27°C (7). If Pasteur had been working in southern France during a torrid Mediterranean summer, rather than in Paris, we may have praised another chemist as being the first to resolve a racemate.

B. Chemical Formation of Diastereomers

The physical properties of enantiomers are identical in an achiral environment. However, chemical reactions that add another asymmetric center create a diastereomeric pair, each of which has physical properties that are not completely the same. Therefore, although an enantiomeric pair cannot be separated by ordinary chromatographic means or fractional recrystallization, the diastereomeric pair can often be separated easily by these means, as is indicated in the chapter by Joseph Gal. After separation, the pure enantiomers can then be regenerated by chemical means. This is today the most fundamental way of resolving a racemate.

Pasteur, in his second lecture, gives the following account, in which the optically active basic alkaloids quinicine or cinchonicine were used to convert the two enantiomeric tartaric acids into diastereomers:

> Thus we find introduced into physiological principles and investigations the idea of the influence of the molecular asymmetry of natural organic products, of this great character which establishes perhaps the only well marked line of demarcation that can at present be drawn between the chemistry of dead matter and the chemistry of living matter.

Later qualified, modified, and generalized by others, Pasteur's new method became applicable to the separation of a number of other racemates (8).

Pasteur then ends his second lecture with the following:

> Such, gentlemen, are in co-ordinated form the investigations which I have been asked to present to you.
>
> You have understood, as we proceeded, why I entitled my exposition, "On the Molecular Asymmetry of Natural Organic Products." It is, in fact, the theory of molecular asymmetry that we have just established, one of the most exalted chapters of the science. It was completely unforeseen, and opens to physiology new horizons, distant, but sure.
>
> I hold this opinion of the results of my own work without allowing any of the vanity of the discoverer to mingle in the expression of my thought. May it please God that personal matters may never be possible at this desk. These are like pages in the history of chemistry which we write successively with that feeling of dignity which the true love of science always inspires.

Although popularly known chiefly for his great work in bacteriology and medicine, Pasteur was by training a chemist, and this work in chemistry alone would have earned him a position as an outstanding scientist.

The development of stereochemical ideas entered a new stage in 1858 when August Kekule (Fig 6) introduced the idea of the valence bond and the pictorial representation of molecules as atoms connected by valence bonds. His main thesis was that the carbon atom is tetravalent, and that a carbon atom can form valence bonds with other carbon atoms to form open chains and that sometimes the carbon chains can be closed to form rings (9). This led directly to his proposal for the structure of benzene. On the occasion of celebrations held in his honor, Kekule in 1890 delivered a speech before the German Chemical Society describing the origin of his idea of the linking of atoms (9).

> During my stay in London I resided for a considerable time in Clapham Road in the neighborhood of the Common. I frequently, however, spent my evenings with my friend Hugo Muller at Islington, at the opposite end of the giant town. . . . One fine summer evening I was returning by the last omnibus, "outside," as usual, through the deserted streets of the metropolis, which are at other times so fully of life. I fell into a reverie and lo, the atoms were gambolling before my eyes! Whenever, hitherto, these diminutive beings had appeared to me, they had always been in motion; but up to that

FIGURE 6 Kekule. [From Japp (9).]

time I had never been able to discern the nature of their motion. Now, however, I saw how, frequently, two smaller atoms united to form a pair; how a larger one embraced two smaller ones; how still larger ones kept hold of three or even four of the smaller; whilst the whole kept whirling in a giddy dance. I saw how the larger ones formed a chain, dragging the

> smaller ones after them, but only at the ends of the chain. . . . The cry of the conductor: "Clapham Road," awakened me from my dreaming; but I spent a part of the night in putting on paper at least sketches of these dream forms. This was the origin of the *Structurtheorie*.

Then he related a similar experience of how the idea for the structure of benzene occurred to him.

> I was sitting writing at my textbook, but the work did not progress; my thoughts were elsewhere. I turned my chair to the fire and dozed. Again the atoms were gambolling before my eyes. This time the smaller groups kept modestly in the background. My mental eye, rendered more acute by repeated visions of this kind, could now distinguish larger structures of manifold conformations; long rows, sometimes more closely fitted together; all twisting and turning in snake-like motion. But look! What was that? One of the snakes had seized hold of its own tail, and the form whirled mockingly before my eyes. As if by a flash of lightning I awoke; and this time also I spent the rest of the night working out the consequences of the hypothesis. Let us learn to dream, gentlemen, and then perhaps we shall find the truth . . . but let us beware of publishing our dreams before they have been put to the proof by the waking understanding.

In speculating on the kind of atomic arrangements that could produce molecular asymmetry, Pasteur, as already indicated, suggested tentatively in 1860 that the atoms of a right-handed compound, for example, might be "arranged in the form of a right-handed spiral, or situated at the corners of an irregular tetrahedron." But he never developed these suggestions. The solution to this problem of what is the cause of molecular asymmetry was presented in the publications of van't Hoff and Le Bel. On September 5, 1874, van't Hoff, while he was still a student at the University of Utrecht and only 22 years of age, published a pamphlet entitled "Proposal for the extension of the structural formulae now in use in chemistry into space, together with a related note on the relation between the optical active power and the chemical constitution of organic compounds" (10). An English translation is presented in van't Hoff (11). Starting with the ideas of August Kekule on the tetravalency of carbon, van't Hoff states, at the beginning of his pamphlet: "It appears more and more that the present constitutional formulas are incapable of explaining certain cases of isomerism; the reason for this is perhaps the fact that we need a more definite statement about the actual positions of the atoms." He then proposed that the four valences of a carbon atom are directed toward the corners of a tetrahedron with the carbon atom at the center, based on the concept of the isomer number, which is illustrated below.

For any atom Y, only one substance of formula CH_3Y has ever been found. For example, chlorination of methane yields only one compound of formula CH_3Cl. Indeed, the same holds true if Y represents, not just an atom, but a group of atoms (unless the group is so complicated that in itself it brings about isomerism); there is only one CH_3OH, and only one CH_3CO_2H. This suggests that every hydrogen atom in methane is equivalent to every other hydrogen atom, so that replacement of any one of them gives rise to the same product. If the hydrogen atoms of methane were not equivalent, then replacement of one would yield a different compound than replacement of another, and isomeric substitution products would be obtained. In what ways can the atoms of methane be arranged so that the four hydrogen atoms are equivalent? There are three such arrangements (Fig. 7): a planar arrangement (I) in which carbon is at the center of a rectangle (or square) and a hydrogen atom is at each corner; a pyramidal arrangement (II) in which carbon is at the apex of a pyramid and a hydrogen atom is at each corner of a square base; a tetrahedral arrangement (III) in which carbon is at the center of a tetrahedron and a hydrogen atom is at each corner. By then comparing the number of isomers that have been prepared for di-, tri- and tetrasubstituted methanes with the number predicted by the above three spatial arrangements, it is possible to decide which one is correct.

For example, with a disubstituted compound CH_2R_2 (Fig. 8); (1) if the molecule is planar, then two isomers are possible. This planar configuration can be either square or rectangular; in each case, there are two isomers only. (2) If the molecule is pyramidal, then two isomers are also possible. There are only two isomers, whether the base is square or rectangular. (3) If the molecule is tetrahedral, then only one form is possible. The carbon atom is at the center of the tetrahedron. In actuality, only one disubstituted isomer is known. Therefore, only the tetrahedral model for a disubstituted methane agrees with the evidence of the isomer number.

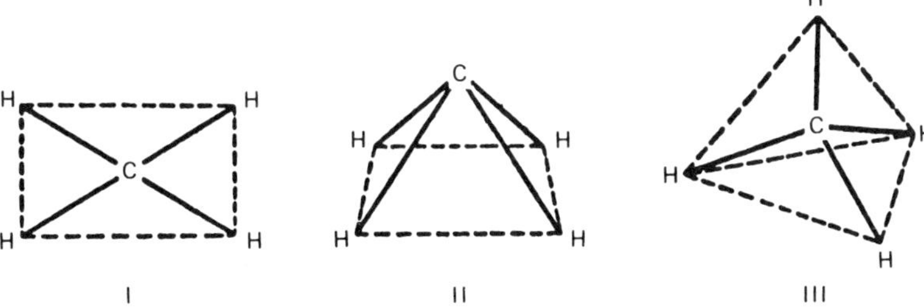

FIGURE 7 Spatial models for methane where the four hydrogen atoms are equivalent. I, planar; II, pyramidal; III, tetrahedral.

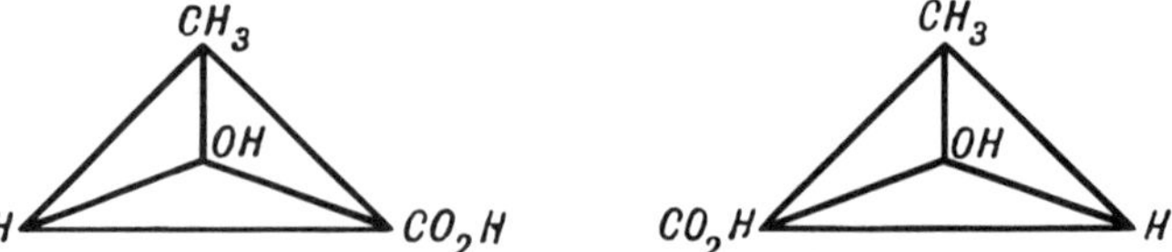

FIGURE 10 Tetrahedral model for lactic acid enantiomers (carbon atom is at the center of the tetrahedron) as envisioned by van't Hoff.

pounds: ethylidene lactic acid (now called α-hydroxypropionic acid), aspartic acid, asparagine, malic acid, glutaric acid, tartaric acid, sugars and glucosides, camphor, borneol, and camphoric acid.

Two compounds from this list are worthy of note: lactic acid (Fig. 10) and tartaric acid (Fig. 11). Wislicenus (Fig. 12) extensively investigated the isomers of lactic acid between 1863 and 1873, and was convinced that the number of isomers exceeded that allowed by the existing structural theory (12). However, due to experimental difficulties in obtaining pure samples of the isomers, in addition to the limits of the structural theory then known to him, he ended up going around in circles. van't Hoff studied the publications of Wislicenus on lactic acids and they led him to his own stereochemical ideas. In fact, lactic acid was the first concrete example of an optically active compound that van't Hoff discussed after his theoretical introduction. He pointed out that ethylidene lactic acid contains an asymmetric carbon. Therefore, it can exist as two pure enantiomers or a racemic mixture, which nicely cleared up the confusion surrounding the lactic acid

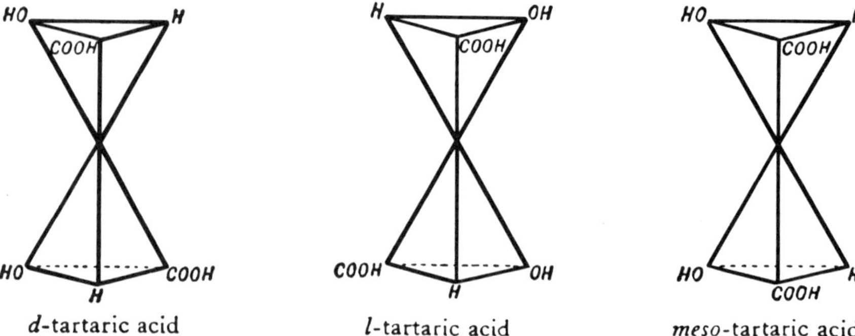

FIGURE 11 Structures for three tartaric acid isomers that are representative of the tetrahedral models used by van't Hoff.

FIGURE 12 Wislicenus.

isomers. In a lecture, held much later in Utrecht on May 16, 1904, van't Hoff said the following:

> Students, let me give you a recipe for making discoveries. In connexion with what has just been said about libraries, I might remark that they have always had a mind-deadening effect on me. When Wislicenus' paper on lactic acid appeared and I was studying it in the Utrecht library, I therefore broke off my study half-way through, to go for a walk; and it was during this walk, under the influence of the fresh air, that the idea of asymmetric carbon first struck me.

These proposals of van't Hoff's came as a breath of fresh air to Wislicenus. No wonder that he was the first to welcome it enthusiastically, or that he sponsored the German translation that made it widely known, or that he was the first to make significant further use of the hypothesis, in his work on geometrical isomers of unsaturated compounds (13).

The other example of note is the optically active tartaric acids (Fig. 11). Tartaric acid contains two asymmetric carbon atoms. The dextro- and levo-tartaric acids are enantiomers. However, a third isomer is possible in which the two rotations due to the two asymmetric carbon atoms compensate and the molecule is optically inactive as a whole. That is, the molecule contains a plane of symmetry. This form, meso-tartaric acid, was also discovered by Pasteur, differs from the two optically active tartaric acids in being internally compensated, and is not resolvable. Thus, the tetrahedral model for carbon and the asymmetric carbon atom proposed by van't Hoff were able to completely explain the observations of Pasteur relating to the three isomers of tartaric acid.

Le Bel published his stereochemical ideas two months later, in November 1874, under the title, "The relations that exist between the atomic formulas of organic compounds and the rotatory power of their solutions" (14). An English translation is presented in Le Bel (15). Le Bel approached the problem from a different direction from van't Hoff. His hypothesis was based on neither the tetrahedral model of the carbon atom nor the concept of fixed valences between the atoms. He proceeded purely from symmetry arguments; he spoke of the asymmetry, not of individual atoms, but of the entire molecule, so that his views would nowadays be classed under the heading of molecular asymmetry. Only once does he mention the tetrahedral carbon atom, which he regarded as not a general principle but a special case. Today, substituted allenes, spiranes, and biphenyls are but a few examples of asymmetric molecules that do not contain any asymmetric carbons, thus confirming Le Bel's views on molecular asymmetry. The reason for the different approaches by van't Hoff and le Bel is easy to understand. van't Hoff came from the camp of structural chemists, and he wished his hypothesis to be understood as an extension of the structural theory to spatial relationships. The tetravalent atomic models used by Kekule in his lectures presumably also prompted his pupil van't Hoff, possibly unconsciously, in the conception of the asymmetric carbon atom. Le Bel, on the other hand, was trained in the tradition of Pasteur (whose investigations he also mentioned expressly in his article), that is, he started out from Pasteur's considerations of the connections between optical rotation and molecular structure.

In 1877 Hermann Kolbe, one of the most distinguished of the older German chemists, published a diatribe in the *Journal für Praktische Chemie* after reading the work of van't Hoff (which had been translated into German by Felix Herrmann at the suggestion of Wislicenus). An English translation of this abusive attack is presented completely in Riddell and Robinson (3). Those individuals interested in seeing an example of the great personal attacks by editors that appeared in journals of the nineteenth century should read this translation. Although defamatory, this criticism served a useful purpose, since it made a decisive contribution to the dissemination of these ideas of van't Hoff. This was fortunate, since van't Hoff soon turned his genius away from stereochemistry to physical chemistry, for which he received the Nobel Prize.

We can now end this historical journey. We have walked through the early days of stereochemistry in the company of giants. In 1949, almost exactly 100 years after the first resolution of *d,l*-tartaric acid by Pasteur, the Dutchman Bijvoet (16), using x-ray diffraction, determined the actual arrangement in space of the atoms of the sodium rubidium salt of (+)-tartaric acid, and thus made the first determination of the absolute configuration about an asymmetric carbon. To further complete the link with the past, Bijvoet did this work while the Director of the van't Hoff Laboratory at the University of Utrecht.

In the intervening years since the first resolution of a racemate by Pasteur, many chromatographic and nonchromatographic methods have been developed for the resolution of racemic compounds. These methods are the subjects of many of the other chapters in this book.

REFERENCES

1. J. Weyer, A hundred years of stereochemistry—The principal development phases in retrospect, *Angew. Chemie. Internat. Ed.*, *13*:591–598 (1974).
2. J. R. Partington, *A History of Chemistry*, Vol. 4, Macmillan and Co., Ltd., London, 1964, pp. 749–764.
3. F. G. Riddell and M. J. T. Robinson, J. H. van't Hoff and J. A. Le Bel—their historical context, *Tetrahedron*, *30*:2001–2007 (1974).

4. L. Pasteur, Researches on the molecular asymmetry of natural organic products. Alembic Club Reprints, No. 14, reissue edition, F. and S. Livingstone, Ltd., Edinburgh, 1948.
5. L. Pasteur, On the asymmetry of naturally occurring organic compounds, *The Foundations of Stereo Chemistry: Memoirs by Pasteur, Van't Hoff, Le Bel, and Wislicenus* (G. M. Richardson, ed.), American Book Co., New York, 1901, pp. 1–33.
6. S. H. Mauskopf, *Crystals and Compounds: Molecular Structure and Composition in Nineteenth-Century French Science*, American Philosophical Society, Philadelphia, 1976, pp. 68–80.
7. J. H. van't Hoff, *The Arrangement of Atoms in Space* (A. Eiloart, transl. ed.), Longmans, Green, and Co., New York, 1898, pp. 34–40.
8. J. H. van't Hoff, *The Arrangement of Atoms in Space* (A. Eiloart, transl. ed.), Longmans, Green, and Co., New York, 1898, pp. 30–33.
9. F. R. Japp, Kekule memorial lecture, *J. Chem. Soc., 73*:97–138 (1989).
10. J. H. van't Hoff, Voorstel tot uitbreiding der tegenwoordig in de scherkunde gebruikte structuur-formules in de ruimte. Greven, Utrecht, 1874.
11. J. H. van't Hoff, A suggestion looking to the extension into space of the structural formulas at present used in chemistry and a note upon the relation between the optical activity and the chemical constitution of organic compounds, *The Foundations of Stereo Chemistry: Memoirs by Pasteur, Van't Hoff, Le Bel, and Wislicenus* (G. M. Richardson, ed.), American Book Co., New York, 1901, pp. 37–46.
12. N. W. Fisher, Wislicenus and lactic acid: The chemical background to van't Hoff's hypothesis, *van't Hoff–Le Bel Centennial* (O. Bertrand Ramsay, ed.), *ACS Symp. Ser., 12*:33–54 (1975).
13. J. Wislicenus, The space arrangement of the atoms in organic molecules and the resulting geometric isomerism in unsaturated compounds, *The Foundations of Stereo Chemistry: Memoirs by Pasteur, Van't Hoff, Le Bel, and Wislicenus* (G. M. Richardson, ed.), American Book Co., New York, 1901, pp. 61–132.
14. J. A. Le Bel, Sur les relations qui existent entre les formules atomiques des corps organiques, et le pouvoir rotatoire de leur dissolutions, *Bull. Soc. Chim. Paris, 22*:337 (1874).
15. J. A. Le Bel, On the relations which exist between the atomic formulas of organic compounds and the rotatory power of their solutions, *The Foundations of Stereo Chemistry: Memoirs by Pasteur, Van't Hoff, Le Bel, and Wislicenus* (G. M. Richardson, ed.), American Book Co., New York, 1901, pp. 49–59.
16. J. M. Bijvoet, A. F. Peerdeman, and A. J. van Bommel, Determination of the absolute configuration of optically active compounds by means of X-rays, *Nature* (Lond.), *168*:271–272 (1951).
17. R. Vallery-Radot, *The Life of Pasteur* (R. L. Devonshire, transl.), Garden City Publishing Co., Inc., New York, 1926.
18. H. A. M. Snelders, J. A. Le Bel's stereochemical ideas compared with those of J. A. van't Hoff, *van't Hoff-Le Bel Centennial* (O. Bertrand Ramsey, ed.), *ACS Symp. Ser., 12*:66–73 (1975).

2

STEREOCHEMICAL TERMS AND CONCEPTS

An Overview

Irving W. Wainer *McGill University and Montreal General Hospital, Montreal, Quebec, Canada*

Alice A. Marcotte *Food and Drug Administration, Washington, D. C.*

I. INTRODUCTION

The terminology used to describe stereochemical relationships is often a maze of capital D's and L's, lowercase *d*'s and *l*'s, mixed in with (R)'s, (S)'s, (+)'s, and (−)'s. This chapter is designed to give the reader a quick overview of stereochemical terms and concepts and to lay a foundation for the more technical discussions to come. It is not meant to be an in-depth treatment of this topic and the interested reader is urged to consult one of the many fine texts on the subject.

II. SYMMETRY AND ASYMMETRY

Symmetry or the lack of it is one of the interesting features of geometric figures with two or more dimensions. We are often confronted with this phenomenon without realizing it. For example, the alphabet contains symmetrical and asymmetrical two-dimensional letters, some of which have different appearances when they are reflected in a mirror. Six of these letters and their mirror images are presented in Fig. 1. There is no difference between each of the symmetrical letters A, H, and M and the corresponding mirror image; however, the mirror images of the asymmetrical letters E, R, and S appear reversed.

The property of nonidentical mirror images also exists for figures with three dimensions. The human body, for example is an asymmetrical three-

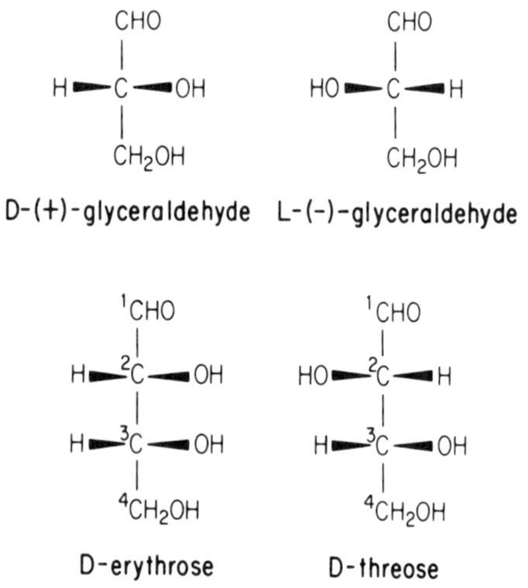

CHO CHO

H—C—OH HO—C—H

CH_2OH CH_2OH

D-(+)-glyceraldehyde L-(-)-glyceraldehyde

^{1}CHO ^{1}CHO

H—^{2}C—OH HO—^{2}C—H

H—^{3}C—OH H—^{3}C—OH

$^4CH_2OH$ $^4CH_2OH$

D-erythrose D-threose

FIGURE 7 The configurations of D-(+) and L-(−)-glyceraldehyde, D-erythrose, and D-threose, according to the Fischer Convention.

hyde. In Fig. 7, the D configuration is assigned to the erythrose and threose series because of the D configuration at C-3.

The Fischer convention is often inexact and difficult to use, especially when complicated chemical transformations are required to convert the molecule under investigation into a molecule of known configuration. In addition, the assigned configuration, D or L, is often confused with the

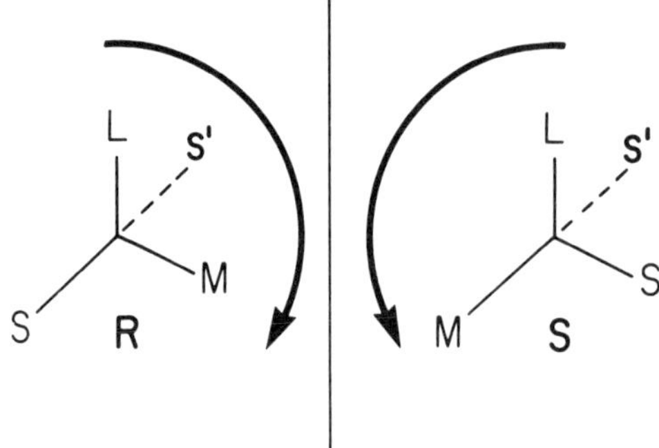

FIGURE 8 The Cahn–Ingold–Prelog sequence rule, where L is the largest group attached to the chiral center, followed in size by M, S, and s′.

TABLE 1 Commonly Used Sterochemical Terms

Stereoisomers	Molecules with the same consitution, but that differ in respect to the spatial arrangement of certain atoms or groups
Enantiomeric molecule	A molecule that is not superimposable on its mirror image
Chiral molecule	A molecule having at least one pair of enantiomers
Enantiomers	Stereoisomers that are related as nonsuperimposable mirror images
Diastereomers	Optical isomers that are not related as an object and its mirror image
Optical activity	A property of a chiral molecule—the ability to rotate a beam of plane polarized light
Optical rotation (α)	The angle that a beam of plane-polarized light is rotated by a chiral molecule
Dextrorotary, *d* or (+) rotation	A clockwise rotation of a beam of plane-polarized light by a stereoisomer, usually used to denote a specific enantiomer of a chiral molecule, that is *d*- or (+)-ephedrine
Levorotatory, *l* or (−) rotation	A counterclockwise rotation of a beam of plane-polarized light by a stereoisomer, usually used to denote a specific enantiomer of a chiral molecule, that is *l*- or (−)- ephedrine
Specific rotation [α]	A quantitative measure of the optical rotation, which is dependent on concentration or density of the chiral substance, length of the light path, temperature, and wavelength of light
Configuration	The description of the spatial arrangement about a chiral atom
Fischer convention (D, L)	The assignment of configuration about a chiral atom by comparison to a standard, (D) − (+) − glyceraldehyde, usually by actual chemical transformation of the molecule under investigation into the standard
Cahn–Ingold–Prelog convention (*R*, *S*)	The assignment of configuration about a chiral atom by designation of the sequence of substituents from the largest (L) to medium (M) to the smallest (S); a clockwise direction of the L-M-S sequence is assigned the *R* configuration and a counterclockwise direction is assigned the *S* configuration

present (24). Obviously, when two or more asymmetric centers are present, the likelihood of being able to devise a nonchiral analog for immunogen formation becomes much smaller (25).

C. Development and Use of Antisera for a Single Enantiomer

1. *Introduction*

In a number of cases, investigators have prepared immunogens from one of two enantiomers and used these for analysis of that particular enantiomer. In some cases, the radioligand was enantiomerically pure; in others, racemic radioligand was used. Usually, this was done because of some specific information regarding the differential pharmacological effect of the enantiomer. In other cases, the drug itself is already administered as a single enantiomer and thus is often readily available for conjugation to protein. Most RIAs for steroids are developed by means of immunogens incorporating the naturally occurring or hormonally effective steroid enantiomer. The voluminous literature on this type of assay will not be further discussed. Likewise, a number of natural products that are generally obtained in the form of a single enantiomer have been subjected to RIA development. Unless there is special information regarding the cross-reaction of enantiomers, this also will not be covered.

2. *Psychoactive Drugs*

A 1974 report described the development of an antiserum for *d*-amphetamine. Inhibition studies were carried out by means of displacement of tritium-labeled *d*-amphetamine, and the cross-reaction for *l*-amphetamine was 4.5% (28). Later an antiserum to *d*-methamphetamine was reported to have no significant cross-reaction with the *l* isomer (29).

RIA for *L*-α-acetyl methadol (LAAM) has been reported (30). The *O*-acetyl group of this compound (Fig. 4 [4a]) was replaced by an *O*-succinyl moiety, which was coupled to bovine thyroglobulin by mediation with a soluble carbodiimide reagent to give an immunogen (Fig. 4 [4b]). Tritium-labeled LAAM was used as a radioligand. A sensitivity of 50 pg was recorded with no serum interference. LAMM has two asymmetric centers, both of the *S* configuration. The diastereoisomer formed by change of the 3-*S* configuration (which bears the acetate residue and was involved in linkage to the protein) to the 3-*R* configuration exhibited a cross-reaction 25% that of LAMM. When the 6 carbon atom in the chain (Fig. 4) was changed from *S* to *R*, the cross-reaction dropped to 0.033% and the 3-*R*, 6-*R* compound (enantiomer of LAAM) had a cross-reaction of 0.17%. Thus, in this case, we see the general trend toward greater differen-

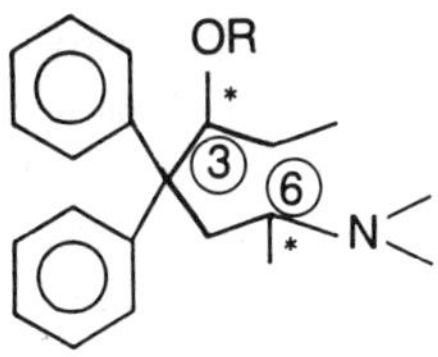

a. R = $-COCH_3$ (LAAM)
b. R = $-COCH_2CH_2CONH$-thyroglobulin

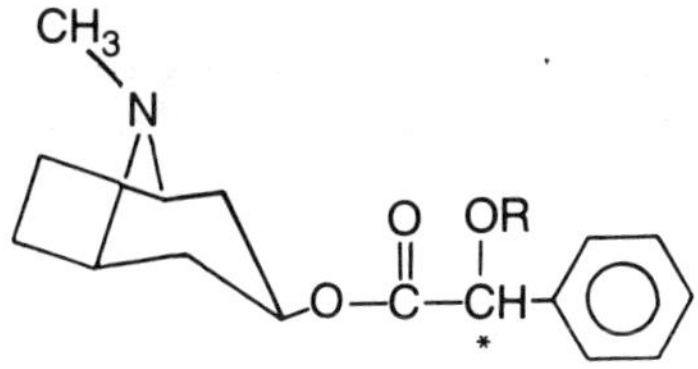

a. R = H (Atropine)
b. R = $-COCH_2CH_2CO-NH-(BSA)_{1/n}$

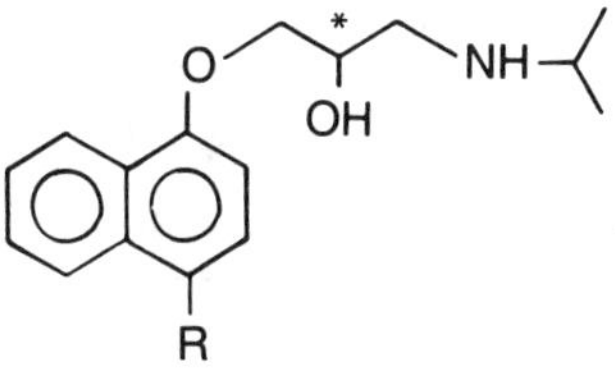

a. R = $-H$ (Propranolol)
b. R = $-OH$ (4-hydroxy propranolol)

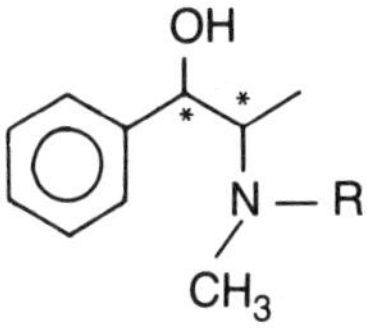

a. R = H (pseudoephedrine)
b. R = $CH_2CH_2CONH(BSA)_{1/n}$

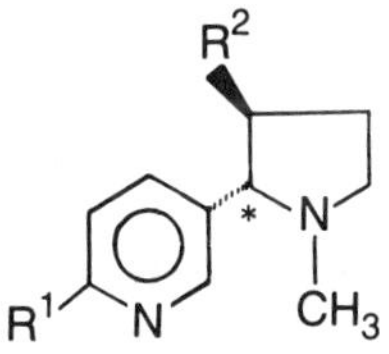

a. $R^1 = R_2$ = H (nicotine)
b. R^1 = $-NHCO-$⟨phenyl⟩$-N=N-(BSA)_{1/n}$, R^2 = H
c. R^1 = H, R^2 = $-CH_2O-COCH_2CH_2CONH(BSA)_{1/n}$

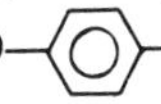

*Chiral center

FIGURE 4 Structures of drugs and immunogens.

tiation as one goes further from the link to protein. This antiserum had good selectivity vs. the isomeric methadols, but exhibited 12% cross-reaction with the *N*-desmethyl metabolite norLAMM. When norLAMM levels were twice those of LAMM, a 25% positive error was introduced. Such ratios can occur as early as 11 hr after LAMM administration (unpublished results) and may be more likely on repeated dosage, due to the longer half-life of the metabolite. This highlights the fact that enantioselectivity alone is not enough. Selectivity vs. metabolites and other closely related compounds must also be taken into consideration.

3. *Atropine*

S-atropine (Fig. 4 [5a]) is markedly more potent than the *R* isomer in a series of tests (31) and binds much more strongly to the muscarinic receptor (32). Indeed, given the propensity of this particular compound for racemization (33) and therefore its lack of complete optical purity, it would appear that essentially all the biological activity resides in the *S* enantiomer.

Various preparations of antisera to atropine have been reported. A racemic hemisuccinate ester was prepared and conjugated to bovine serum albumin by the carbodiimide technique. Antisera formed to the original immunogen selectively bound the *R* isomer (34), but a later antiserum prepared by this approach was reported to bind both *R* and *S* forms with "equal efficiency" (35). *R,S*-atropine was treated with diazotized *p*-aminobenzoic acid, and the resulting compound (which was not further characterized) was used for conjugation to bovine serum albumin by means of a carbodiimide-mediated reaction. Antisera resulting from use of this material were quite selective for the *R* isomer, with a cross-reaction of only about 2% for the *S* isomer (36). Virtanen et al. followed this procedure with *S*-atropine. Their antiserum bound equally to *S*- and *R,S*-atropine, as measured by displacement of tritium-labeled *R,S*-atropine (37). In another study (31), both racemic atropine and the *S* isomer were coupled to human serum albumin by the technique of Wurtzburger et al. (36). Antisera were obtained that were selective for both the *R* and *S* isomers (33).

The work with atropine presents a confusing picture. Use of racemic radioligand in these studies may contribute to the observed cross-reactivity picture. The rather facile enolization and hence racemization of atropine-type structures is another problem. Finally, the structure of the product of atropine and diazotized *p*-aminobenzoic acid has never been established conclusively. For these reasons, the radioreceptor assay for this compound (see below) is generally preferable.

4. *Propranolol*

The importance of propranolol (Fig. 4 [6a]) has led to the development of a number of immunoassays for it. Most of the workers dealt with the racemic compound. However, Kawashima et al., in an early demonstration of the utility of enantioselective assays, developed a procedure for analysis of the *l* isomer (38). These workers made a hemisuccinyl derivative of *l*-propranolol and coupled it to bovine serum albumin by a mixed anhydride technique. Antisera were generated in rabbits, and the radioligand was racemic tritiated propranolol. With this antiserum, this group measured plasma concentrations of the *l* isomer in rat blood (38), as well as in mice (39). Because they also had generated an antiserum against the racemic material that had little capability for distinguishing *d*-, *l*-, or *d,l*-propranolol, they also determined total propranolol and by difference could obtain an estimate of the *d* isomer as well. In the rat they found a higher initial concentration of *d* than *l* isomer, but much higher amounts of the *l* isomer in the heart. This was attributed to selective uptake of the *l* isomer by receptors. Because of the relatively high (7%) cross-reaction of the *d* isomer with the *l* antiserum, concentrations of the *l* isomer were corrected. The actual concentration of *l* was equal to [(measured concentration of *l* isomer) − 0.07(measured concentration of *d,l*-propranolol)]/0.93.

5. *Pseudoephedrine*

Findlay and co-workers developed an RIA for *d*-pseudoephedrine (Fig. 4 [7a]) (40). The compound was allowed to add to methyl acrylate. After hydrolysis, the resulting carboxylic acid was conjugated to bovine serum albumin by means of a carbodiimide procedure to form an immunogen (Fig. 4 [7b]). The radioligand for the binding studies was either tritium-labeled *d,l*-pseudoephedrine (5 Ci/mmol) or a conjugate of the *d*-pseudoephedrine-methyl acrylate adduct (analogous to [7b]) with tyrosine methyl ester labeled with ^{125}I. Charcoal or polyethylene glycol was used to separate free and bound ligand. With the tritium-labeled radioligand, serum could be analyzed directly, but plasma had to be extracted if the ^{125}I radioligand was used. The plasma extract was treated with methyl acrylate to convert the *d*-pseudoephedrine to a compound more closely resembling the original immunogen. By this procedure it was possible to achieve sensitivities down to 0.2 ng/mL. When the iodinated radioligand was used, the cross-reaction with *l*-pseudoephedrine was less than 0.003%. With the tritium-labeled *d,l* radioligand, the cross-reaction with the *l* isomer was 0.01–0.05%. *l*-Ephedrine, a diastereoisomer, cross-reacted to the extent of 0.13–0.32%.

6. *S-Bioallethrin*

S-Bioallethrin (Fig. 5 [9a]) is a synthetic pyrethroid insecticide with three asymmetric centers. The *S* designation refers to the 1-*R*, 3-*R*, 4′-*S* isomer and is not a true designation of stereochemistry. Conversion of the terminal allylic group to $—CH_2CH_2CH_2OH$, formation of hemisuccinate,

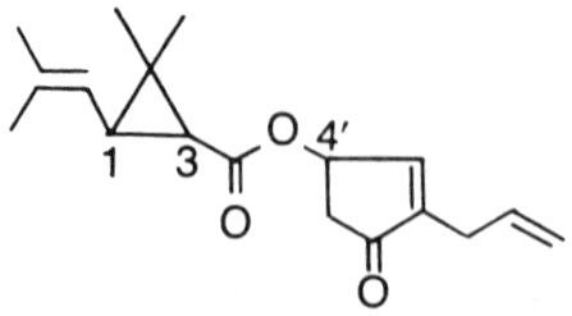

[9]

a. 1–R, 3–R, 4′–S
b – h. See Table 1.

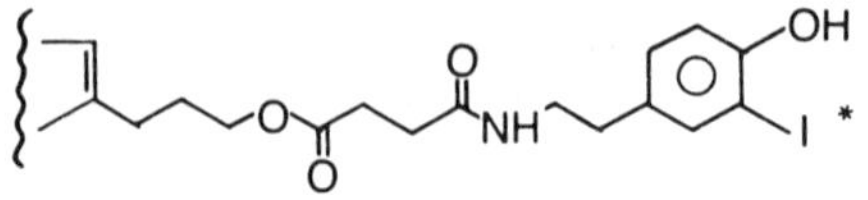

[10]

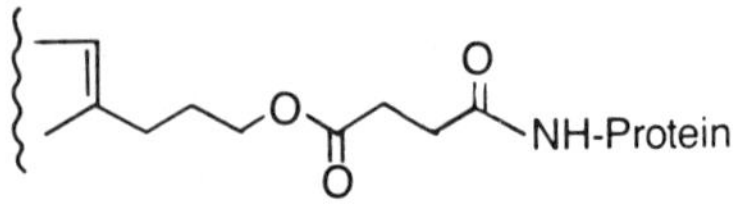

[11]

FIGURE 5 Bioallethrin structures.

and conjugation to protein led to an immunogen (Fig. 5 [11]) by means of which stereoselective antibodies were obtained. A ^{125}I-labeled tyramine conjugate (Fig. 5 [10]) was used as the radioligand. All eight isomers were studied for their ability to inhibit radioligand binding. As Table 1 shows, the enantiomer [9b] of compound [9a] (Fig. 5) had less than 1% cross-reaction. Changing only the asymmetric center at the 4′ position is less critical for binding than changing the orientation of the groups on the

TABLE 1 Stereochemistry and Antibody Binding in the Bioallethrin Series[a]

Compound	Cyclopropane substituent orientation	Absolute configuration 1	3	4′	Percent cross-reaction at 50% inhibition
[9a]	*Trans*	*R*	*R*	*S*	100
[9b]	*Trans*	*S*	*S*	*R*	0.8
[9c]	*Trans*	*R*	*R*	*R*	57
[9d]	*Trans*	*S*	*S*	*S*	5
[9e]	*Cis*	*R*	*S*	*S*	<0.1
[9f]	*Cis*	*S*	*R*	*S*	2
[9g]	*Cis*	*R*	*S*	*R*	3
[9h]	*Cis*	*S*	*R*	*R*	≪0.1

[a]The antiserum was generated from a conjugate of an analog of compound Fig. 5[9a].
Source: Wing and Hammock (41).

cyclopropane ring (compare Fig. 5 [9c] vs. [9d] or [9c] vs. [9e] or [9f]). Although the authors attribute this to the closer proximity of the 4′ position to the conjugation link (41), the 4′ position is well removed from the link. An alternative explanation would be the flexibility of rotation around the 4′ *O*—*C* bond compared to the rigidity of the cyclopropane ring.

7. Nicotine

Although *l*-nicotine is the predominant natural isomer, the occurrence of some *d*-nicotine makes it of interest to develop enantioselective assays for this compound. A review of nicotine and cotinine assays has been given (42). An early report of an RIA for *l*-nicotine described an immunogen prepared from 6-aminonicotine. This compound was converted to 6-(*p*-aminobenzamido)nicotine and conjugated to albumin (Fig. 4 [8b]) by diazotization (43). The enantiomeric purity of the 6-aminonicotine was not discussed, and the synthesis of 6-aminonicotine is reported to cause racemization (44). However, the antiserum eventually obtained showed only 6% cross-reaction with *d*-nicotine (43).

A racemic nicotine analog, *trans*-3-hydroxymethyl-nicotine, was converted to the hemisuccinate, which was conjugated to protein to form an immunogen (Fig. 4 [8c]). The resulting antiserum was used with tritiated *l*-nicotine as radioligand. With this radioligand the assay was highly selective for *l*-nicotine, with less than 0.01% cross-reaction with *d*-nicotine. Similar enantioselectivity is claimed for *l*-cotinine (42).

Hybridomas producing monoclonal antibodies to *S*-nicotine were obtained from mice immunized with conjugated racemic 3′hydroxymethyl-nicotine. Affinity constants were around 10^8 M^{-1} with 4% cross-reaction to *R*-nicotine (67).

D. Development and Use of Antisera for Both Enantiomers

1. Introduction

In not all instances in which racemic drug is administered is only one enantiomer of interest. Even when the major portion of the activity is associated with a single enantiomer, knowledge of concentrations of the other enantiomer may be of interest because of potential for toxicity or other side effects. Furthermore, it would be anticipated that the best selectivity would be achieved if both immunogen and labeled ligand were enantiomerically equivalent and optically pure. The enantiomeric immunogen would favor formation of enantioselective antibodies, and the enantiomerically pure labeled ligand would further enhance enantioselectivity by allowing the analyst to observe and use those antibodies with highest affinity for the enantiomer. A few examples of this approach have been published.

2. Warfarin

A method for the determination of both warfarin enantiomers based on enantioselective immunoassay has been reported (45). Warfarin analogs (Fig. 6) containing a 4′-carboxyethyl group (as the methyl ester) were converted to their diastereoisomeric camphor-sulfonates (Fig. 6 [12b]). These were separated chromatographically. Base-catalyzed hydrolysis then removed the camphor-sulfonyl group to yield the pure enantiomeric warfarin analogs. These were individually conjugated to bovine serum albumin by a mixed anhydride procedure to yield an immunogen (Fig. 6 [12c]). For radioligands, halogenated warfarin derivatives were resolved in a similar manner and then reduced with tritium gas to yield radioligands (Fig. 6 [12d]) of high optical purity (25–31 Ci/mmol). To demonstrate that racemization had not occurred under the catalytic conditions of the reduction, the tritiated *S* enantiomer was mixed with unlabeled *R*,*S*-warfarin. The *d*-10-camphor-sulfonates were synthesized and separated chromatographically, and it was shown that the tritium in the diastereoisomer containing the *R*-warfarin enantiomer was less than 1% that in the diastereoisomer containing the *S*-warfarin enantiomer.

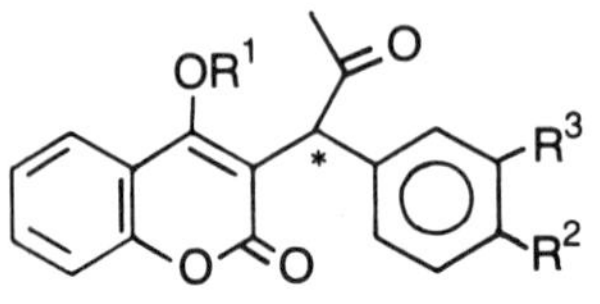

OR^1 O R^3 * O O R^2

[12]

a. $R^1 = R^2 = R^3 = H$ (warfarin)
b. R^1 = Camphorsulfonyl
$R^2 = CH_2CH_2COOMe$
$R^3 = H$
c. $R^1 = R^3 = H$
$R^2 = CH_2CH_2CONH(BSA)_{1/n}$
d. $R^1 = R^2 = H$
$R^3 = T$
*Chiral center

FIGURE 6 Warfarin, analogs, and immunogen.

R-warfarin exhibited a cross-reaction of only about 0.3% with the *S*-antiserum and *S* radioligand. *S*-warfarin cross-reacted to the extent of 3% with the *R*-warfarin antiserum and *R* radioligand. Various warfarin metabolites were shown to have low cross-reactions, with the exception of 4′-hydroxy warfarin. This latter compound is not a human metabolite, although it is found in rats in low concentration after a single dose of warfarin. The method could be used to determine the half-life of the individual enantiomers in rats given racemic drug; the resulting ratios of half-lives were in accord with those previously reported after administration of individual enantiomers. This enantioselective immunoassay was used in humans to demonstrate that vaccination against influenza did not change the relative pharmacokinetics of warfarin enantiomers (46).

The development of the warfarin immunoassays illustrates several points that are of value in development and use of enantioselective assays. In assays of this type, not only must enantioselectivity be considered, but also the cross-reaction with metabolites is still of importance. As in any RIA, high-specific-activity radioligand is required for the best sensitivity. The use of optically pure radioligand is a further advantage in enantioselectivity. The standard samples used for competitive binding assays must also be essentially optically pure. Otherwise, misleadingly high cross-reactions may be observed.

Two rabbits were challenged with each immunogen, and all four rabbits gave highly enantioselective antisera, thus indicating that the use of enantiomerically pure immunogens can be relied on to generally produce antisera of high enantioselectivity.

Unpublished results of immunoadsorption studies indicate that the low cross-reactions between enantiomers are probably not due to the presence of small amounts of antiserum selective for the opposite enantiomer. Such antibodies might be generated if an animal responds strongly to a very small amount of the opposite enantiomer present as an impurity. It was observed that even a low cross-reaction may still result in some inaccuracies, particularly in instances in which one enantiomer is present in much higher concentrations than the others. Also, cross-reaction curves are not generally parallel. Thus, in the assay of *R*- and *S*-warfarin in rat plasma, it was necessary to set up procedures for correction of the observed cross-reaction. In the rat, the *R* isomer, which yields the highest cross-reaction with the opposite enantiomer, is also the one with higher plasma concentration and longer half-life. The observation that corrections are necessary, and the earlier problems discussed with antisera to racemic immunogens constitute a strong argument for preparing and using antisera to both enantiomers of a drug.

radioligand. Cross-reactions of antisera with the opposite enantiomer were less than 2%. When plasma from a subject given *d,l*-ephedrine was analyzed by RIA, the sum of the two enantiomer concentrations agreed closely with the total ephedrine concentration determined by GLC-electron capture detection (57).

7. Monoclonal Antibodies to Soman

With the advent of hybridoma technology, the question arises as to whether, given a racemic immunogen, one would be able to generate hybridomas producing monoclonal antibodies to both isomers with high enantioselectivity. This is supported by the observed enantioselectivity of polyclonal antisera to racemic immunogens discussed earlier and the results obtained with nicotine (67). The selectivity of such antibodies was less clear cut with a diastereoisomeric compound. Monoclonal antibodies were developed to the nerve agent soman (Fig. 8 [15a]) by immunizing mice with a phenyldiazonium analog of soman conjugated to keyhold limpet hemocyanin or bovine serum albumin (Fig. 8 [15b]). Soman has two asymmetric centers: one on carbon [C(±)] and one on phosphorus [P(±)]. One antibody-producing clone was obtained from each protein. A competitive inhibition enzyme assay was used to test binding affinity. The first monoclonal antibody had the following order of affinity: C(+)P(+) ⩽ C(−)P(+) < C(−)P(−) < C(+)P(−). The second antibody also exhibited a preference for the more toxic P(−) diastereoisomers. Unfortunately, the observed affinity constants were very low (covering a range from 5×10^3 L/mol to $< 10^3$ L/mol for the first antibody and from 5.9×10^5 to 5.2×10^4 for the second antibody). Other selectivity observations indicated that only two loci on the soman molecule (the P=O oxygen and *t*-butyl group)

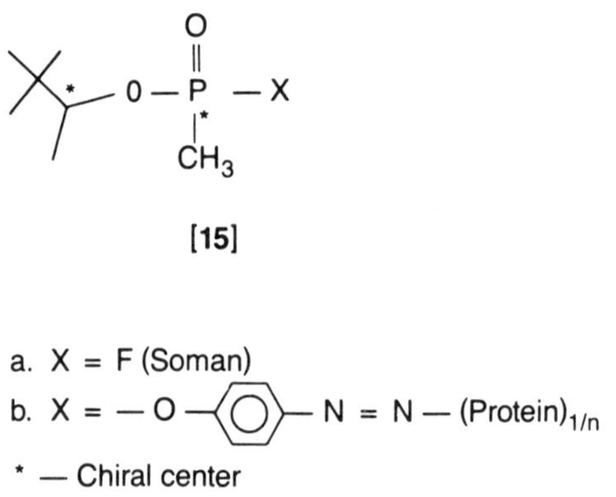

[15]

a. X = F (Soman)

b. X = —O—⟨phenyl⟩—N = N—$(\text{Protein})_{1/n}$

* — Chiral center

FIGURE 8 Soman and immunogen based on soman.

provided the major source of interaction with the antibodies. Because three loci are needed for enantioselectivity, the rather poor range of enantioselectivities may be explained. The immunogen used for this coupling contains a highly immunodominant azophenyl residue, and it is not surprising that the resulting antibodies have low affinity for the actual soman molecule itself (56).

IV. RECEPTOR BINDING ASSAYS

A. Introduction

Receptors for drug molecules generally exhibit stereoselectivity, including enantioselectivity. Thus, receptors are in many ways ideally suited for the development of enantioselective binding assays. A property of receptors that often makes them different from antibodies is the fact that receptors will tend to bind to a series of drugs in a manner correlated with their pharmacological activity. Pharmacologically active metabolites may be therefore analyzed in a receptor binding assay. Results from such assays are therefore best quoted in terms of drug equivalents unless it has been established for certain that only one analyte is involved. This property has the disadvantage that the actual analyte may be a poorly defined mixture of substances. On the other hand, it can be argued that the analysis result may have a better correlation with drug effects than a more precise measurement of a single substance, which may not represent all the pharmacologically active material present. Brief reviews of radioreceptor assays have been given (58,59).

B. Antimuscarinic Drugs

A good example of the potential advantages of radioreceptor assay (RRA) is illustrated by the case of atropine (Fig. 4 [5a]). As has been previously discussed, the generation of enantioselective antisera for this particular compound has presented problems; the drug has often been analyzed using a nonenantioselective immunoassay, even though most, if not all, of the biological activity resides in the *S* isomer. Metcalfe (60) isolated a fraction of porcine brain that contained muscarinic binding sites and had high affinity for tritium-labeled quinuclidinyl benzilate, which was used as the radioligand (40 Ci/mmol). The receptor fraction could be kept up to 1 yr at −20°C without significant loss of activity. Assay sensitivity was increased by using sequential addition (allowing the receptor to react first with the atropine and then with the radioligand). Radioligand bound to receptor was separated from unbound material by filtration through glass fiber disks, and the radioactivity in the disk was measured. Nonspecific binding (a difficulty sometimes associated with RRA) was less than 5%.

Plasma proteins were precipitated with methanol, and the methanol extract was evaporated before analysis. The sensitivity of the assay was 300 fmol in the assay tube (equivalent to 87 pg of atropine), and plasma samples could be measured down to 1.4 ng/mL (60). The assay in general responded to muscarinic antagonists and was reportedly considerably more sensitive to (−)-hyoscyamine (*S*-atropine) (60) than an RIA procedure (36).

Aaltonen et al. (61) compared RRA and RIA for atropine. These workers obtained preparations of receptor from rat brain and lyophilized them to a stable, dry form. They used the tritium-labeled quinuclidinyl benzilate at 35 Ci/mmol. The affinity constant was 0.48 nM, and by analysis of 25-μL serum samples they could obtain a sensitivity down to 1.25 ng/mL in serum. Nonspecific binding was again quite reasonable (4%) and a filtration-type separation was used. The *d* isomer of an atropine did not bind, and therefore, the cross-reaction of the *d,l* compound was 50% that of the *l* isomer. For comparison they used RIA developed by the method of Virtanen et al. (37). The immunogen was an *l*-hyoscyamine-bovine serum albumin conjugate, but the antiserum was sensitive to both *d,l* and *l* isomers. Racemic tritium-labeled atropine was used as the radioligand.

When plasma samples were analyzed by both techniques, the RIA showed much greater concentrations in plasma than the RRA. The AUC calculated from RIA was 104 μg·h/L but only 28.9 μg·h/L when calculated from RRA. The RRA results fit a three-compartment model and the volume of distribution by RRA was two times that found by RIA. Half-lives calculated from the two methods were about the same, but the clearance by the RRA was about three times that calculated from the RIA results (61). Clearly, in this case, more valuable results were obtained by an enantioselective RRA than a poorly selective or nonselective RIA.

C. β-Adrenergic Antagonists

The beta blocker drugs often contain an asymmetric center, since many of them are based on an aminodihydroxypropane structure. Nahorski et al. (62) obtained membranes from bovine lung tissue that contained beta receptors. In combination with tritium-labeled (−)-dihydroalprenolol (48–58 Ci/mmol), these membrane fragments were used to develop an assay applicable to several beta blocking drugs. *l*-Propranolol (Fig. 4 [6a]) competed very well with the radioligand for binding sites (K_i 0.8 nM vs. k_d of 0.95 nM for dihydroalprenolol). *d*-Propranolol competed only about 1.3% as well, with a K_i of 60 nM. A series of other beta blocking drugs were examined. Although in several instances, the *l* isomer was tested (timolol, isoprenoline, adrenaline, and noradrenaline), the *d* isomer was not similarly examined. In general, however, the compounds tested displaced the radioligand with affinities that matched their pharmacological potency. The assay was used to measure *l*-propranolol in volunteers who took 40 mg of *d,l*-propranolol. The lack of interference of the *d* isomer was established by analyzing plasma from volunteers given 40 mg of *d*-propranolol. In this case, the apparent concentration of propranolol was essentially indistinguishable from zero. Proteins did not have significant effects on the analysis of *l*-propranolol. This suggests that dissociation from plasma proteins occurs during the assay, since propranolol is highly protein bound.

4-Hydroxypropanolol (Fig. 4 [6b]) is an active metabolite that contributes to the beta blocking effect of propranolol. Its affinity (as the racemate) for the beta receptor in this instance was only 5% that of propranolol. If we assume that its enantioselectivity was similar to that of the parent drug, this would be equivalent to about 10% cross-reactivity for the *l* metabolite. It would appear that 4-hydroxypropanolol would contribute little to the apparent propranolol equivalents in plasma unless its concentration was markedly greater than that of the parent drug.

Another radioreceptor assay for propranolol used an ^{125}I radioligand (iodohydroxybenzylpindolol) and β-receptor sites were obtained from turkey erythrocyte membranes (63). The membranes were stable indefinitely when stored at −80°C in buffer and could be repeatedly frozen and thawed. Glass fiber filters were used to separate free and bound radioligand. Serum samples inhibited binding to the membrane very readily, and it was necessary to extract propranolol from serum before assay. Indications were that some constituent of serum destroyed the ability of the radioligand to bind to the β-receptors. Most of the interference could be avoided by heating serum to 95°C. The sensitivity limit of the assay was about 0.25–0.5 ng/mL. The assay was selective for the *l*-isomer, although high concentrations of *d*-propranolol could slightly inhibit binding. It was more sensitive to propranolol than to its metabolites 4-hydroxypropanolol or desisopropylpropanolol. The conditions used for treatment of plasma and extraction appeared to get rid of these compounds, so the authors concluded that mainly *l*-propranolol was being analyzed (63).

The method correlated well with both a high-performance liquid chromatography and an RIA procedure, with a slope approximating 1 in each case. The HPLC procedure at least measures both enantiomers. The most probable explanation of this result is that the enantiomers differ little in concentration in human plasma under the conditions of the experiment. Since racemic propranolol was used for generating the standard displacement curve, inability to detect the *d* isomer would be compensated for.

This procedure was modified to measure total β-adrenergic blocking activity in plasma as well as the separate contributions of propranolol and 4-hydroxypropanolol (64). The cross-reaction of 4-hydroxypropanolol with this receptor protein is about 25% that of propranolol itself (probably 50% with the *l* enantiomer, since racemic 4-hydroxypropanolol was used as a standard). Pharmacologically inactive metabolites of propranolol do not react with the receptor. 4-Hydroxypropanolol was stabilized in plasma by addition of sodium bisulfite and could then be recovered quantitatively by extraction. Alternatively, oxidation of plasma with dilute hydrogen peroxide destroyed 4-hydroxypropanolol without affecting propranolol concentration. Thus, by use of both of these techniques, it was possible to obtain 4-hydroxypropanolol concentrations by difference.

In spite of the demonstration that at least the receptor assays using receptor from bovine lung measure very specifically *l*-propranolol (62), most other workers appear to have used the racemic material as a standard and to have recorded assay results in terms of racemic substance. That this procedure can give usable results is supported by the generally good correlation between the receptor assay methodology and other nonenantioselective techniques such as GLC or fluorometry. However, it would seem that given the enantioselective properties of the receptor system, it would behoove researchers to analyze and report *l*-propranolol or *l*-propranolol equivalents whenever possible.

D. Miscellaneous

Calcium channel antagonist drugs are a relatively new class of therapeutic agents, all of which influence the binding of tritium-labeled nitrendipine to receptor sites. Many of these drugs are not asymmetric but others, such as verapamil, are. RRA are available for these compounds (65,66).

Other drug classes for which RRA have been reported include benzodiazepines, neuroleptics, and tricyclic antidepressants. Many of the drugs in these classes are not asymmetric and information on enantioselectivity is sparse. Barnett and Nahovski (59) give references to these assays.

E. Conclusions

As more receptors are identified and isolated, the importance of RRA will likely increase, and they will be applied to enantiomer determination. As with enantioselective immunoassays, careful attention must be paid to the optical purity of standards and radioligands. The potential for interference by the inactive enantiomer should be considered. When possible, results should be obtained by use of a standard curve of the enantiomer and reported as concentration of the enantiomer. A disadvantage of RRA is its lack of ability to measure the pharmacologically inactive enantiomer, which may be important for toxicological reasons. The ability of RRA to measure drug equivalents, including active metabolites, may help or hurt, depending on the purpose of the assay.

REFERENCES

1. C. E. Cook, C. R. Tallent, E. Amerson, G. Taylor, and J. A. Kepler, Quinidine radioimmunoassay: Development and characterization of antiserum, *Pharmacologist*, *17*:A241 (1975).
2. R. S. Yalow and S. A. Berson, Assay of plasma insulin in human subjects by immunological methods, *Nature* (London), *184*:1648–1649 (1959).
3. *Methods in Enzymology* (S. P. Colonick and N. O. Kaplan, eds.), *Immunochemical Techniques, Parts A-E* (J. J. Langone and H. V. Vanakis, eds.), Vols. 70, 73, 74, 84, and 92, Academic Press, New York, 1980–1983.
4. A. J. Moss, G. V. Dalrymple, and Charles M. Boyd, *Practical Radioimmunoassay*, C. V. Mosby Co., St. Louis, Mo., 1976.
5. D. S. Skelley, L. P. Brown, and P. K. Besch, Radioimmunoassay, *Clin. Chem.*, *19*:146–186 (1973).
6. V. P. Butler, Jr., The immunological assay of drugs, *Pharmacol. Rev.*, *29*:104–184 (1978).
7. International Atomic Energy Agency, *Radioimmunoassay and Related Procedures in Medicine*, Vols. I and II, New York, 1974 (unpublished).
8. R. P. Ekins, G. B. Newman, and J. L. H. O'Riordan, Theoretical aspects of "saturation" and radioimmunoassay, *Radioisotopes in Medicine: In Vitro Studies* (R. L. Hayes, F. A. Gorewitz, and B. E. P. Murphy, eds.), U. S. Atomic Energy Commission Conference 67111, 1968, pp. 59–100.
9. D. Rodbard, Statistical quality control and routine data processing for radioimmunoassays and immunoradiometric assays, *Clin. Chem.*, *20*:1255–1270 (1974)
10. D. Rodbard, R. H. Lenox, H. L. Wray, and D. Ramseth, Statistical characterization of the random errors in the radioimmunoassay dose-response variable, *Clin. Chem.*, *22*:350–358 (1976).
11. B. F. Erlanger, Principles and methods for the preparation of drug-protein conjugates for immunological studies, *Pharmacol. Rev.*, *25*:271–280 (1973).
12. B. F. Erlanger, The preparation of antigenic hapten-carrier conjugates: A survey, *Methods Enzymol.*, *70*:85–104 (1980).
13. M. Ollerich, Enzyme immunoassay: A review, *J. Clin. Chem. Clin. Biochem.*, *22*:895–904 (1984).
14. C. Blake and B. J. Gould, Use of enzymes in immunoassay techniques, a review, *Analyst*, *109*:533–547 (1984).
15. D. S. Smith, M. H. H. Al-Hakiem, and J. Landon, A review of fluoroimmunoassay and immunofluorometric assay, *Ann. Clin. Biochem.*, *18*:253–274 (1981).
16. R. F. Schall, Jr. and H. J. Tenoso, Alternatives to radioimmunoassay: Labels and methods, *Clin. Chem.*, *27*:164–167 (1981).
17. M. Pourfarzaneh, R. S. Kamel, J. Landon, and C. C. Dawes, The use of magnetizable particles in solid-phase immunoassay, *Methods. Biochem. Anal.*, *28*: 267–295 (1982).

18. E. A. Kabat, Basic principles of antigen/antibody reactions, *Methods Enzymol.*, *70*:3–49 (1980).
19. K. Landsteiner, *The Specificity of Serological Reactions* (revised ed.), Dover Press, New York, 1962.
20. G. Kohler and C. Milstein, Continuous cultures of fused cells secreting antibody of predetermined specificity, *Nature* (London), *256*:495–497 (1975).
21. R. P. Ekins and S. Dakubu, The development of high-sensitivity pulsed-light, time-resolved fluoroimmunoassays, *Pure Appl. Chem.*, *57*:473–482 (1985).
22. C. C. Harris, R. H. Yolken, H. Krokan, and I. C. Hsu, Ultrasensitive enzymatic radioimmunoassay: Application to detection of cholera toxin and rotavirus, *Proc. Natl. Acad. Sci., USA*, *76*:5336–5339 (1979).
23. M. Maeda and A. Tsuji, Radioimmunoassay of cyclazocine and stereospecificity of antibody, *J. Pharm. Dyn.*, *4*:167–174 (1981).
24. C. E. Cook, T. P. Seltzman, C. R. Tallent, and J. D. Wooten, III. Immunoassay of racemic drugs: A problem of enantioselective antisera and a solution, *J. Pharmacol. Exp. Therap.*, *220*:568–573 (1982).
25. K. L. Rominger and H. J. Albert, Radiological determination of fenoterol. Part I. Theoretical fundamentals, *Arzneim. Forsch./Drug Res.*, *35*:415–420 (1985).
26. A. DeLean, P. J. Munson, and D. Rodbard, Simultaneous analysis of families of sigmoidal curves: Application to bioassay, radioligand assay and physiological dose-response curves, *Am. J. Physiol.*, *235*:E97–E102 (1978).
27. J. J. Pratt, M. G. Woldring, R. Boonman, and J. Kittikool, Specificity of immunoassays. II. Heterogeneity of specificity of antibodies and antisera used for steroid immunoassay and the selective blocking of less specific antibodies, including a new method for the measurement of immunoassay specificity, *Eur. J. Nuc. Med.*, *4*:159–170 (1979).
28. S. J. Gross and J. R. Soares, Hapten determinants and purity—the key to immunologic specificity, *Immunoassays for Drugs Subject to Abuse* (S. J. Mulé, I. Sunshine, M. Braude, and R. E. Willette, eds.), CRC Press, Cleveland, Ohio, 1974, pp. 3–11.
29. T. Niwaguchi, Y. Kanda, T. Kishi, and T. Inoue, Determination of *d*-methamphetamine in urine after administration of *d*- or *d,l*-methamphetamine to rats by radioimmunoassay using optically sensitive antiserum, *J. Forensic Sci.*, *27*:592–597 (1982).
30. K. L. McGilliard and G. D. Olsen, Stereospecific radioimmunoassay of α-*l*-acetylmethadol (LAAM), *J. Pharmacol. Exp. Therap.*, *215*:205–212 (1980).
31. R. B. Barlow, *Introduction to Chemical Pharmacology*, Wiley, New York, 1963, p. 211.
32. P. B. Marshall, *Br. J. Pharmacol.*, *10*:270 (1955).
33. A. Huhtikangas, T. Lehtola, R. Virtanen, and P. Peura, Application of radioimmunoassay to racemization studies, *Finn. Chem. Lett.*, 63–66 (1982).
34. A. Fasth, J. Sollenberg, and B. Sörbo, Production and characterization of antibodies to atropine, *Acta Pharm. Suec.*, *12*:311–322 (1975).
35. L. Berghem, U. Berghem, B. Schildt, and B. Sörbo, Plasma atropine concentrations determined by radioimmunoassay after single-dose IV and IM administration, *Br. J. Anaesth.*, *52*:597–601 (1980).

36. R. J. Wurzburger, R. L. Miller, H. G. Boxenbaum, and S. Spector, Radioimmunoassay of atropine in plasma, *J. Pharmacol. Exp. Therap.*, *203*:435–441 (1977).
37. R. Virtanen, J. Kanto, and E. Iisalo, Radioimmunoassay for atropine and *l*-hyoscyamine, *Acta Pharmacol. Toxicol.*, *47*:208–212 (1980).
38. K. Kawashima, A. Levy, and S. Spector, Stereospecific radioimmunoassay for propranolol isomers, *J. Pharmacol. Exp. Therap.*, *196*:517–523 (1976).
39. A. Levy, S. H. Ngai, A. D. Finck, K. Kawashima, and S. Spector, Disposition of propranolol isomers in mice, *Eur. J. Pharmacol.*, *40*:93–100 (1976).
40. J. W. A. Findlay, J. T. Warren, J. A. Hill, and R. M. Welch, Stereospecific radioimmunoassays for *d*-pseudoephedrine in human plasma and their application to bioavailability studies, *J. Pharm. Sci.*, *70*:624–631 (1981).
41. K. D. Wing and B. D. Hammock, Stereoselectivity of a radioimmunoassay for the insecticide *S*-bioallethrin, *Experientia*, *35*:1619–1620 (1979).
42. J. J. Langone and H. Van Vanakis, Radioimmunoassay of nicotine and its metabolites, *Methods*, *84*:628–640 (1982).
43. S. Matsukura, N. Sakamoto, H. Imura, H. Matsuyama, T. Tamada, T. Ishiguro, and H. Muranaka, Radioimmunoassay of nicotine, *Biochem. Biophys. Res. Commun.*, *64*:574–580 (1975).
44. J. J. Langone, H. B. Gjika, and H. A. Vaunakis, Nicotine and metabolites, radioimmunoassays for nicotine and cotinine, *Biochemistry*, *12*:5025–5030 (1973).
45. C. E. Cook, N. H. Ballentine, T. B. Seltzman, and C. R. Tallent, Warfarin enantiomer disposition: Determination by stereoselective radioimmunoassay, *J. Pharmacol. Exp. Therap.*, *210*:391–398 (1979).
46. P. Kramer, M. Tsuru, C. E. Cook, C. J. McClain, and J. L. Holtzman, Effect of influenza vaccine on warfarin anticoagulation, *Clin. Pharmacol. Therap.*, *35*:416–418 (1984).
47. L. C. Mark, L. Brand, J. M. Perel, and F. I. Carroll, Barbiturate stereoisomers: Direction for the future?, *Excerpta Med. Int. Congr. Ser.*, *399*:143–146 (1976).
48. H. D. Christensen and I. S. Lee, Anesthetic potency and acute toxicity of optically active disubstituted barbituric acids, *Toxicol. Appl. Pharmacol.*, *26*:496–503 (1973).
49. C. E. Cook and C. R. Tallent, Synthesis of (*R*)-5-(2′-pentyl)-barbituric acid derivatives of high optical purity, *J. Heterocycl. Chem.*, *6*:203–206 (1969).
50. F. I. Carroll and R. Meck, Synthesis and optical rotatory dispersion studies of (*S*)-5-(2′-pentyl)barbituric acid derivatives, *J. Org. Chem.*, *34*:2676–2680 (1969).
51. C. E. Cook, C. R. Tallent, and T. B. Seltzman, Immunoassay as a tool for analysis of enantiomers: Stereoselective radioimmunoassay for *R*- and *S*-pentobarbital and analogs, *7th Internat. Congr. Pharmacol.*, A424 (1987).
52. C. E. Cook, M. A. Myers, C. R. Tallent, T. Seltzman, and A. R. Jeffcoat, Secobarbital enantiomers: Stereoselective radioimmunoassay in human saliva and plasma, *Fed. Proc.*, *38*:A2713 (1979).
53. C. E. Cook, Stereoselective drug analysis, *Topics in Pharmaceutical Sciences 1983* (D. D. Breimer and P. Speiser, eds.), Elsevier, New York, 1983, pp. 87–98.

54. F. Bartos, D. Bartos, G. D. Olsen, B. Anderson, and G. D. Daves, Jr., Stereospecific antibodies to methadone. II. Synthesis of *d*- and *l*-methadone antigens, *Res. Commun. Chem. Path. Pharmacol.*, *20*:157–164 (1978).
55. K. L. McGilliard, J. E. Wilson, G. D. Olsen, and F. Bartos, Stereospecific radioimmunoassay of *d*- and *l*-methadone, *Proc. West. Pharmacol. Soc.*, *22*:463–466 (1979).
56. A. A. Brimfield, K. W. Hunter, Jr., D. E. Lenz, H. P. Benschop, C. VanDijk, and L. P. A. deJong, Structural and stereochemical specificity of mouse monoclonal antibodies to the organophosphorus cholinesterase inhibitor soman, *Molec. Pharmacol.*, *28*:32–39 (1985).
57. K. K. Midha, J. W. Hubbard, J. K. Cooper, and C. Mackonka, Stereospecific radioimmunoassays for *l*-ephedrine and *d*-ephedrine in human plasma, *J. Pharm. Sci.*, *72*:736–739 (1983).
58. K. Ensing and R. A. deZeeuw, Radioreceptor assay—A tool for the bioanalysis of drugs, *Trends Anal. Chem.*, *3*:102–106 (1984).
59. D. B. Barnett and S. R. Nahovski, Drug assays in plasma by radioreceptor techniques, *Trends Pharmacol. Sci.*, *4*:407–409 (1983.
60. R. F. Metcalfe, A sensitive radioreceptor assay for atropine in plasma, *Biochem. Pharmacol.*, *30*:209–212 (1981).
61. L. Aaltonen, J. Kanto, E. Iisalo, and K. Pihlajamäki, Comparison of radioreceptor assays and radioimmunoassay for atropine: Pharmacokinetic application, *Eur. J. Clin. Pharmacol.*, *26*:613–617 (1984).
62. S. R. Nahorski, M. I. Batta, and D. B. Barnett, Measurement of β-adrenoceptor antagonists in biological fluids using a radioreceptor assay, *Eur. J. Pharmacol.*, *52*:393–396 (1978).
63. J. P. Bilezikian, D. E. Gammon, C. L. Rochester, and D. G. Shand, A radioreceptor assay for propranolol, *Clin. Pharmacol. Therap.*, *26*:173–180 (1979).
64. C. L. Rochester, D. E. Gammon, E. Shane, and J. P. Bilezikian, A radioreceptor assay for propranolol and 4-hydroxypropranolol, *Clin. Pharmacol. Therap.*, *28*:32–39 (1980).
65. H. R. Lee, W. R. Roeske, and H. Y. Yamamura, The measurement of free nitrendipine in human serum by an equilibrium dialysis-radioreceptor assay, *Life Sci.*, *33*:1821–1829 (1983).
66. R. J. Gould, K. M. M. Murphy, and S. H. Snyder, A simple sensitive radioreceptor assay for calcium antagonist drugs, *Life Sci.*, *33*:2665–2672 (1983).
67. R. J. Bjercke, G. Cook, N. Rychlik, H. B. Gjika, H. Van Vunakis, and J. J. Langone, Stereospecific monoclonal antibodies to nicotine and cotinine and their use in enzyme-linked immunosorbent assays, *J. Immunol. Meth.*, *90*:203–213 (1986).
68. C. E. Cook, T. B. Seltzman, C. R. Tallent, B. Lorenzo, and D. E. Drayer, Pharmacokinetics of pentobarbital enantiomers as determined by enantioselective radioimmunoassay after administration of racemate to humans and rabbits, *J. Pharmacol. Exp. Therap.*, *241*:779–785 (1987).

4

Indirect Methods for the Chromatographic Resolution of Drug Enantiomers

Synthesis and Separation of Diastereomeric Derivatives

Joseph Gal *University of Colorado School of Medicine, Denver, Colorado*

I. INTRODUCTION

When two chiral compounds, racemic A and racemic B, react to form covalent adducts in a reaction not affecting the asymmetric centers, the stereochemical course of the reaction may be depicted as follows:

$$(\pm)\text{-A} + (\pm)\text{-B} \rightarrow \underset{[1]}{[+\text{A}+\text{B}]} + \underset{[2]}{[+\text{A}-\text{B}]} + \underset{[3]}{[-\text{A}+\text{B}]} + \underset{[4]}{[-\text{A}-\text{B}]}$$

Four different products, [1–4], are possible. Derivatives [1] and [4] are enantiomers of each other, as are [2] and [3]. Any two of the derivatives that are not enantiomerically related, for example, [1] and [3], are diastereomeric pairs. Nonchiral chromatographic systems can separate diastereomers, but not enantiomers. Thus, chromatographic analysis of a mixture of [1–4] in a nonchiral chromatographic system can yield, in principle, two peaks: one consisting of [1] and [4], and the other of [2] and [3].

The above forms the basis for the chromatographic separation of enantiomers via diastereomeric derivatives. A mixture of the enantiomers of A is reacted, before chromatographic analysis, with a pure enantiomer of B, for example, (−)-B, to produce, in this case, diastereomers [2] and [4]. In this theoretical example, A may represent a sample of a drug whose enantiomeric composition is to be determined, and B is the chiral derivatizing agent (CDA) used to convert the enantiomers of A to diastereomeric derivatives. The diastereomeric products are then separated using a nonchiral chromatographic system. The ratio of diastereomeric peak areas (if we assume equal response to the diastereomers by the detector used) is the enantiomeric ratio of A in the sample analyzed. This approach to enantiomeric analysis is often termed the indirect method, inasmuch as the enantiomers are converted to diastereomers, and it is the diastereomeric species, rather than the enantiomers per se, that are separated chromatographically.

The indirect chromatographic separation of enantiomers has been used for many years (1–7). However, the spectacular developments during the past 10 years in the synthesis and application of chiral chromatographic stationary phases have revolutionized the field of enantiospecific analysis and provided a wide array of alternatives to the indirect method. Nevertheless, as predicted several years ago (7), the indirect approach continues to enjoy wide popularity in enantiospecific drug analysis. There are several reasons for this popularity of indirect chromatographic separation methods. Thus, despite the increased availability of direct methods—based primarily on chiral stationary phases—many separations have been achieved only via an indirect method; in many cases, the increased sensitivity in detection that may be provided by the CDA is essential; some of the new chiral high-performance liquid chromatography (LC) columns, although providing the desired enantioselectivity for a given separation problem, may not offer adequate chemical selectivity, thereby requiring additional purification steps (e.g., column-switching, etc.); some of the chiral columns have a relatively short lifetime and/or are expensive; in many cases, the use of a chiral column still requires derivatization (with a nonchiral reagent) in order to obtain chiral recognition, good chromatographic behavior, or sufficient sensitivity; a large variety of CDAs are available; in many cases, both enantiomers of the CDA are available, an advantage not shared by most chiral LC and gas-liquid-chromatography (GLC) columns. In summary, then, while it is unquestionable that the direct separation of enantiomers is conceptually superior to the indirect approach, the latter has a well-deserved niche in the armamentarium of the biomedical analyst.

In this chapter, the use of derivatization with CDAs will be reviewed, with emphasis on pharmaceutical and pharmacological applications. The intent here is not the exhaustive listing of publications in this field but rather the illustration of principles and applications, including discussion of recent and interesting examples.

II. GENERAL CONSIDERATIONS

There are several important requirements that must be met for the successful use of the indirect chromatographic separation of enantiomers.

A. The Derivatization Reaction

Derivatization with a CDA results in the chemical modification of the analytes. It is important, therefore, to ascertain that the expected products are obtained. Thus, when a new derivatization reaction is carried out or a new CDA is first used, it is essential to confirm rigorously the structure of the derivatives using appropriate analytical techniques (mass spectrometry, nuclear magnetic resonance, elemental analysis, etc.). This is particularly important in complex cases, for example, when more than one functional group in the analyte may react with the CDA.

The derivatization reaction should proceed under sufficiently mild conditions to avoid significant degradation of the reactants. Most important, under the reaction conditions used, racemization or epimerization of the chiral components, that is, the drug to be derivatized, the chiral reagent, or the derivatives, must not occur. Ideally, the reaction of the analyte enantiomers with the CDA should proceed at the same rate. The derivatization reaction should proceed until completion, that is, until all the compound to be derivatized has reacted with the CDA. This will obviate any concerns about kinetic resolution, that is, unequal rates of reaction of the enantiomers with the CDA. In practice, this requirement usually means use of excess CDA. This, in turn, may have significant implications for the chromatographic separation of the excess CDA from the derivatives formed. It is important, when developing a derivatization reaction, to study the time course of the reaction, that is, to demonstrate that under the conditions used the reaction proceeds to completion.

In rare instances, the derivatization with a CDA was carried out enzymatically (see below). Clearly, the requirements for optimization of such reactions are different from those of nonenzymatic chemical transformations and should be carefully examined before an enzymatic derivatization reaction is routinely applied for quantification.

B. The Chiral Derivatizing Agent

The CDA must also meet some important requirements. It should be of high chemical purity and stable under the derivatization conditions used, as discussed above. Furthermore, the CDA should also be chemically and stereochemically stable under the storage conditions used. In addition, the CDA should react with the target functional group readily and selectively. It is highly desirable that the derivatives formed should produce an equal response by the detection system used.

A critically important question in using a CDA is its enantiomeric purity. If we refer to the original theoretical example of the equation above, it is clear that if CDA (−)-B were contaminated with a small amount of (+)-B, the products of the reaction with a mixture of the enantiomers of A would include, in addition to [2] and [4] derived from (−)-B, compounds [1] and [3] derived from (+)-B. As pointed out earlier, in nonchiral chromatographic systems [1] is not separable from [4], and [2] is not separable from [3]. Thus, the presence of contaminant (+)-B in (−)-B will cause a contamination of the peaks of [2] and [4] with [3] and [1], respectively. Obviously, this will result in a false value for the enantiomeric composition of A.

It may be argued that if the actual extent of enantiomeric contamination of a CDA is known accurately, the reagent may be safely used, because the appropriate correction in diastereomeric peak ratios can be made. An objection (5) to this argument is that if the enantiomerically impure CDA is present in excess, differences, if any, between the CDA enantiomers in their reaction rates with the analyte enantiomers (i.e., diastereoselective kinetics) will still result in an error in the determination of the enantiomeric ratio. In practice, however, such kinetic differences are usually negligible. A more precise but cumbersome solution to this problem is to separate the four stereoisomeric derivatives using chiral chromatographic conditions, for example, a chiral stationary phase. Under such conditions, four distinct peaks are obtainable as a matter of principle (whether the four stereoisomers are actually resolved depends, of course, on the chromatographic conditions chosen). A review of the literature indicates that small (1–2%) enantiomeric contamination of a CDA may not necessarily render the CDA useless in many applications. It is clear, nevertheless, that the CDA used should be enantiomerically pure whenever possible. This simplifies the analysis and eliminates any uncertainty associated with enantiomeric contamination. There is, in fact, an application in which enantiomerically impure CDAs cannot be used safely: the determination of trace enantiomeric impurity in an analyte. If the CDA used is itself enantiomerically contaminated, the accurate determination of the extent of trace enantiomeric contamination of the analyte may be difficult if not impossible.

Although the danger posed by enantiomeric contamination of the CDA has been pointed out frequently, it must be emphasized that many CDAs are available in optically pure form. For example, CDAs derived from natural products, such as those based on D-glucose, (−)-methanol, etc. (see below), are often available in enantiomerically pure form.

The enantiomeric purity of a CDA is best determined by chromatographic means. This may be accomplished (1) by chromatographic separation of the enantiomers of the CDA using direct methods, that is, a chiral stationary phase or chiral mobile phase; for this method, both enantiomers of the CDA should be available to demonstrate reliably their separability; or (2) indirectly, by derivatizing the CDA with a compound known to be enantiomerically pure. Here also, both enantiomers of the CDA, or alternatively, those of the compound used in the derivatization, should be available to demonstrate chromatographic separation and thereby detectability of enantiomeric contamination.

There are other requirements the CDA must meet. The derivatives produced by the CDA must have suitable chromatographic properties. In GLC applications, for example, the derivatives must have adequate volatility. Because derivatizations with CDAs frequently involve acylation, esterification, etc., of such polar centers as hydroxyl, carboxyl, amino, etc., groups of the analytes, the resulting derivatives, often display improved chromatographic behavior relative to the parent compounds. In some cases, however, the diastereomeric derivatives formed still contain functional groups that interfere with the chromatographic separation, for example, via insufficient volatility, peak tailing, inadequate resolution of diastereomers, etc. Such additional functional groups must also be derivatized or masked if the diastereomers are to be successfully separated. This usually involves masking the troublesome polar groups via derivatization with nonchiral agents. Thus, carboxyl groups may be esterified with diazomethane, hydroxyl groups silylated, etc. Clearly, however, such additional derivatization requirements increase the complexity of the analytical method.

It may also be necessary, especially in biological applications, for the CDA to increase the detectability of the analyte. Some CDAs have, in fact, been designed with sensitive detection in mind, for example, to produce electron-capturing derivatives for GLC or fluorescent derivatives for LC. Sensitive and selective detection is of utmost importance in applications in drug metabolism and pharmacokinetics, where drugs present at low concentrations in complex biological fluids are to be identified and quantified.

Finally, a significant consideration in the use of a particular CDA is its availability. A CDA that must be synthesized via an elaborate route and resolved into its enantiomers before it can be used clearly is not a practical candidate in many laboratories. Fortunately, however, many of the CDAs described in the literature are available from commercial suppliers. Others, while not commercially available, are easily prepared from commercially available resolved starting materials.

C. Types of Applications

Chromatographic resolution of enantiomers, including the indirect approach, can be used in several different types of applications. The recent surge of interest (8–12) in the stereochemical aspects of drug action and disposition has been accompanied by a rapidly increasing need for synthetic and analytical methods for drug stereoisomers. Regarding analytical methods, chromatographic separation of enantiomers, directly or indirectly, has clearly become the technique of choice. As for the preparation of optically active compounds, although asymmetric synthesis is now a practical possibility in many cases (13), preparative LC separation of enantiomers is rapidly becoming an important alternative to chemical resolution and asymmetric synthesis, at least on a laboratory scale (3).

1. *Determination of Enantiomeric Composition*

The enantiomeric purity or composition of drugs, synthetic intermediates, analogs, etc., in bulk samples can be determined via chiral derivatization in laboratory research, manufacturing processes, dosage forms, quality control, etc. In these applications, limitations on the amount of material available are usually nonexistent, allowing for convenient derivatization and ready detection. In such derivatizations, 0.1–1.0 mg is typically the amount of analyte derivatized.

Another major role for the chromatographic analysis of enantiomeric composition is in the areas of drug metabolism and pharmacokinetics. Many drugs, in research or in clinical use, are applied in their racemic form; chiral or nonchiral drugs may be metabolized to chiral metabolites, etc. Thus, there is a need for the stereospecific detection, identification, and quantification of the individual enantiomers of drugs and their metabolites in various biological media. The concentration of these analytes can often be as low as a few nanograms per milliliter in complex biological fluids containing many other substances. Sensitivity and specificity are therefore critical factors in such analyses. Derivatization with a CDA, in addition to permitting stereochemical analysis, can also improve the chromatographic behavior and detectability of the analytes, an important consideration in the analysis of xenobiotics in biological media. It is important to recognize, however, that derivatization with a CDA in such trace analysis procedures can be very different from derivatizing, for example, a milligram of a pure standard. In fact, the limitations of chiral derivatization in this respect are in many ways very similar to those encountered in nonchiral chemical derivatization for chromatographic trace analysis, which have been discussed (14,15).

2. *Determination of Absolute Configuration*

Determination of absolute configuration is another important application of the chromatographic separation of diastereomeric derivatives of enantiomers. This can be accomplished in several different ways. Obviously, if a standard of known configuration is available for comparison, the configuration of a compound of unknown stereochemical identity in a given sample can be readily assigned. If an authentic standard of known configuration is not available, the absolute configuration of the analyte may be determined by separating and isolating the derivatives using LC, followed by determination of the configuration using physical measurements such as circular dichroism and nuclear magnetic resonance (NMR).

Still another approach to the determination of absolute configuration is based on the elution order of the diastereomeric derivatives. When a series of closely related chiral compounds is derivatized with the same CDA, it is often observed that the order of elution of the two diastereomers is the same for all members of the series, that is, the elution order is correlated with configuration. Indeed, this phenomenon was recognized early (16), and empirical correlations of elution order with configuration are frequently used to assign absolute configurations (17,18). Caution must be exercised, however, in using this approach to assign configurations, since occasionally an "unexpected" reversal in the elution order of the derivatives is observed (19). Indeed, Pirkle and Finn have pointed out that in order to assign with confidence the absolute configurations on the basis of elution order of the diastereomers, one must know the absolute configuration of the CDA and the mechanism of chromatographic separation for the relevant stereoisomers (3).

3. *Preparative Chromatography*

The recovery of useful amounts of the chromatographically separated components is a capability provided by a variety of chromatographic techniques, including GLC, thin-layer chromatography (TLC), and LC. It is the latter technique, however, that can most readily serve in the preparative mode for the isolation of relatively large amounts of material. There is no doubt, however, that the requirement in the indirect approach to cleave chemically the derivatizing moiety from the derivatives after chromatographic separation is a significant disadvantage and deterrent to the preparative use of this technique. Nevertheless, such preparative separations have been described in the literature and in many instances can provide useful alternatives to preparative direct separations, especially when one considers the lack of availability of preparative-size versions of many chiral LC columns or the often prohibitively high price of those that are available.

There are several important considerations in the design of preparative separations via derivatization with CDAs. First, the CDAs should produce diastereomeric derivatives that, after the chromatographic separation, can be cleaved chemically to yield the original drug enantiomers. The components resulting from the reaction used to cleave the derivatives should be separable without undue difficulty. An important factor is the enantiomeric purity of the CDA, since enantiomeric contamination of the CDA will result in enantiomeric contamination of the drug stereoisomer(s) retrieved. Depending on the scale used, the cost of the CDA may become a consideration.

III. SEPARATIONS

A. General Considerations

Two different measures of the extent of chromatographic separation of two peaks are used in characterizing stereoisomer separations. The separation factor α is the ratio of the retention times corrected for the retention time of an unretained substance, whereas the peak resolution R is a measure of the extent of overlap of the two peak areas (20). The two parameters α and R contain different information (6,21). If two equal-sized peaks have $R >$ 1.5, the two peaks are essentially completely separated (22). When developing a chromatographic separation for the enantiomers of a given drug, one should determine R under the chromatographic conditions used to be able to judge the extent of peak separation. Ideally, R should be > 1.5 for the racemate if the chromatographic peaks are to be used in quantification.

The chromatographic separation of enantiomers as diastereomers was first developed using GLC (1). Subsequently, many separations using GLC were reported, but modern LC has dominated the field of enantiospecific drug analysis in recent years. Nevertheless, the arrival of high-resolution capillary GLC has revived interest in the use of indirect enantiomer separation via this type of chromatography, and today GLC remains important in the analytical separation of enantiomers after derivatization with CDAs. The availability of sensitive detection methods, for example, mass spectrometry, electron capture, etc., enhances the applicability of GLC in indirect enantiospecific drug analysis.

The advent of modern column LC in the 1970s rapidly led to the use of this chromatographic technique in the separation of enantiomers as diastereomeric derivatives. Today, most of the reported new developments in indirect enantiomer resolutions use LC, and LC is particularly important in the resolution of chiral pharmaceuticals.

TLC has also been used for the separation of diastereomeric derivatives of enantiomers, but this form of chromatography has not attained widespread use in indirect resolutions. Other chromatographic techniques, for example, supercritical fluid chromatography, capillary electrophoresis, countercurrent chromatography, etc., have not received much attention in indirect enantioseparation.

A variety of compounds have been resolved via the indirect method, and a large variety of chromatographic systems and conditions have been used in the separations. Relatively few of these chromatographic resolutions have been examined in detail to elucidate the physiochemical mechanism responsible for the separation of the diastereomers. Indeed, many of the pharmaceutical applications of indirect resolution were developed empirically, without available rational information on suitable CDAs and chromatographic conditions that would likely produce optimal separation. Much of the pioneering work aimed at elucidating the mechanisms responsible for the separation of diastereomers in GLC was carried out by Karger and his associates (23,24). In LC resolutions, Pirkle et al. expended considerable effort in developing models that explain the separation of diastereomeric amides, carbamates, and ureas on normal-phase columns (3,19,21,25). Other investigators have also studied the factors underlying the separability of diastereomers, and the reader is referred to recent detailed discussions of the subject (3–5,26). Additional studies on separation mechanisms, particularly in reversed-phase (RP) LC, could provide further useful information for the rational selection of conditions for the indirect resolution of pharmacologically active agents.

B. Examples of Separations

In this section, applications of the indirect-resolution approach to compounds of primarily pharmaceutical, pharmacological, or toxicological interest will be reviewed. Because a fundamental aspect of the method is the chemical reaction between the analyte and the CDA, it is convenient to divide the applications according to the drug functional groups involved in the derivatization.

1. Resolutions via Derivatization of Amino Groups

The amino group is present in a wide variety of asymmetric drugs. Furthermore, this functional group undergoes a variety of chemical reactions, many of which have been exploited in derivatizing with CDAs.

Formation of Amides. Carboxamide bond formation (Eq. 1, CDA is first reactant) appears to be the most common reaction used in chiral derivatization of primary and secondary amino groups.

$$\text{R-CO-X} + \text{HNR'R''} \rightarrow \text{R-CO-NR'R''} \quad (1)$$

This is not surprising, since amide-forming reactions between amines and (activated) carboxylic acids proceed readily, and many optically active carboxylic acids are readily available. Furthermore, the diastereomeric amides produced are often well-resolved by LC or GLC. A variety of leaving groups (X in Eq. 1) have been used in the acylation reaction, for example, chloride, imidazole, carboxylate, etc.

N-perfluoroacyl derivatives of L-proline, such as (*S*)-*N*-trifluoroacetylprolyl chloride (corresponding to L-proline), [5] (see structures at end of chapter), have been used as CDAs (27–35). The use of some of these reagents illustrates several of the potential problems associated with the indirect method. CDA [5] is well known (29,34–36) to be contaminated with the *R* enantiomer formed via racemization during synthesis and storage of the reagent. Commercial samples of [5] have been reported to contain varying amounts, sometimes as much as 15% (35), of the enantiomeric contaminant. Some investigators found that CDAs such as [5] racemized during the derivatization reaction if triethylamine was present as the catalyst (32,33). Lim et al., however, were able to develop aqueous reaction conditions under which racemization of the *N*-heptafluorobutyryl analog of [5] was negligible (33).

Other attempts to circumvent the enantiomeric contamination of CDAs such [5] include the work of Silber and Riegelman, who carefully synthesized [5] in > 98% enantiomeric purity and found that derivatization at −78°C resulted in <2% racemization (35). Others, on the other hand, found that the *N*-heptafluorobutyryl analog of [5] racemized during the derivatization reaction even at −78°C (33). Liu et al. used [5] to derivatize racemic methamphetamine and separated the four derivatives on a chiral GLC column (30). It was concluded that an achiral column is adequate for the determination of the methamphetamine enantiomers, provided that the extent of enantiomeric contamination of [5] is known (30). Attempts to use leaving groups other than chloride in [5] have been disappointing, leading to reagents that reacted sluggishly and stereoselectively with the analyte enantiomers and were still not free from racemization problems (29). More recently, however, the *N*-heptafluorobutyryl analog of [5] has been successfully used in the resolution of several primary and secondary amines. Thus, several amphetamine derivatives and ephedrines were resolved (37,38), and the determination of the concentration of methoxyphenamine and several of its metabolites by GLC and ECD were reported (39). It would seem, nevertheless, that the possibility of racemization of these reagents must be carefully checked during their use. That the problems are specific to structures such as [5] is illustrated by observations that other L-proline-based CDAs, such as [6] (40) and [7] (41), are free from racemization problems.

Reagent [6] was developed for the determination of the enantiomeric composition of amphetamines in pharmaceutical dosage forms using LC separation. The derivatives show strong UV-light absorption, but baseline separation of the diastereomers was not achieved (40). Compound [7], containing the pentafluorophenyl group, produces derivatives that elicit a high response from the electron capture detector (41) and has proven valuable in drug metabolism studies (42,43). For example, the absolute configuration of α-methyldopamine formed from α-methyl DOPA in a hypertensive patient could be determined using [7]. The metabolite, isolated from urine, was derivatized with [7], and its GLC retention time was shown to correspond exactly to that of the derivative of an authentic sample of known absolute configuration (42).

Krull and his co-workers developed a new and promising approach in derivatizations of amino groups with CDAs (44). The CDA, an amino acid derivative, was covalently linked via esterification of the carboxyl group to an insoluble, structurally rigid organic polymer backbone. The amino group of the CDA moiety was linked to a fluorescent group, for example, fluorenylmethoxycarbonyl, in order to provide sensitive detectability to the derivatives. In the derivatization reaction (aminolysis of an ester), the amino group of the analyte reacted at the labile-activated ester link between the CDA and the polymer support, severing the link with release of the derivatives, in which the analyte was now covalently attached to the CDA via the newly formed amide bond. The derivatization can be carried out on-line or off-line with respect to a LC system used for the separation of the diastereomeric derivatives. The authors carefully evaluated the nature of the CDA, its immobilization on the polymer backbone, the conditions of derivatization, etc. Such an approach has several potential advantages: The absence of excess reagent in the analyte mixture is highly advantageous, especially when trace amounts of the analyte are measured; the analysis is amenable to automation; the CDA includes a built-in moiety for sensitive detectability. Clearly, this technique has considerable promise.

α-Methoxy-α-trifluoromethylphenylacetyl chloride, [8], is a widely used and highly useful CDA. Both enantiomers of the parent, [9], are commercially available and can be converted in a simple one-step procedure to [8] (36,45). Lacking a hydrogen α to the carboxyl group, [8] is stereochemically highly stable. Applications of this CDA for the resolution of amines using packed-column GLC include the resolution of amphetamine and several related compounds (36) and DOPA and α-methylDOPA (46), studies on the stereochemical course of the metabolism of amphetamines in vivo (36,47) and in vitro (48,49), and investigations of the stereoselective disposition of the antiarrhythmic agent tocainide in animals (50) and humans (51–53). More recently, the reagent has been used in conjunction with capillary GLC. Changchit et al. derivatized 3-amino-1-phenylbutane, a metabolite of labetalol, with [8] and separated the diastereomeric derivatives on capillary GLC columns with detection by mass spectrometry (54).

The reagent has also been evaluated for the resolution of amphetamines by RP LC, but for several compounds incomplete separation of the diastereomers was obtained (55). On the other hand, *N*-acylation of praziquantel, a chiral anthelmintic agent, with [8] produced diastereomers that were well resolved (α = 1.18) using normal-phase LC in a study of the absolute configuration of the drug (56).

Coleman (57) synthesized the anhydride of acid (*R*)-[9] and used it to determine the enantiomeric purity of oxfenicine [(*S*)-(+)-*p*-hydroxyphenylglycine], a therapeutic agent that promotes carbohydrate oxidation. The anhydride had to be used in the derivatization since the use of acid chloride [8] led to racemization of the drug due to the presence of the required base catalyst. The diastereomers were separated by silica gel LC and detected at 254 nm (57).

The above-mentioned antiarrhythmic agent tocainide has also been resolved with (*R*)-(−)-*O*-methylmandelic acid chloride, [10], using normal-phase LC in a study of the stereoselective disposition of the drug (58). In the assay, CDA [10] was dissolved in the organic solvent used for the extraction of tocainide from urine (58). The *S* form of this CDA was used by Helmchen and Strubert in early studies on indirect resolutions of chiral amines using LC (59). Both enantiomers of the parent acid of [10] are commercially available.

(−)-Camphanic acid chloride, [11], was used in the RP LC determination of the enantiomeric purity of ^{11}C-labeled D- and L-methionine synthesized for positron-emission tomography studies of the brain (60). CDA [11] has received limited attention despite its commercial availability, but interest in its use appears to be increasing. Nichols et al. used [11] and capillary GLC to determine the enantiomeric purity of resolved samples of the hallucinogenic compound 1-(3,4-methylenedioxyphenyl)-2-aminopropane (MDA) and several of its homologs (61). CDA [11] has also been applied and to the determination of the enantiomeric purity of nicotine, the stereochemical aspects of which are of interest. Jacob et al. (62) developed an elegant procedure in which nicotine is *N*-demethylated to nornicotine, and the latter in turn is derivatized with [11]. The resulting diastereomers are conveniently separated on a capillary GLC column, thus providing a method for the determination of the enantiomeric composition of both nicotine and nornicotine (62). CDA [11] is a readily available agent of high enantiomeric purity, stereochemical stability, and high reactivity toward amines (and alcohols, see below), which should prove useful in the chromatographic resolution of a variety of chiral compounds.

Another readily available and potentially useful chiral reagent that has received very little attention is τ-butyloxycarbonyl-L-leucine-*N*-hyroxysuccinimide ester, [12]. This reagent was used to acylate the amino group of the chiral drug γ-vinyl-γ-aminobutyric acid, followed by trifluoroacetic acid-catalyzed removal of the t-butyloxycarbonyl group (63). The diastereomeric derivatives were resolved on a C8 column, with an α value as high as 5.75 (63). A related CDA, γ-butyloxycarbonyl-L-leucine anhydride was described by Hermansson (64,65). Acylation with this reagent produces the same derivatives as those produced by [12], the difference between the two CDAs being only in the identity of the leaving group. Hermansson's reagent was used in the determination of the enantiomers of propranolol and several related β-adrenergic antagonists in blood plasma (64,65). Interest in this area is intense as a result of the stereoselective pharmacology and disposition of these drugs. The assay procedure described by Hermansson permitted the simultaneous quantification of the propranolol enantiomers in plasma at a concentration as low as 1.0 ng/mL using LC with fluorometric detection, the fluorescence being due to the drug moiety in the derivatives rather than the CDA (64).

An interesting fluorescent CDA was developed by Weber et al. (66). These workers converted racemic benoxaprofen, [13], to the corresponding *N*-(*R*)-α-methylbenzylamide diastereomers, which were preparatively separated on silica gel LC columns. The separated diastereomers were acid-hydrolyzed to retrieve the individual enantiomers of benoxaprofen. Either enantiomer could then serve as CDA for the resolution and determination of chiral amines. The diastereomeric derivatives of several amines were highly fluorescent and well resolved on silica gel columns or by TLC. One disadvantage of the procedure was that the CDAs had slightly less than 100% enantiomeric purity.

Goto et al. (67) synthesized the succinimidyl ester [14] of (−)-α-methoxy-α-methyl-1-naphthaleneacetic acid for the normal-phase LC resolution of chiral amines. The reagent permitted the determination of the enantiomers of an amphetamine derivative in blood plasma after administration of racemic drug to rabbits. With detection at 280 nm, the lower limit of sensitivity was 5 ng/mL for each enantiomer (67). Several chiral acids from the "profen" group of nonsteroidal antiinflammatory drugs have been adapted as CDAs. One of these, naproxen, [15], is the *S* enantiomer and is commercially available as the resolved acid; several of these acids have the advantage of providing fluorescent derivatives (68,69).

Demian and Gripshover evaluated several CDAs for the LC resolution of 3-aminoquinuclidine, an intermediate in the synthesis of several pharmaceuticals (70). It was found that the best resolution was provided by *O*,*O*-dibenzoyltartaric anhydride, [16], a CDA originally developed for the derivatization of alcohols by Lindner and his co-workers (see below). Interestingly, this reagent, although commercially available as the (*R*,*R*) and also the (*S*,*S*) enantiomer, has not received much attention for amines.

Formation of Carbamates. The reaction of amines with chloroformates bears some similarity to amide formation and produces carbamates (Eq. 2)

$$\text{RO-CO-Cl} + \text{HNR}'\text{R}'' \rightarrow \text{RO-CO-NR}'\text{R}'' \quad (2)$$

Although some of the early work on the separation of diastereomers by GLC (71) or LC (19) involved carbamates, derivatization of chiral amines with chloroformate CDAs has received relatively little attention until recently in pharmaceutical/pharmacological applications. This is surprising, inasmuch as chiral chloroformates can be readily synthesized from chiral alcohols, many of which are commercially available. There are indications, however, that interest in the use of chloroformates is increasing, and several reports on such reagents have appeared in recent years.

(−)-Menthyl chloroformate, [17], is commercially available and has been used in several applications. For example, Seeman et al. (72) carried out the preparative chromatographic resolution of nornicotine using [17] as a CDA. The diastereomeric carbamates formed via the reaction of racemic nornicotine with [17] were separated by preparative silica gel LC, and the pure enantiomers of nornicotine were liberated by acid-catalyzed hydrolysis of the carbamates (72). Several reports have described the use of [17] in the enantiospecific determination of drug concentrations in biological fluids; thus, the reagent has been used in the analysis of encainide and some of its metabolites (73), flecainide (74), and propranolol (75), and other drugs (76).

An interesting chloroformate CDA that appeared recently is (+)-1-(9-fluorenyl)ethyl chloroformate, [18]. This reagent is suitable for the separation of the enantiomers of amino acids and amines via RP LC (77) and has already been applied to the determination of the β-adrenergic antagonists propranolol (78) and atenolol (79) in biological fluids. In these drugs, there is a hydroxyl group in addition to the secondary amino functionality; it was shown that only the amino group reacted when propranolol was derivatized with [18] (78). A significant advantage of [18] is that it produces fluorescent derivatives. A disadvantage, on the other hand, is that the reagent is rather expensive.

A unique property of chloroformates is their ability to *N*-dealkylate tertiary amines to produce the carbamate of a corresponding secondary amine (Eq. 3)

$$\text{RO-CO-Cl} + \text{NR}'_3 \rightarrow \text{RO-CO-NR}'_2 \quad (3)$$

Depending on the nature of the substituents on nitrogen, this reaction may give a complex product mixture, but in some cases such derivatization could be the basis of a useful indirect enantiomer separation. When one or two of the three *N*-substituents are methyl, for example, demethylation may be favored, and the reaction may be usable as a precolumn derivatization for chromatographic resolution. This approach was used in the analysis of encainide, a tertiary-amine antiarrhythmic drug (73).

Formation of Ureas. Primary and secondary amino groups undergo facile reaction with isocyanates to give the corresponding ureas (Eq. 4), and this transformation has been exploited for the derivatization of amines

$$\text{R-N=C=O} + \text{HNR}'\text{R}'' \rightarrow \text{RNH-CO-NR}'\text{R}'' \qquad (4)$$

Two commercially available optically active isocyanates have been extensively used as CDAs in LC: 1-phenylethyl isocyanate ([19], also known as α-methylbenzyl isocyanate) and 1-(1-naphthyl)ethyl isocyanate, [20]. Both enantiomers of each of these CDAs are commercially available. Thompson et al. examined the products of the reaction of propranolol, an amino-alcohol, with [19], and found that the amino group of the drug was derivatized, whereas the hydroxyl group did not react under the reaction conditions used (80). There are many reports on the use of CDA [19] and its naphthalene analog [20] for the resolution of amine drugs by LC; many of these describe applications to β-adrenergic antagonists or related drugs that are β-amino alcohols (81–89), and many of the applications involve determination of the enantiomers in biological fluids. In addition to the above β-adrenergic antagonists, several other amine drugs have been derivatized with one or the other enantiomer of [19] or [20] and the derivatives resolved on analytical and/or preparative LC columns, for example, flecainide (90), cathinone (91), noreseroline (92), mecamylamine (93), tocainide (94), fluoxetine and its metabolite norfluoxetine (95), and others (96). It is noteworthy that the naphthalene moiety in [20] provides fluorescent derivatives, an advantage over [19]. Other chiral isocyanates have also been evaluated as CDAs for amines (97).

It was shown by Shonenberger et al. that thermal decomposition of the separated diastereomeric (α-methylbenzyl) urea derivatives produced by the reaction of [19] and chiral secondary amines is a convenient technique for the retrieval of the optically pure amines (93).

Isocyanates are readily synthesized from primary amines, and many such amines are available in resolved form. It is somewhat surprising therefore that only relatively few chiral isocyanates have been used as CDAs. On the other hand, such lack of interest in other isocyanates may be the reflection of the considerable success afforded by [19] and [20].

Formation of Thioureas. If the isocyanate in Eq. 4 is replaced by an isothiocyanate, the products formed are thioureas (Eq. 5).

$$\text{R-N=C=S} + \text{HNR}'\text{R}'' \rightarrow \text{RNH-CS-NR}'\text{R}'' \qquad (5)$$

This reaction is selective for primary and secondary amino groups, and hydroxyl groups do not react under conditions in which the amine is readily derivatized. A significant difference between isothiocyanates and isocyanates is that the former react with water or alcohols much more slowly than the latter. This difference is an advantage when the derivatization of amino acids, catecholamines, and other highly polar compounds is considered: Isothiocyanate CDAs are compatible with aqueous reaction media, in which such polar analytes have greater solubility.

Nambara et al. described in 1978 the synthesis of two optically active terpene isothiocyanates and their use in the LC resolution of amino acids (98). These CDAs did not receive further attention, but in 1980 Nimura et al. reported on the use of 2,3,4,6-tetra-*O*-acetyl-β-D-glucopyranosyl isothiocyanate, [21], for the LC resolution of amino acid esters (99). The use of this CDA was then extended to amino acids (100) and catecholamines (101). Concerning the latter, [21] was also shown to be suitable for the determination of the enantiomeric purity of (−)-epinephrine in cardiac and dental dosage forms (102,103). CDA [21] is commercially available, and in recent years its use in indirect resolutions by LC has greatly expanded (104). Many β-adrenergic antagonists (105), other amino alcohols (106–109), several antiarrhythmic primary and secondary amines (110), and amphetamines (55) have been resolved via derivatization with CDA [21]. Many of the separations were achieved on RP LC columns, and the reagent has been shown to be applicable to problems of stereoselective drug metabolism and pharmacokinetics (104). Of particular interest in this regard is the report by Walle et al. (111) on the enantiomeric composition of the sulfate conjugates of 4-hydroxypropranolol. The intact sulfate conjugates could be derivatized with [21], and the derivatives were well resolved by RP LC. With detection at 313 nm, the minimum detectable amount of each enantiomer of the conjugate was 20 ng. By using this method, the plasma concentration of the enantiomeric sulfate conjugates of 4-hydroxypropranolol in a patient on chronic propranolol therapy could be determined (111). The technique could also be applied to in vitro studies of the conjugation of 4-hydroxypropranolol (112). Clearly, this is a promising method for the investigation of the stereoselectivity of the formation and disposition of those drug conjugates that are amenable to derivatization with [21].

Particularly good separation of the diastereomers derived from amino alcohol structures was achieved with [21] (104–110). In the derivatization of such compounds, only the amino group reacts with the isothiocyanate moiety. It is interesting to compare the resolution of the amino alcohol ephedrine (*R,S*/*S,R* racemate), [22], with that of methamphetamine, [23], which lacks the hydroxy group. Derivatization of racemic ephedrine with the isothiocyanate [21] produces derivatives that are well resolved (α = 1.19) by RP LC (106), whereas the derivatives of methamphetamine are minimally resolved under similar conditions (no α given) (113). An analogous comparison can be made for norephedrine and amphetamine (55,106), which are the *N*-demethyl analogs of [22] and [23], respectively.

Although a detailed separation mechanism that would account for these differences has not been proposed, it is clear that the hydroxyl group in amino alcohols plays a crucial role in producing differences between the diastereomers in their partition characteristics. A likely mechanism could be the formation of an intramolecular hydrogen bond involving the hydroxyl group, which would result in the molecule favoring a more rigid conformation. Such reduction in conformational mobility by intramolecular binding via various forces (hydrogen bonding, dipole-dipole, etc.) is thought to be important in maximizing physiochemical differences between diastereomers and appears to explain a variety of diastereomer separations (3,5). It should be noted that the considerable success of the isocyanate reagents in the resolution of aminoalcohols discussed in the previous section above may also be due to similar intramolecular hydrogen-bonding mechanisms.

The successful use of [21] in a variety of resolutions, as outlined above, has stimulated the search for other optically active isothiocyanate CDAs, partly because [21] failed to resolve some racemates. Several promising alternatives to [21] have emerged, for example, (*R*)-α-methylbenzyl isothiocyanate, [24] (114), (*S*)-1-(1-naphthyl)ethyl isothiocyanate [25] (110,115), (*R*)-1-(2-naphthyl)ethyl isothiocyanate [26] (110), and (*S*,*S*)-2,2-dimethyl-4-phenyl-5-isothiocyanato-1,3-dioxane [27] (54); these CDAs are suitable for the LC resolution of a variety of amines using RP conditions. Interestingly, significant differences were observed between the regioisomeric α-naphthyl and β-naphthyl CDAs [25] and [26] in their abilities to resolve some of the amines (110,116). One of the new isothiocyanates, reagent [24], is commercially available (114).

It appears, then, that isocyanate and isothiocyanate CDAs are useful in LC resolutions of amines. Several of these agents are commercially available, and many others can be readily prepared from a variety of commercially available optically active primary amines. The stereochemical stability of these CDAs is excellent, their reaction with amines is rapid and convenient, and the derivatives are readily and sensitively detected. These reagents are likely, therefore, to enjoy continued popularity.

Formation of Isoindoles. In 1971 Roth described a sensitive analytical method for amino acids based on their reaction with *o*-phthaldialdehyde (OPA) and a thiol, 2-mercaptoethanol (117). The reaction produces an intensely fluorescent isoindole derivative of the amino acid and is specific for the primary amino group (118,119). Subsequently, the OPA method was extended to the enantiospecific LC analysis of amino acids by substituting an optically active (single enantiomer) thiol for 2-mercaptoethanol in the derivatization reaction (120–122). This results in the formation of two

paring three optically active acid chlorides, [8], [30], and [31] in the chromatographic resolution of a variety of chiral alcohols (129). Five capillary GLC columns and one LC column were screened. The data obtained provided a good starting point for the selection of the appropriate chromatographic system and conditions for a particular resolution problem (129).

The earlier-cited CDA [8] has been used successfully and extensively in the chromatographic resolution of chiral alcohols. In addition to its value in the GLC resolution of alcohols (130) and hydroxy-fatty acids (131), the reagent is also useful in the LC resolution of hydroxyl compounds, as shown, for example, by the work of Doolittle and Heath (129) and others (132). In this connection, the applications of [8] in studies of the chemistry and biology of carcinogenic polycyclic aromatic hydrocarbons (PAHs) are worthy of consideration. These hydrocarbons, for example, benzo[a]-pyrene, undergo biotransformation to toxic metabolites; furthermore, the biotransformations and toxicity of the metabolites can be highly stereoselective. Among the metabolites involved are arene epoxides that undergo rapid conversion to vicinal dihydroxy dihydro derivatives of the aromatic hydrocarbons. Derivatization of the diols with [8] and chromatographic separation of the resulting diastereomeric bis-esters have been used in the determination of the enantioselectivity of biotransformations (133,134) and the preparation of the pure enantiomers (135–137). The latter applications involved preparative LC separation of the diastereomers, followed by hydrolysis of the bis esters (135–137). The separations were usually carried out on silica gel LC columns, although in some cases TLC separation of the diastereomeric esters was also feasible (137). Bromohydrin analogs of the diols, useful in the synthesis of some of the chiral metabolites, have been similarly resolved (138).

Another CDA that has been similarly applied (139–143) in the separation of dihydrodiol derivatives of PAHs is the acid chloride [32] of (−)-menthoxyacetic acid. This acid is commercially available from several sources. CDA [32] has also been used in the separation of the isomers of the new antihypertensive agent nipradilol. The chemical structure of this drug includes two asymmetric centers, and therefore the drug is a mixture of four stereoisomers, that is, two diastereomeric racemates. Derivatization of nipradilol, an amino alcohol, with [32] produced the four *N,O*-bis-menthoxyacetyl derivatives, which could be cleanly separated using silica gel LC, with detection at 275 nm (144).

The important anticoagulant drug warfarin is chiral and is administered clinically as the racemic mixture. Significant pharmacodynamic and pharmacokinetic differences between the enantiomers have generated considerable interest in the stereochemical aspects of this drug (145).

Banfield and Rowland developed an elegant procedure for the chromatographic resolution of the warfarin enantiomers (146). The enol hydroxyl group in the molecule was esterified with carbobenzoxy-L-proline, [33], a commercially available agent, and the derivatives were separated on silica LC columns. The procedure could be adapted to pharmacokinetic studies of warfarin using a UV detector at 313 nm (146). The original procedure was subsequently modified to use fluorescence detection to achieve greater sensitivity and permit monitoring of metabolites of the drug (147). The modification involved postcolumn on-line reaction of the eluted derivatives to generate fluorescent species. The procedure was recently adapted to the stereospecific analysis of a related chiral drug, nicoumalone (148,149) and large pharmacokinetic differences between its enantiomers were found (149).

Camphanic acid chloride, [11], described above in the resolution of amines, has also been used in the derivatization and separation of hydroxyl compounds. The bronchodilator proxyphylline was resolved after derivatization with [11] using RP LC in a study of the optical purity of synthetic samples of the enantiomers (150).

Other optically active acid chlorides have also been used in chromatographic separations of hydroxyl-containing drug enantiomers. Shimizu et al. (151) synthesized and resolved 2-(2-naphthyl)-propionic acid and used the acid chloride [34] of the dextrorotatory acid to acylate the hydrolyzed derivative of the calcium antagonist drug diltiazem. The derivatives were separated using RP LC. The purpose of the procedure was the determination of the enantiomeric purity of diltiazem, since this drug is used clinically as a pure enantiomer. The disadvantage of the procedure is the requirement to synthesize and resolve the parent acid of the CDA. Indeed, the same workers recently published a new procedure for the same purpose: The new CDA used, [35], was the (*S*)-*N*-1-(2-naphthylsulfonyl) derivative of L-proline. This CDA is readily prepared in a simple procedure from the resolved and inexpensive precursor,' L-proline. No racemization occurred during the synthesis of the CDA and its optical purity was high. The CDA afforded derivatives of diltiazem that were well separated by silica (152). Reagent [35] merits further attention.

Several derivatives of mandelic acid have been used as CDAs for the resolution of chiral alcohols as the corresponding esters. Thus, in a study of the steric course of the cytochrome P-450-mediated hydroxylation of phenylethane (ethylbenzene), White et al. derivatized the enzymatically formed 1-phenylethanols with (*S*)-*O*-propionylmandelyl chloride, [36], and separated the diastereomeric derivatives by GLC (153). Comber and Brouillette synthesized the enantiomers of carnitine via derivatization of a racemic precursor alcohol with (*R*)- or (*S*)-α-methoxyphenylacetic [(*O*)-

methylmandelic] acid (see acid chloride [10]). After adding a required methyl group to the derivatives, the diastereomeric esters were separated via flash chromatography, followed by hydrolysis of the individual diastereomeric esters to remove the esterifying CDA moiety and thereby obtain the desired individual enantiomers (154).

Several lactic acid derivatives were used by Gessner et al. for the determination of the enantiomeric purity of flavor substances such as chiral alcohols from natural sources. Diastereomeric *O*-acetyl-, propionyl-, and hexanoyllactic acid esters of the chiral alcohols were separated by GLC (155). A report from the same laboratories described characterization of several chiral aroma substances that are δ-lactones. The lactones were hydrolyzed to the corresponding hydroxy acids, and the acid moiety was esterified to the isopropyl ester. The remaining hydroxyl group was esterified with (*R*)-2-phenylpropionic acid chloride or [30], and the diastereomeric derivatives were separated using preparative silica gel LC. The derivatives were also separated on an analytical scale by GLC (156).

A new type of CDA for the normal-phase LC resolution of hydroxyl compounds was developed by Goto et al. (157). The reagent ([37], both enantiomers were prepared) is chiral by virtue of restricted rotation around the binaphthyl bond. Hydroxyl groups react with the carbonyl nitrile moiety of the CDA to give the esters. The resolution of hydroxyl acids (157) and β-adrenergic antagonists (158) was carried out with [37] using normal-phase LC. The ester derivatives formed are highly fluorescent, whereas the reagent itself is nonfluorescent, a great advantage indeed.

A series of CDAs was reported by Lindner et al. (159). The reagents, optically pure symmetrically *O,O*-disubstituted (*R,R*)- or (*S,S*)-tartaric acid anhydrides, were developed for the LC separation of amino alcohols. Under the derivatization conditions used, only the hydroxyl group reacts with the anhydride to produce a monoester. The derivatives were well resolved by RP LC, and the zwitterionic nature (amino and carboxyl groups) of the derivatives was thought to be important to their separability. The procedure was applied to a series of β-adrenergic antagonists, and the presence of 0.2% of enantiomeric contamination of either enantiomer of propranolol could be determined, illustrating the power of the indirect method, provided an essentially 100% optically pure CDA is used (159). In an application of the method, Demian and Gripshover used (*R,R*)-dibenzoyltartaric anhydride to derivatize 1-methyl-3-pyrrolidinol at the hydroxyl group and separated the diastereomeric esters on a C_8 RP LC column. The procedure was used in the determination of the enantiomeric purity of the substrate, an intermediate in the synthesis of several drugs (160).

The earlier-mentioned chiral acid (*S*)-naproxen [15] has also been used, as the acid chloride derived from the *S* acid, to form diastereomeric esters of 2-pentanol, 2-hexanol, 2-heptanol, 2-octanol, and 2-nonanol, and the diastereomers were separated on packed OV-17 GLC columns (69). Walther et al. recently described a new CDA for the resolution of enantiomeric alcohols. The reagent, now available commercially as (*S*)-Trolox™ methyl ether [38], appears superior to several other CDAs for the same purpose in that unlike the latter, [38] provided enantiomer separations by capillary GLC of several primary alcohols, that is, alcohols in which the stereogenic (asymmetric) center was not the hydroxyl-bearing carbon (161).

Formation of Carbonates. The reaction of an alcohol with a chloroformate produces the corresponding carbonate (Eq. 7)

$$\text{RO-CO-Cl} + \text{HOR}' \rightarrow \text{RO-CO-OR}' \qquad (7)$$

This reaction was among the earliest used for the GLC resolution of enantiomeric alcohols: Studies 20 years ago showed that the reaction of (−)-menthyl chloroformate, [17], with chiral alcohols produced diastereomeric carbonates that could be resolved on packed columns (71,162). Despite its commercial availability, this CDA has been little used for the resolution of alcohols. Prelusky et al. (163) used [17] in a study of enantioselective metabolic ketone reductions. The metabolite alcohols were derivatized with [17] and the derivatives were separated by capillary GLC. Good resolution was obtained under the chromatographic conditions used (163). The derivatization of warfarin with [17] and LC of the derivatives was described, but the peaks were not fully separated, even with retention times of 80–100 min (164). Brash et al. used CDA [17] to determine the enantiomeric composition of several enantiomeric pairs of hydroperoxyeicosatetraenoic acids using silica LC, but some of the derivatives could not be resolved (165). It appears that [17] may be useful in the resolution of hydroxyl compounds and the evaluation of capillary GLC in the separations deserves attention.

Formation of Carbamates. Chiral isocyanates were discussed above as CDAs for the derivatization of amines, but these reagents have also proven useful for hydroxyl compounds. The reaction (Eq. 8) of an isocyanate with an alcohol yields a carbamate (sometimes also referred to as a urethane)

$$\text{RN}{=}\text{C}{=}\text{O} + \text{HOR}' \rightarrow \text{RHN-CO-OR}' \qquad (8)$$

In pioneering work by Pereira et al., the derivatization of racemic alcohols with (+)-[19] was carried out to produce diastereomeric carbamates that could be resolved on stainless steel capillary GLC columns coated with Carbowax 20M or OV-225 stationary phase (166). The retention times, however, ranged between 28 and 90 min, rendering the technique somewhat inconvenient. Gal et al. used capillary GLC columns to separate carbamate derivatives of (−)-[19] with retention times of 15 min or less and applied the procedure to the investigation of the stereoselectivity of ketone reduction by rat- and rabbit-liver cytosol (50,167). The CDA was also used in a study of the pharmacokinetics of misonidazole, using RP LC to separate the carbamate derivatives (168). The naphthalene analog, [20], has also been used in the separation of enantiomeric alcohols (169–173). Pirkle and his associates carried out extensive studies on the LC separation of diastereomeric carbamate derivatives using adsorption chromatography (3). Sakaki and Hirata (174) studied the separation by supercritical fluid chromatography of diastereomeric derivatives of several chiral secondary alcohols after derivatization with [20]. The role played by the stationary phase, mobile phase, and structure of the analyte was studied, and it was shown that useful resolutions could be obtained for several alcohols (174).

Dehydroabietyl isocyanate, [39], was used by Falck et al. to determine the absolute configuration of hydroxyeicosatetraenoic acid methyl esters in a study of the enzymatic epoxidation of arachidonic acid (175). In another study of similar metabolites, the stereochemical identity of 12-hydroxy-5,8,10,14-eicosatetraenoic acid derived from the lesional scale of patients with psoriasis was studied via separation of the enantiomers after derivatization with [39] (176). In these studies (175,176), LC separation of the derivatives was used. CDA [39] can be prepared from commercially available resolved dehydroabietylamine. It would be worthwhile to examine the applicability of this CDA to the resolution of other compounds.

Formation of Other Derivatives. The reactivity of isothiocyanates toward alcohols is less than that of isocyanates (177), and it appears that chiral isothiocyanates have not received attention as CDAs for the resolution of hydroxyl compounds.

A common metabolic biotransformation of alcohols is conjugation with D-glucuronic acid at the hydroxyl group, and it has been often reported that when the substrate alcohol is chiral, the diastereomeric glucuronide conjugates of the two enantiomeric alcohols are separable by LC. It could be expected, therefore, that this "natural CDA method" would be exploited for indirect resolutions; however, little work has appeared on enzymatic derivatizations for indirect resolutions. One reason for this paucity of interest may be that enzymatic reactions are more complex than nonenzymatic derivatizations. Gerding et al. described (178) the derivatization of compound [40], a D-2 dopamine agonist, with uridine-5′-diphosphoglucuronic acid (UDPGA), the natural co-factor involved in the conjugation of alcohols with glucuronic acid. The enzyme required, glucuronyl transferase, was obtained from fresh bovine livers.

The diastereomeric conjugates were well separated by RP LC, and the procedure could be used to determine the enantiomeric purity of the drug enantiomers in the range of 99.84–99.98%, illustrating the utility of this approach. Furthermore, this work is another exception to the generalization that trace-enantiomeric-impurity determinations cannot generally be performed reliably with the indirect method.

Meyers et al. (179) described a procedure for the determination of the enantiomeric composition of 2-substituted-1,2-glycols. The glycols were reacted with optically pure *S*-(+)-2-propylcyclohexanone, [41], to form diastereomeric ketals. A new asymmetric center is formed in the reaction, and therefore, a racemic glycol gives, upon reaction with [41], four diastereomers. For several glycols the four derivatives could be separated by LC, and therefore, the enantiomeric composition could be determined (179). Because some drugs are metabolized to glycols (180), this procedure may be of interest in drug disposition studies.

3. Resolutions via Derivatization of Carboxyl Groups

Many chiral drug molecules contain the carboxyl group. This functional moiety can also be generated metabolically from other structures via such biotransformations as hydrolysis of esters, oxidation of alkyl groups, alcohols, aldehydes, amines, etc. Of particular interest in recent years has been the stereoselectivity in the action and disposition of a group of nonsteroidal antiinflammatory drugs (NSAIDs), for example, ibuprofen, fenoprofen, flurbiprofen, etc. (181). These drugs have the 2-arylpropionic acid moiety in their structure and are chiral by virtue of the asymmetric center α to the carboxyl group. Most such drugs are used clinically in the racemic form, even though it is known that the desired activity resides primarily in the *S* isomers, the *R* enantiomers possessing low or no activity (181). Furthermore, it has been shown that many of these NSAIDs undergo in vivo metabolic inversion of the chiral center of the inactive *R* enantiomer to the active *S* antipode (181), and this unusual and interesting biotransformation has been the subject of many investigations (181). As a result of such interest, much of the recently published work on chiral derivatization of carboxylic acids deals with arylpropionic acid NSAIDs.

Formation of Esters. One frequently used approach to the formation of diastereomeric derivatives of carboxylic acids is esterification with an optically active alcohol (Eq. 9)

$$\text{ROH} + \text{HOOCR}' \rightarrow \text{RO-OC-R}' \qquad (9)$$

A variety of alcohols, for example, 2-octanol (182,183), (−)-2-butanol (184), (+)-3-methyl-2-butanol (185), (−)-menthol (186), (*S*)-ethyl lactate (187), (*R*)-1-phenylethanol (188), etc., have been used. Several different sets

toxification pathway, inasmuch as epoxides are often highly reactive compounds that can react with various cellular nucleophiles to produce, ultimately, serious toxic effects (211).

In recent years, there has been considerable interest in the stereochemical aspects of metabolic epoxidation, prompted by the recognition that the toxicity, metabolic formation, and further metabolism of epoxides can be highly stereoselective (211). As discussed earlier, when the epoxide formed is an arene oxide (i.e., epoxidation of an aromatic double bond), it is often rapidly converted to dihydrodiols, and such hydroxyl compounds have been analyzed by the indirect resolution approach. However, when the metabolite epoxide is more stable, it is often possible to examine its stereochemistry. For this purpose, several methods for the chromatographic separation of enantiomeric epoxides have been developed, including some indirect methods.

The thiol nucleophile glutathione is known to react with many epoxides in a biotransformation catalyzed by glutathione-*S*-transferases (GST). In the reaction, the thiol nucleophile reacts at one and/or the other of the epoxide-ring carbons to give ring-opened derivatives, and such conjugations can be markedly regio- and enantioselective (212,213). Under appropriate conditions, the ring-opening reaction can also take place nonenzymatically. Such reactions form the basis of derivatizations of epoxides, with glutathione serving as the CDA. Glutathione is a chiral compound of high enantiomeric purity that has been found to be suitable for use in enzymatic and nonenzymatic reactions as a CDA for epoxides. Armstrong et al. (214) and van Bladeren et al. (215) used this approach to study the stereoselectivity of the cytochrome P-450-catalyzed epoxidation of PAHs. The enzymatically formed epoxides were reacted nonenzymatically with glutathione, and the derivatives were separated by RP LC. The CDA was found to add to the epoxides in *trans* fashion, but often in a nonregiospecific manner, each enantiomer yielding two positionally isomeric adducts, crating a challenging chromatographic separation problem (214,215). In subsequent investigations, glutathione was replaced by *N*-acetylcysteine as the thiol nucleophile, but regioisomerism in the reaction of the CDA with the epoxides was still a complication, and complete chromatographic separation of all the adducts was not possible in some cases (216,217). Nevertheless, the thiol-CDA approach to the derivatization of arene oxides was used successfully in studies of the stereochemical course and nature of enzymatic epoxidations (214–217).

Foureman et al. took a somewhat different approach in using glutathione as the thiol CDA in an in vitro study of the stereoselectivity of the oxidation of styrene catalyzed by purified cytochrome P-450 and by microsomes from rat liver (218). After incubation of styrene with the P-450 preparation, glutathione and a preparation containing GST were added, and the epoxide metabolite was converted enzymatically to the glutathione adducts. Here again, attack by the thiol occurred nonregiospecifically, and therefore, each enantiomer of the epoxide gave two adducts. Nevertheless, the analytical method was suitable for the determination of the stereoselectivity of the epoxidation of styrene (219).

A different approach was taken by Panthananickal et al., who studied the steric course of enzymatic epoxidations of PAHs (220,221). These investigators chose a chiral amino compound, polyguanylic acid, to serve as the CDA. In the reaction, the exocyclic amino group of the guanine moiety in the CDA reacted regio-specifically (i.e,. at only one of the two ring carbons) with the epoxides, but both *cis* and *trans* addition occurred, so that each epoxide enantiomer yielded two adducts. The initial adducts were hydrolyzed with potassium hydroxide and then digested with alkaline phosphatase to guanosine derivatives, which could be separated by RP LC (220,221).

Gal developed a procedure for the determination of the enantiomeric composition of several types of epoxides (222). Nonchiral simple aklylamines, for example, isopropylamine, were used in the ring-opening reaction to produce enantiomeric amino alcohols, which were then derivatized with the optically active isothiocyanate [21], and the derivatives were resolved by RP LC (222). Despite the two-step derivatization, the procedure is simple and practical to carry out and could be adapted to the determination of the enantioselectivity of the rat-liver-microsomal epoxidation of an alkene (223).

5. *Resolutions via Derivatization of Other Functional Groups*

Other functional groups have been derivatized with CDAs for the purpose of analytical or preparative chromatographic resolution. Isocyanates and isothiocyanates, for example, can be derivatized with chiral amines serving as CDAs (114). Similarly, the enantiomeric purity of isocyanates may be determined via derivatization with an optically pure alcohol (167). Isocyanates may also be derivatized with chiral 2-oxazolidones [21]. Pirkle et al. described derivatization and resolution of lactams using chiral isocyanates such as [19] and [20] (224).

The natural product gossypol, [45], a polyphenolic binaphthyl, is a male antifertility agent of considerable current interest. The available evidence suggests that the antifertility action of [45] may reside only in the (−) enantiomer (225). Procedures have been devised for the chromatographic separation of the enantiomers of [45] as diastereomeric derivatives of optically active β-amino alcohols (226,227). L-Phenylalaninol was used in one procedure (227), whereas (−)-norepinephrine was employed in another (226). The latter CDA produced the bis-Schiff base upon reaction with the aldehyde groups of gossypol (226), whereas the former CDA was thought to yield the bis-oxazolidine derivative (227). This difference in the reaction course of [45] with the two CDAs may be the result of a difference between the two CDAs in steric hindrance around the hydroxyl groups.

Reid et al. developed an interesting albeit complex derivatization scheme for the enantiospecific analysis of cyclophosphamide (228). Cyclophosphamide [46] is an antineoplastic drug that is chiral by virtue of the stereogenic phosphorus center. In the derivatization scheme, the drug is reacted with chloral to produce the secondary alcohol derivative [47]. In this reaction a new asymmetric center is created, and thus, two diastereomeric products are possible from each enantiomer of cyclophosphamide. However, the authors found that either only one of the two diastereomers was formed or the two diastereomers were not separable under the chromatographic conditions used, as only a single product peak was obtained regardless of the stereochemical identity of the drug enantiomer. The secondary alcohols were then reacted with the acid chloride of (+)-naproxen, and the resulting diastereomeric esters were separated by normal-phase or RP LC. The technique could be applied to the determination of the enantiomeric composition of cyclophosphamide in human serum (228). It should be recognized that although this derivatization scheme is indeed complex, the chemistry of cyclophosphamide precludes simple one-step derivatizations.

Chiral olefins were reacted with optically platinum complexes, and the diastereomeric derivatives formed were separated by LC (229); thiols may be derivatized with optically active acetals for chromatographic resolution (230). The chromatographic separation of enantiomeric sugars, a rather specialized field, has been reviewed (231).

IV. SUMMARY AND CONCLUSIONS

The enantiomers of a large number of compounds of pharmaceutical, pharmacological, or toxicological interest have been resolved as diastereomeric derivatives. A variety of functional groups can be derivatized, and a large variety of chiral derivatizing agents are available. The indirect method can be used to determine enantiomeric composition or purity, assign absolute configuration, and isolate the enantiomers on a preparative scale. GLC, TLC, and LC have been used in the separations; the majority of new applications use LC, although capillary GLC is also gaining popularity.

Some CDAs suffer from racemization problems and should therefore be abandoned, but most CDAs described have demonstrated excellent

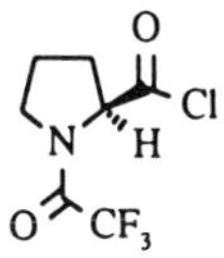

O
Cl
N
H
O
CF$_3$

[5]

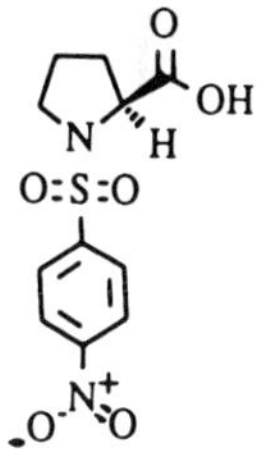

O
OH
N
H
O:S:O
N$^+$
O$^-$ O

[6]

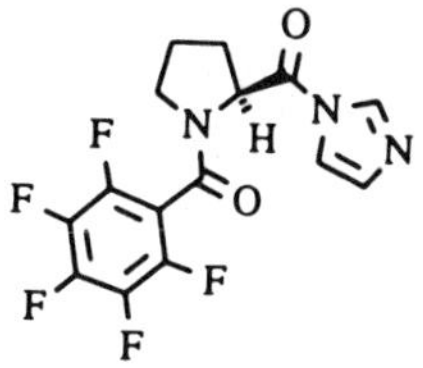

O
N
H
N
N
F
F
O
F
F
F

[7]

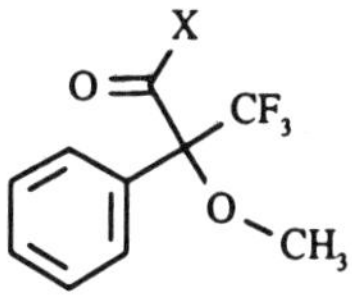

X
O
CF$_3$
O
CH$_3$

[8] X = Cl
[9] X = OH

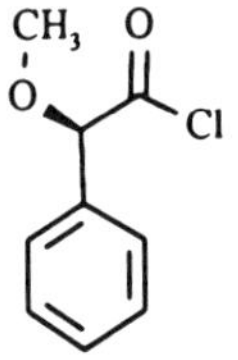

CH$_3$
O
O
Cl

[10]

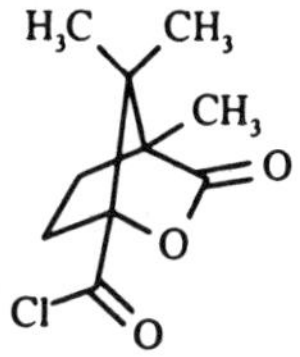

H$_3$C CH$_3$
CH$_3$
O
O
Cl
O

[11]

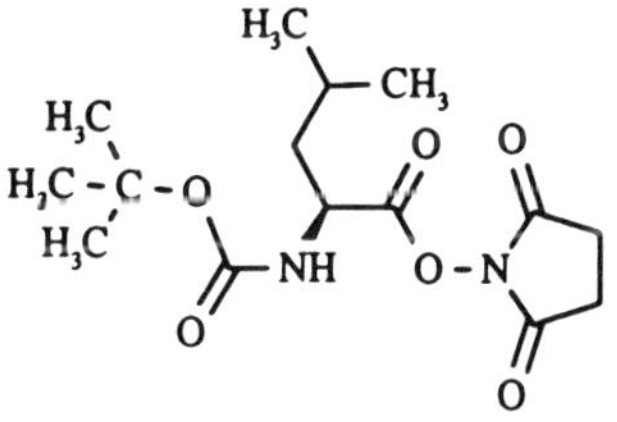

H$_3$C
CH$_3$
H$_3$C
H$_3$C - C - O
H$_3$C
O
O
NH
O - N
O
O

[12]

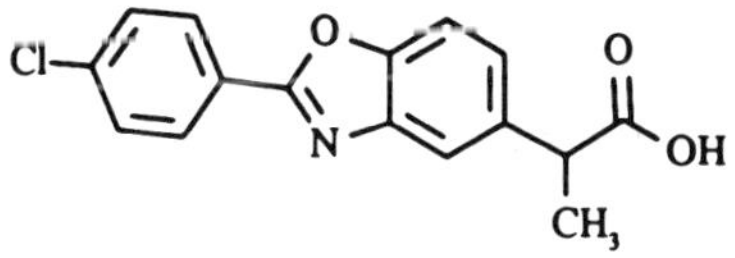

Cl
O
N
O
OH
CH$_3$

[13]

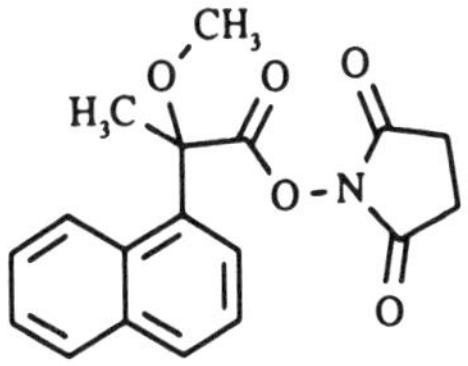

CH$_3$
O
O
O
H$_3$C
O - N
O

[14]

O
OH
CH$_3$
H$_3$C
O

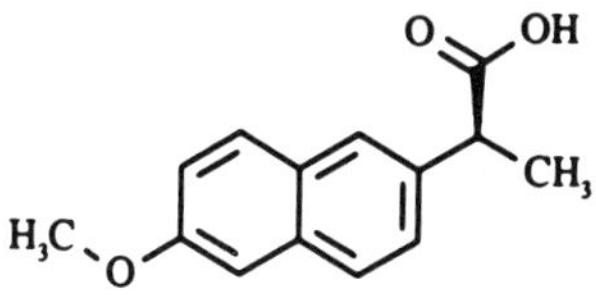

[15]

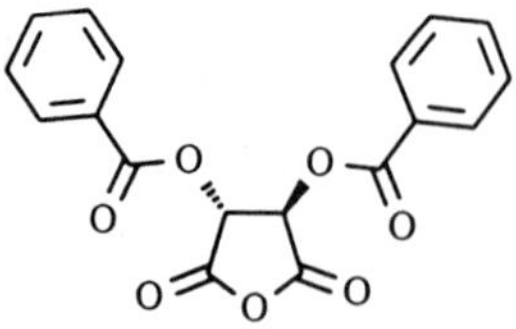

[16] 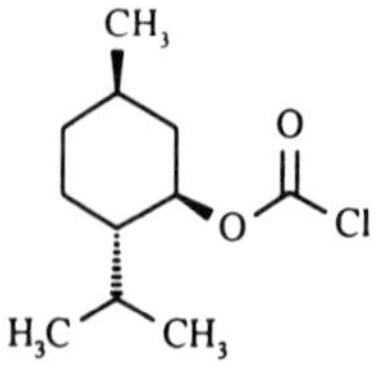 [17] 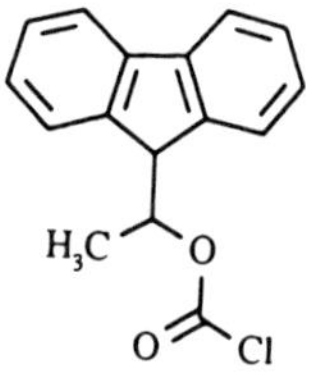 [18]

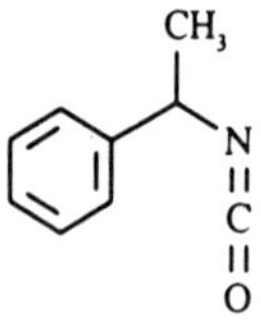

[19] 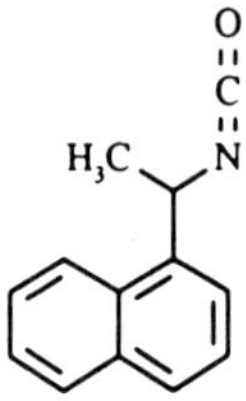 [20] 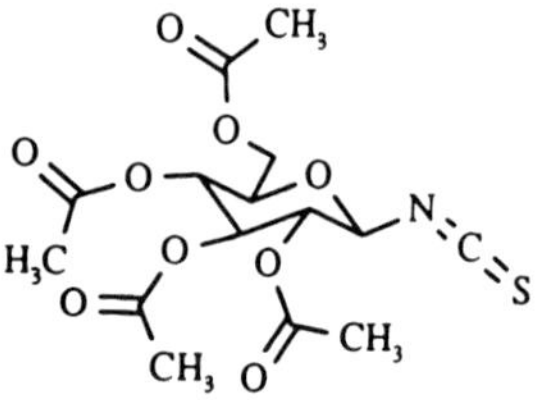 [21]

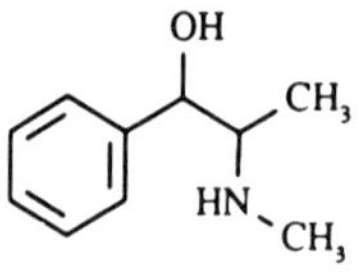

[22] 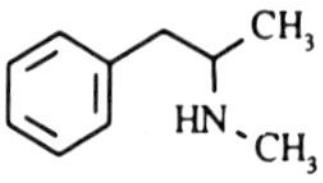 [23] 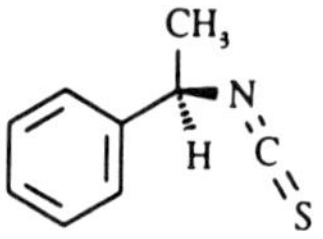 [24]

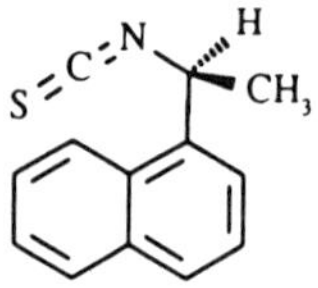

[25] 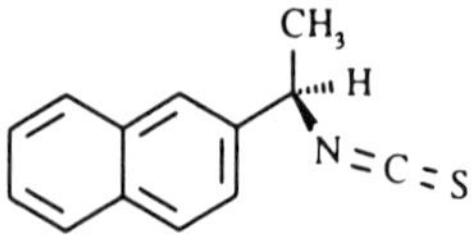 [26] 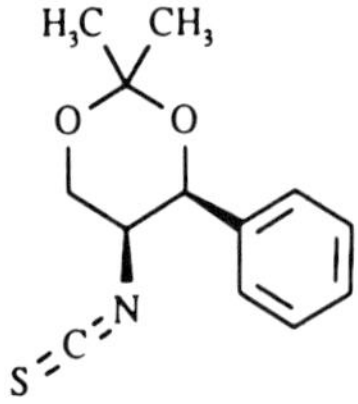 [27]

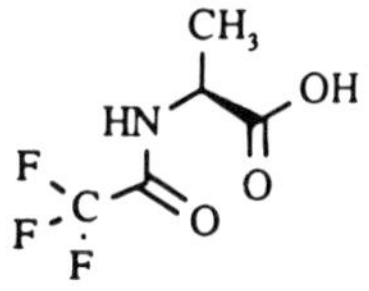

[28]

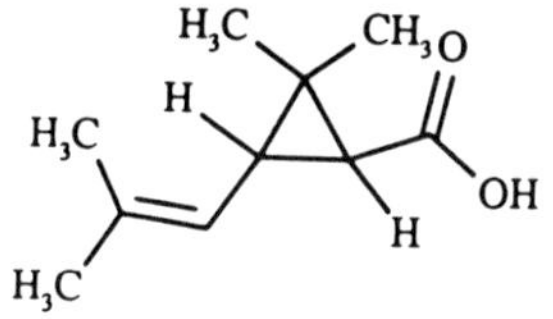

[29]

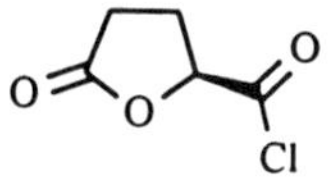

[30]

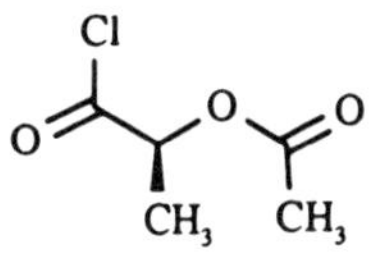

[31]

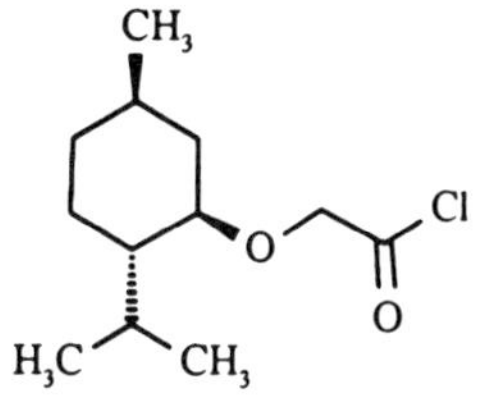

[32]

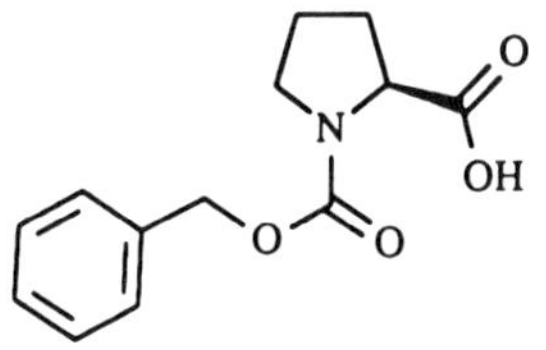

[33]

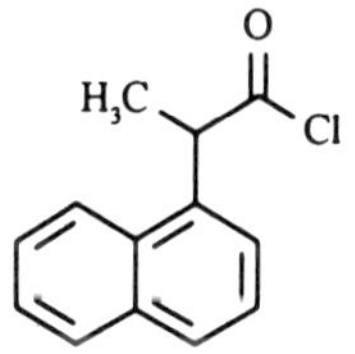

[34]

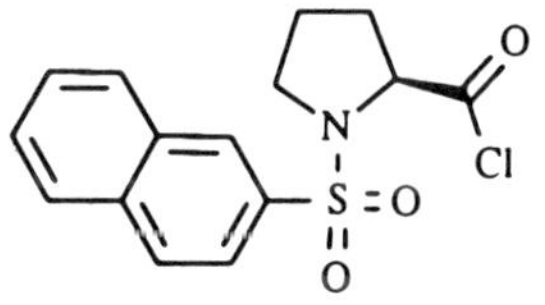

[35]

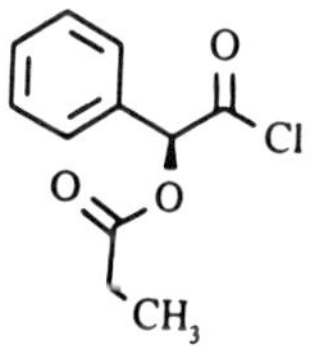

[36]

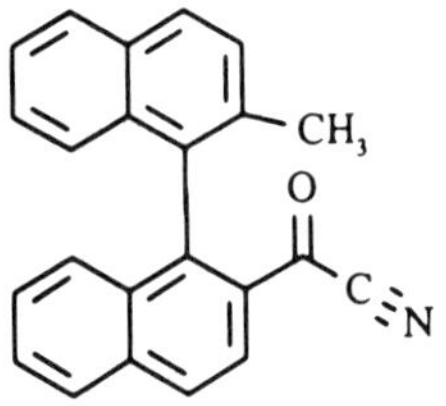

[37]

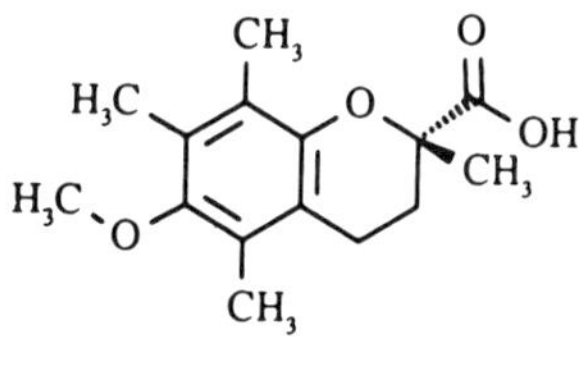

[38]

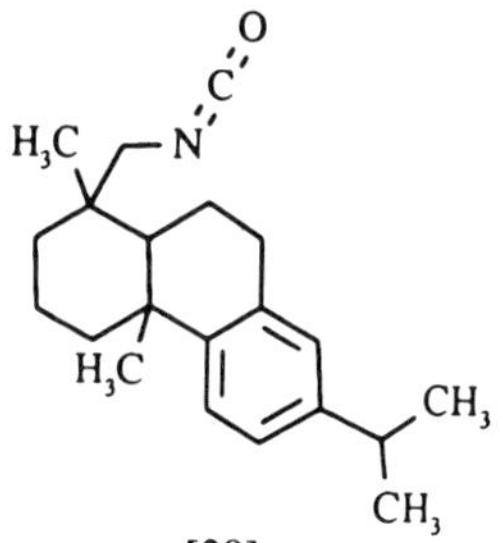

[39]

144. M. Yoneda, M. Shiratsuchi, M. Yoshimura, Y. Ohkawa, and T. Muramatsu, *Chem. Pharm. Bull.*, *33*:2735 (1985).
145. C. Banfield, R. O'Reilly, E. Chan, and M. Rowland, *Br. J. Pharmacol.*, *21*:564P (1986).
146. C. Banfield and M. Rowland, *J. Pharm. Sci.*, *72*:921 (1983).
147. C. Banfield and M. Rowland, *J. Pharm. Sci.*, *73*:1392 (1984).
148. T. S. Gill, K. J. Hopkins, and M. Rowland, *Br. J. Clin. Pharmacol.*, *21*:564P (1986).
149. T. S. Gill, J. K. Hopkins, and M. Rowland, *J. Pharm. Pharmacol.*, *36* (Supplement):12P (1984).
150. M. Ruud-Christensena and B. Salvesen, *J. Chromatogr.*, *303*:433 (1984).
151. R. Shimizu, K. Ishii, N. Tsumagari, M. Tanigawa, M. Matsumoto, and I. I. Harrison, *J. Chromatogr.*, *253*:101 (1982).
152. R. Shimizu, T. Kakimoto, K. Ishii, Y. Fujimoto, H. Nishi, and N. Tsumagari, *J. Chromatogr.*, *357*:119 (1986).
153. R. E. White, J. P. Miller, L. V. Favreau, and A. Bhattacharyya, *J. Am. Chem. Soc.*, *108*:6024 (1986).
154. R. N. Comber and W. J. Brouillette, *J. Org. Chem.*, *52*:2311 (1987).
155. M. Gessner, W. Deger, and A. Mosandl, *Z. Lebensm. Unters. Forsch.*, *186*:417 (1988).
156. A. Mosandl and M. Gessner, *Z. Lebensm. Unters. Forsch.*, *187*:40 (1988).
157. J. Goto, N. Goto, and T. Nambara, *Chem. Pharm. Bull.*, *30*:4597 (1982).
158. J. Goto, M. Ito, N. Goto, and T. Nambara, in *Proceedings of the 10th International Symposium on Column Liquid Chromatography*, San Francisco, Calif., 1986.
159. W. Lindner, Ch. Leitner, and G. Uray, *J. Chromatogr.*, *316*:605 (1984).
160. J. Demian and D. F. Gripshover, *J. Chromatogr.*, *387*:532–535 (1987).
161. W. Walther, W. Vetter, M. Vecchi, H. Schneider, R. K. Muller, and T. Netscher, *Chimia*, *45*:121 (1991).
162. M. W. Anders and M. J. Cooper, *Anal. Chem.*, *43*:1093 (1971).
163. D. B. Prelusky, R. T. Coutts, and F. M. Pasutto, *J. Pharm. Sci.*, *71*:1390 (1982).
164. G. L. Jeyaraj and W. R. Porter, *J. Chromatogr.*, *315*:378 (1984).
165. A. R. Brash, A. T. Porter, and R. L. Maas, *J. Biol. Chem.*, *260*:4210 (1985).
166. W. Pereira, V. A. Bacon, W. Patton, B. Halpern, and G. E. Pollok, *Anal. Letters*, *3*:23 (1970).
167. J. Gal, D. DeVito, and T. W. Harper, *Drug Metab. Disp.*, *9*:557 (1981).
168. K. M. Williams, *Clin. Pharmacol. Ther.*, *36*:817 (1984).
169. Y. Yamazaki and H. Maeda, *Tetrahedron Lett.*, *26*:4775 (1985).
170. Y. Yamazaki and H. Maeda, *Agric. Biol. Chem.*, *50*:79 (1986).
171. Y. Yamazaki and H. Maeda, *Agric. Biol. Chem.*, *49*:3202 (1985).
172. P. Michelsen, E. Aronsson, G. Odham, and B. Akesson, *J. Chromatogr.*, *350*:417 (1985).
173. P. E. Sonnet, R. L. Dudley, S. Osman, P. Pfeffer, and D. Schwartz, *J. Chromatogr.*, *586*:255 (1991).
174. K. Sakaki and H. Hirata, *J. Chromatogr.*, *585*:117 (1991).

175. J. R. Falck, S. Manna, H. R. Jacobson, R. W. Estabrook, N. Chacos, and J. Capdevila, *J. Am. Chem. Soc.*, *106*:3334 (1984).
176. P. M. Wollard, *Biochem. Biophys. Res. Commun.*, *136*:169 (1986).
177. J. March, *Advanced Organic Chemistry: Reactions, Mechanisms, and Structure*, McGraw-Hill, New York, 1968.
178. T. K. Gerding, B. F. H. Drenth, V. J. M. Van de Grampel, N. R. Niemeijer, R. A. De Zeeuw, P. G. Tepper, and A. S. Horn, *J. Chromatogr.*, *487*:125 (1989).
179. A. I. Meyers, S. K. White, and L. M. Fuentes, *Tetrahedron Lett.*, *24*:3551 (1983).
180. M. H. Holshouser and M. Kolb, *J. Pharm. Sci.*, *75*:619 (1986).
181. A. J. Hutt and J. Caldwell, *Clin. Pharmacokinet.*, *9*:371 (1984).
182. D. M. Johnson, A. Reuter, J. M. Collins, and G. F. Thompson, *J. Pharm. Sci.*, *68*:112 (1979).
183. P. L. Anelli, C. Tomba, and F. Uggeri, *J. Chromatogr.*, *589*:346 (1992).
184. J. P. Kamerling, M. Duran, G. J. Gerwig, D. Ketting, L. Bruinvis, J. F. G. Vliegenthart, and S. K. Wadman, *J. Chromatogr.*, *222*:276 (1981).
185. W. A. Konig and I. Benecke, *J. Chromatogr.*, *195*:292 (1980).
186. M. Hasegawa and I. Matsubara, *Anal. Biochem.*, *63*:308 (1975).
187. P. Mohr, N. Waespe-Sarcevic, C. Tamm, K. Gawronska, and J. K. Gawronski, *Helv. Chim. Acta*, *66*:2501 (1984).
188. T. Kaneda, *J. Chromatogr.*, *366*:217 (1986).
189. K. D. Ballard, T. D. Eller, and D. R. Knapp, *J. Chromatogr.*, *275*:161 (1983).
190. J. P. Guette and A. Horeau, *Tetrahedron Lett.*, 3049 (1965).
191. E. J. D. Lee, K. M. Williams, G. G. Graham, R. O. Day, and G. D. Champion, *J. Pharm. Sci.*, *73*:1542 (1984).
192. E. J. D. Lee, K. Williams, R. Day, G. Graham, and D. Champion, *Br. J. Clin. Pharmacol.*, *19*:669 (1985).
193. T. Yamaguchi and Y. Nakamura, *Drug. Metab. Disp.*, *13*:614 (1985).
194. A. Rubin, M. P. Knadler, P. P. K. Ho, L. D. Bechtol, and R. L. Wolen, *J. Pharm. Sci.*, *74*:82 (1985).
195. N. N. Singh, F. M. Pasutto, R. T. Coutts, and F. Jamali, *J. Chromatogr.*, *378*:125 (1986).
196. N. N. Singh, F. Jamali, F. M. Pasutto, A. S. Russell, R. T. Coutts, and K. S. Drader, *J. Pharm. Sci.*, *75*:439 (1986).
197. H. Nagashima, Y. Tanaka, and R. Hayashi, *J. Chromatogr.*, *345*:373 (1985).
198. H. Nagashima, Y. Tanaka, H. Watanabe, R. Hayashi, and K. Kawada, *Chem. Pharm. Bull.*, *32*:251 (1984).
199. S. Bjorkman, *J. Chromatogr.*, *339*:339 (1985).
200. K. C. Rice, *J. Org. Chem.*, *47*:3617 (1982).
201. M. Jiang and D. M. Soderlund, *J. Chromatogr.*, *248*:143 (1982).
202. M. Hirama, T. Noda, and S. Ito, *J. Org. Chem.*, *50*:127 (1985).
203. A. Abas and P. J. Meffin, *J. Pharmacol. Exp. Ther.*, *240*:637 (1987).
204. Avgerinos and A. J. Hutt, *J. Chromatogr.*, *415*:75 (1987).
205. B. Blessington, N. Crabb, S. Karkee, and A. Northage, *J. Chromatogr.*, *469*:183 (1989).

206. N. N. Singh, F. Jamali, F. M. Pasutto, and R. T. Coutts, *J. Chromatogr.*, *382*:331 (1986).
207. J. Goto, M. Ito, S. Katsuki, N. Saito, and T. Nambara, *J. Liq. Chromatogr.*, *9*: 683 (1986).
208. K. Shimada, E. Haniuda, T. Oe, and T. Nambara, *J. Liq. Chromatogr.*, *10*:3161 (1987).
209. E. L. Palylyk and F. Jamali, *J. Chromatogr.*, *568*:187 (1991).
210. A. Carlson and O. Gyllenhaal, *J. Chromatogr.*, *508*:333 (1990).
211. O. Pelkonen and D. W. Nebert, *Pharmacol. Rev.*, *34*:189 (1982).
212. T. Watabe, A. Hiratsuka, and T. Tsurumori, *Biochem. Biophys. Res. Comm.*, *130*:65 (1985).
213. I. G. C. Robertson, H. Jensson, B. Mannervik, and B. Jernstrom, *Carcinogenesis*, *7*:295 (1986).
214. R. N. Armstrong, W. Levin, D. E. Ryan, P. E. Thomas, H. D. Mah, and D. M. Jerina, *Biochem. Biophys. Res. Commun.*, *106*:602 (1982).
215. P. J. van Bladeren, R. N. Armstrong, D. Cobb, D. R. Thakker, D. E. Ryan, P. E. Thomas, N. D. Sharma, D. R. Boyd, W. Levin, and D. M. Jerina, *Biochem. Biophys. Res. Comm.*, *106*:602 (1982).
216. P. J. van Bladeren, J. M. Sayer, D. E. Ryan, P. E. Thomas, W. Levin, and D. M. Jerina, *J. Biol. Chem.*, *260*:10226 (1985).
217. P. J. van Bladeren, K. P. Vyas, J. M. Sayer, D. E. Ryan, P. E. Thomas, W. Levin, and D. M. Jerina, *J. Biol. Chem.*, *259*:8966 (1984).
218. G. L. Foureman, C. Harris, F. P. Guengerich, and J. R. Bend, *J. Pharmacol. Exp. Ther.*, *248*:492 (1989).
219. E. Caspi, S. Shapiro, and J. U. Piper, *Tetrahedron*, *37*:3535 (1981).
220. A. Panthananickal and L. J. Marnett, *Chem.-Biol. Interactions*, *33*:239 (1981).
221. A. Panthananickal, P. Weller, and L. J. Marnett, *J. Biol. Chem.*, *258*:4411 (1983).
222. J. Gal, *J. Chromatogr.*, *331*:349 (1985).
223. D. Burnett and J. Gal, *The Pharmacologist*, *27*:249 (1985).
224. W. H. Pirkle, M. R. Robertson, and M. H. Hyun, *J. Org. Chem.*, *49*:2433 (1984).
225. Z. D. Kai, S. Y. Kang, M. J. Ke, Z. Jin, and H. Liang, *J. Chem. Soc., Chem. Comm.*, 168 (1985).
226. D. S. Sampath and P. Balaram, *Biochim. Biophys. Acta*, *882*:183 (1986).
227. D. S. Sampath and P. Balaram, *J. Chem. Soc., Chem. Comm.*, 649 (1986).
228. J. M. Reid, J. F. Stobaugh, and L. A. Sternson, *Anal. Chem.*, *61*:441 (1989).
229. J. Kohler, A. Deege, and G. Schomburg, *Chromatographia*, *18*:119 (1984).
230. C. R. Noe, *Chem. Ber.*, *115*:1591 (1982).
231. M. R. Little, *J. Biochem. Biophys. Meth.*, *11*:195 (1985).

5

The Direct Resolution of Enantiomeric Drugs by Chiral-Phase Gas Chromatography

Wilfried A. König *Institut für Organische Chemie, Universität Hamburg, Hamburg, Germany*

I. INTRODUCTION

Similar to many other cases of biologically active compounds, stereochemistry influences the pharmacological effect of a chiral drug. This can be explained by the fact that there is only one energetically favorable (specific) interaction of an active molecule with its receptor, both being chiral structures. Qualitative and quantitative differences are caused by different receptor affinities as demonstrated in Fig. 1 (1). The metabolism (biotransformation) of drugs is mainly caused by enzymes, which are chiral macromolecules and discriminate between substrate molecules of different stereochemistry. This may result in metabolites of different activity and in different pharmacokinetics, resorption, and excretion. Therefore, racemic drugs should be looked on as a 1:1 mixture of two different compounds.

About 40% of synthetic pharmaceuticals are chiral, but only approximately 10% of them are used as pure stereoisomers, whereas the rest are applied as racemic mixtures. A disastrous accident with racemic thalidomide combining the sedative action of the *R* enantiomer with the highly teratogenic properties of the *S* enantiomer has brought the importance of stereochemistry to public attention.

There are many less dramatic cases of different activities of stereoisomers. Even in cases where only one of the enantiomers is active, its inactive optical antipode should be avoided and considered as "enantio-

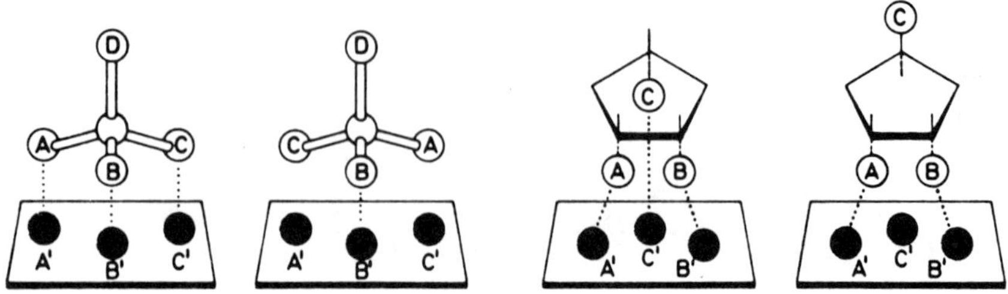

OPTICAL ISOMERS CIS-TRANS-ISOMERS

FIGURE 1 Schematic representation of the molecular interaction of a receptor with different stereoisomers of drugs. (From Schunack, Mayer, and Haake [1]).

meric waste," which needs to be metabolized by the organism of a patient and ultimately contaminates the environment. There is also a possibility that negative side effects may not be obvious and sometimes hard to recognize. In addition, findings exist of synergistic effects of two enantiomers or of different although favorable effects. Recently, toxicity studies on both enantiomers of a racemic drug have begun to be demanded by the regulatory authorities of most countries before release of a racemic drug.

Growing awareness of the relevance of drug stereochemistry has not only greatly stimulated the development of methods for asymmetric synthesis of enantiomerically pure drugs as well as the preparative separation of racemic pharmaceuticals, but also initiated the development of methods for precise and sensitive determination of enantiomeric proportions. On the other hand, access to pure stereoisomers has enabled scientists to study physiological activity and stereoselective metabolism of enantiomerically pure drugs.

II. ENANTIOSELECTIVE GAS CHROMATOGRAPHY

A. Chiral Diamide Phases

Capillary gas chromatography with optically active stationary phases became a well-established technique for stereochemical analysis after the pioneering work of Gil-Av and his associates in the mid–1960s (2). Thermal stability of the initially low-molecular-weight amino acid and peptide derivatives was greatly improved after the introduction of chiral polysiloxanes by Frank, Nicholson, and Bayer (Chirasil-val [1], 1977) (3) and of

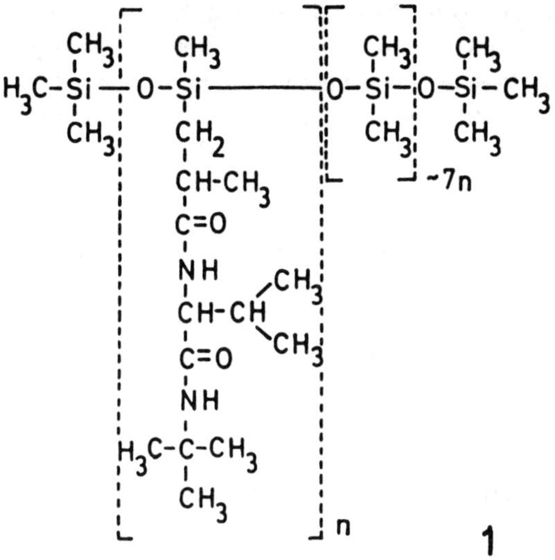

1

similar optically active polymer diamide phases (e.g., XE-60-L-valine-(*S*)- and (*R*)-α-phenylethylamide, [2], 1981) by König et al. (4). As in all chiral diamide phases, the stereoselective interaction of the enantiomers with the chiral selector molecules (diasteromeric interaction) is mainly based on hydrogen-bonding forces. The ability to separate enantiomers using this type of phase is, therefore, with only few exceptions restricted to sub-

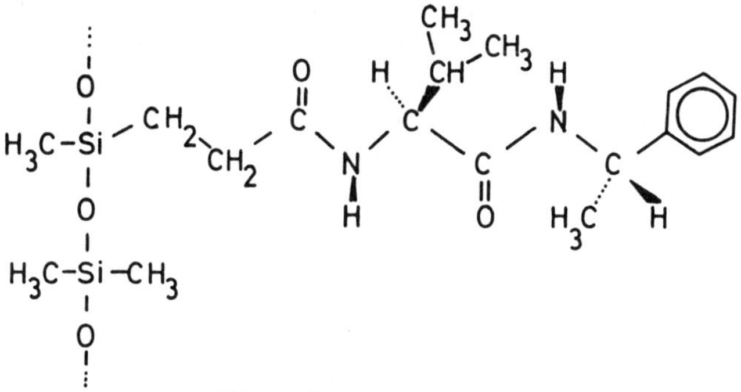

XE-60-L-VAL-(S)-α-PEA

2

strates with hydrogen donor or acceptor sites. Gil-Av et al. have proposed the formation of transient five- and seven-membered cyclic association complexes for explaining enantioselective selector-selectand interaction (5). By special derivatization procedures (formation of carbamates, urethanes, cyclic carbonates, oxazolidin-2-ones, or oximes), the range of compound classes susceptible to enantiomeric resolution could be enlarged (6–8). The results obtained with diamide-type chiral stationary phases were summarized in a monograph (9). A large number of chiral sympathomimetic and β-receptor-blocking drugs could be resolved after conversion of the amino alcohols to cyclic oxazolidin-2-ones with phosgene or phosgene-substituting reagents (10,11). A comprehensive summary of these studies was published in the previous edition of this work (12).

B. Cyclodextrin Derivatives

In 1983 Koscielski et al. reported on gas chromatographic enantiomer separations with cyclodextrins (cyclic glucose oligomers with 6, 7, or 8 α-1→4-linked glucose units, Fig. 2) dissolved in formamide as chiral stationary phases in packed-column gas chromatography (13). In the fundamental studies of Smolkova-Keulemansova (14,15), the unique "topological" selectivity of cyclodextrins toward guest molecules that could be included in the macrocyclic cavity (Fig. 2, formation of inclusion complexes) had already been indicated, although separation efficiency of these columns was rather poor. Enantioselectivity of the complex formation of guest molecules with cyclodextrins was demonstrated by Cramer who separated enantiomers by fractionated crystallization as early as 1952 (16). Szejtli et al. were able to improve gas chromatographic efficiency by using per-*O*-methylated cyclodextrins and capillary columns (17,18). Due to the high melting points of these cyclodextrin derivatives, the operation temperature had only a very limited range. The great potential of cyclodextrins was again demonstrated by Armstrong et al. who applied cyclodextrins bonded to silica gel as chiral support in liquid chromatography and succeeded in the separation of chiral pharmaceuticals with mainly aromatic substituents (19).

A breakthrough in the application of cyclodextrins in capillary gas chromatography was achieved in 1988 by substitution of the sugar hydroxylic groups with alkyl, for example, pentyl, residues and by preparing selectively alkylated/acylated derivatives (20,21). The resulting hydrophobic compounds are liquid at room temperature, soluble in unpolar solvents, temperature-stable up to 400°C, and highly enantioselective. At the same time, it was shown that the rather hydrophilic per-*O*-methylated cyclodextrins could be used over a temperature range of about 70–220°C after dilution with a polysiloxane, particularly OV 1701 (22). With both

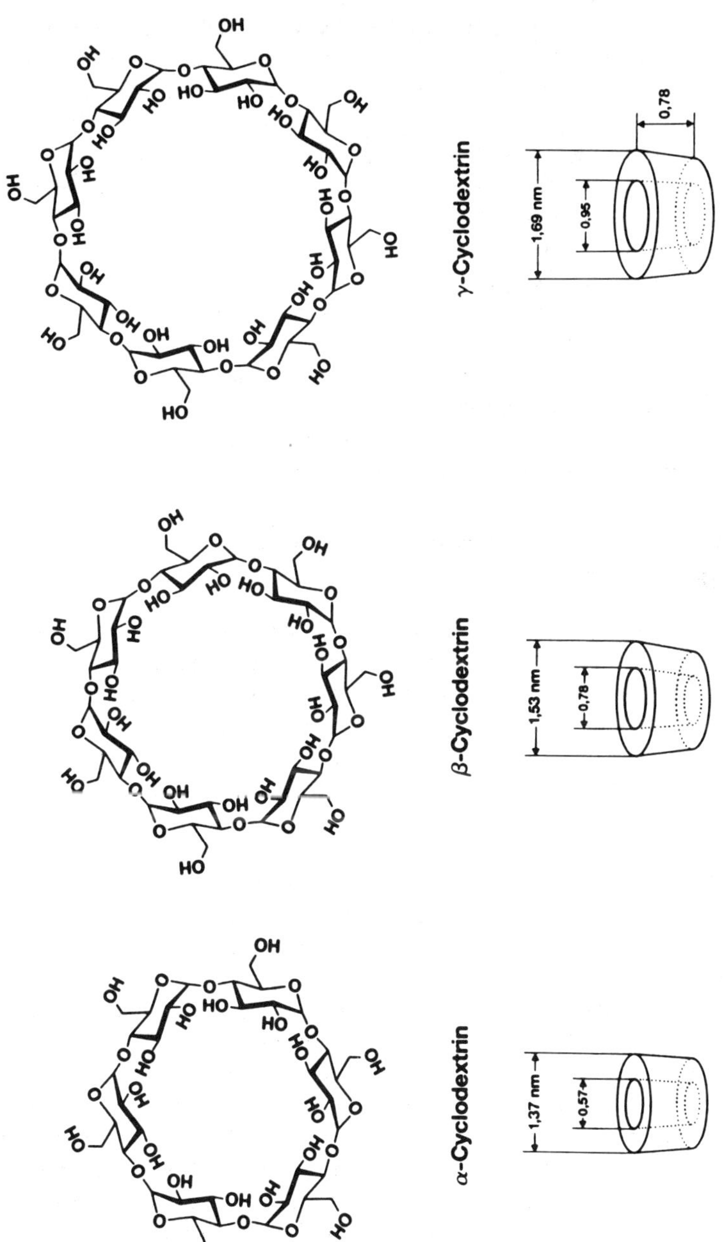

FIGURE 2 Chemical structure and molecular dimensions of α-, β-, and γ-cyclodextrin.

approaches, enantiomeric resolutions of a wide variety of racemic compounds could be achieved, including enantiomers of many compound classes, which had never been resolved by any chromatographic method before (23,24). Somewhat later several other research groups, including those of Armstrong, Li, and Pitha (25), Mosandl et al. (26), Fischer et al. (27), and others, joined in research activities on cyclodextrin phases and an increasing number of papers in this field are published.

Today mainly fused silica tubing is used to prepare highly efficient and well-deactivated capillary columns, although Pyrex glass proved to be equally suited, except that handling of glass capillaries is not common knowledge any more. The most recent advancement was reported by Schurig, who chemically immobilized per-*O*-methylated cyclodextrins and obtained capillary columns that can be used in supercritical fluid chromatography (28), which may be particularly useful for the investigation of drugs with volatilities insufficient for gas chromatography.

III. PREPARATION OF HYDROPHOBIC CYCLODEXTRIN DERIVATIVES

Hydroxy groups of cyclodextrins can be regioselectively alkylated in positions 2 and 6 of the glucose residues due to the high reactivity of these hydroxy groups. The 3-positions are much less reactive and can be alkylated or acylated under more drastic conditions. The preparation of pure per-*O*-alkylated cyclodextrin derivatives, which are mainly used for the separation of unpolar compounds (alkanes, alkenes, spiroacetals, alkyl halides, etc.) as well as a method for characterization of these phases have been described in detail (29).

Particularly useful for enantiomer separation of more polar compounds are 2,6-di-*O*-pentylated/3-*O*-acylated (acetyl, butyryl) cyclodextrin derivatives. They are prepared by acylation of the 2,6-di-*O*-pentylated intermediates (30–32). Substitution of hydroxy groups of cyclodextrins not only causes hydrophobicity and solubility in unpolar solvents and a decrease in the melting point, but also changes the conformation of the macrocyclic system, which again has a great influence on the selectivity toward the structure of a guest molecule. We found that all the selectively different-position di-pentylated/methylated cyclodextrin derivatives have quite different properties and are generally very useful chiral stationary phases. Cyclodextrin derivatives with selective alkylation in the 6- or 2-position can be prepared by using removable protecting groups (33,34). An example is shown in Scheme 1. Some of the hydrophobic cyclodextrin derivatives that were prepared and evaluated in our group are shown in

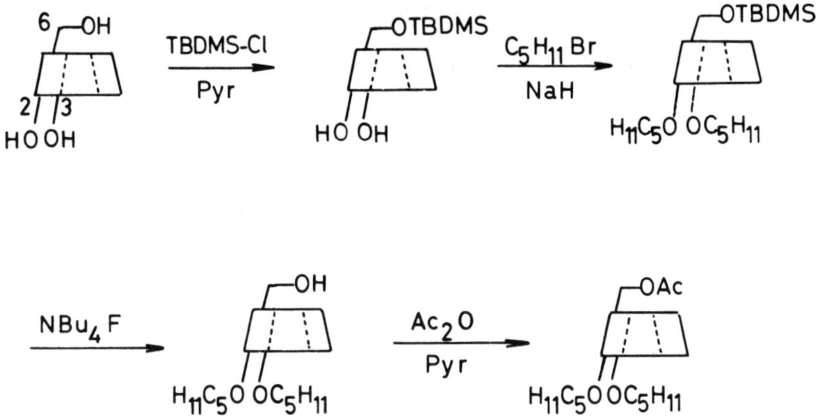

SCHEME 1 Reaction pathway for the preparation of hydrophobic cyclodextrin derivatives with "inverse" substituion pattern. (TBDMS-Cl = tert-butyldimethylsilyl chloride, NBu_4F = tetra-*n*-butylammonium fluoride.) [From König et al. (33).]

Fig. 3. It should be emphasized that a careful examination of the cyclodextrin derivatives must be performed after the final purification. Several authors have indicated that incomplete derivatization may be the reason for irreproducible results. Except for thorough NMR investigations (^{1}H- and ^{13}C-NMR), the "reductive cleavage" of cyclodextrin derivatives and subsequent quantitative gas chromatographic analysis of the 1,5-anhydrosorbitol derivatives as developed by Mischnick-Lübbecke and Krebber (35) have proved a sensitive and reliable method to detect even traces of under- or over-alkylated derivatives.

IV. APPLICATION OF ENANTIOSELECTIVE GAS CHROMATOGRAPHY TO PHARMACEUTICALS

The introduction of "inclusion" as a new principle of selectivity has considerably extended the range of applications of enantioselective gas chromatography in enantiomer separation of drugs, although systematic investigations have not yet been published. In many cases, chiral pharmaceuticals suffer from insufficient volatility, however, in some cases, derivative formation may be a way to enhance volatility to the degree that gas chromatography becomes applicable. This is particularly true for polyfunctional compounds as amino acids (e.g., L-DOPA), amino alcohols (e.g., adrenergic drugs), amines (e.g., amphetamine), and diols (*trans-*

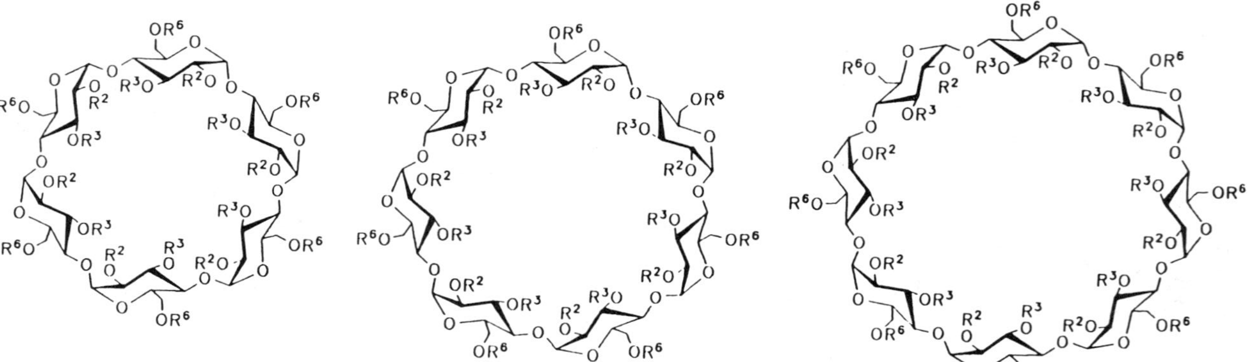

FIGURE 3 Structures of hydrophobic α-, β-, and γ-cyclodextrin derivatives:
$R^2 = R^3 = R^6 =$ *n*-pentyl
$R^2 = R^6 =$ *n*-pentyl, $R^3 =$ methyl
$R^2 = R^3 =$ *n*-pentyl, $R^6 =$ methyl
$R^2 = R^6 =$ *n*-pentyl, $R^3 =$ acetyl
$R^2 = R^3 =$ *n*-pentyl, $R^6 =$ acetyl
$R^2 = R^6 =$ *n*-pentyl, $R^3 =$ butyryl
$R^2 = R^3 =$ *n*-pentyl, $R^6 =$ butyryl

sobrerol). For other compounds missing sites for derivatization, for example, compounds with heterocyclic structures (hydantoins, succinimides, barbiturates), enantioselective gas chromatography using cyclodextrin derivatives has proved to be superior to the formerly used chiral polysiloxane stationary phases.

Although *high-pressure liquid chromatography* (HPLC) with chiral stationary phases is certainly more popular in stereochemical drug analysis, enantioselective gas chromatography may be beneficial in certain cases where sensitivity and precision are major issues. Also, once the separation conditions are elaborated, gas chromatography will provide highly reproducible results, whereas HLPC conditions for optimal resolutions may be more sensitive to operational parameters. As will be shown, in certain cases enantioselective gas chromatography is the only way to separate a racemic pharmaceutical.

A. Amino Alcohol-Type Structures and Sympathomimetics

A great number of compounds of this type, including β-receptor blocking agents and sympathomimetic drugs, have been resolved on capillary columns with XE-60-L-valine-(*R*)-α-phenylethylamide (10,11). For the separation of most of the important β-blockers meanwhile, efficient HPLC-methods have been established and gas chromatography has lost some of its attractiveness because of the necessity to derivatize samples. It was shown, however, that ephedrine, pseudoephedrine, and norephedrine can be easily resolved using 3-*O*-acylated-2,6-di-*O*-pentylated β-cyclodextrin (Lipodex D®; Fig. 4, (36)).* Probably this chiral stationary phase would be suitable to separate many more compounds of this type. Results of systematic investigations have not yet been published.

B. Amine-Type Drugs

Lipodex D is the chiral phase of choice to investigate amine-type pharmaceuticals (30). Figure 5 demonstrates the separation of some trifluoroacetylated compounds. Amphetamine is a nervous system stimulant. Its *S*-enantiomer is three to four times more potent than the *R*-enantiomer (38). Mexiletin is used as an antiarrhythmic drug, pholedrine is a sympathomimetic agent, which could not be separated on chiral diamide phases due to the lack of a hydrogen-bonding donor site. An extraordinarily large separation effect is observed for the enantiomers of tranylcypromine, a drug used as an antidepressant. In this case, the different

---

*Capillary columns with this phase are available from Macherey-Nagel, D-5160 Düren, Germany.

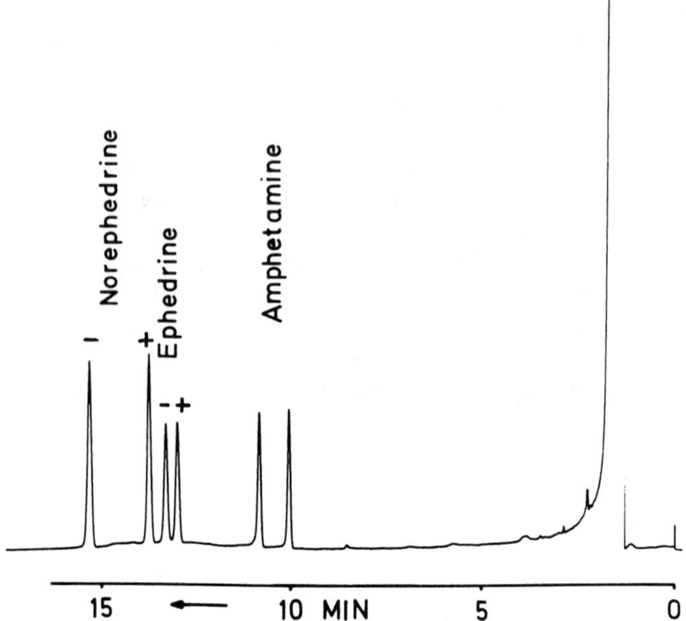

FIGURE 4 Enantiomer separation of (trifluoroacetylated) amphetamine, ephedrine, and norephedrine on a 40-m Pyrex glass capillary with heptakis(3-*O*-acetyl-2,6-di-*O*-pentyl)-β-cyclodextrin at 150°C. Carrier gas (as in all other figures) is hydrogen. [Trifluoroacetylation with trifluoroacetic acid anhydride in dichloromethane (1:4, v/v) for 30 min at 100°C.]

physiological activities of the individual enantiomers have been well documented (39).

The potency of enantioselective gas chromatography became apparent in the pharmacokinetic studies of flecainide [3] and two of its chiral metabolites as carried out by Fischer et al. (40). Trace amounts of these compounds in plasma samples were screened for enantiomeric discrimination.

3

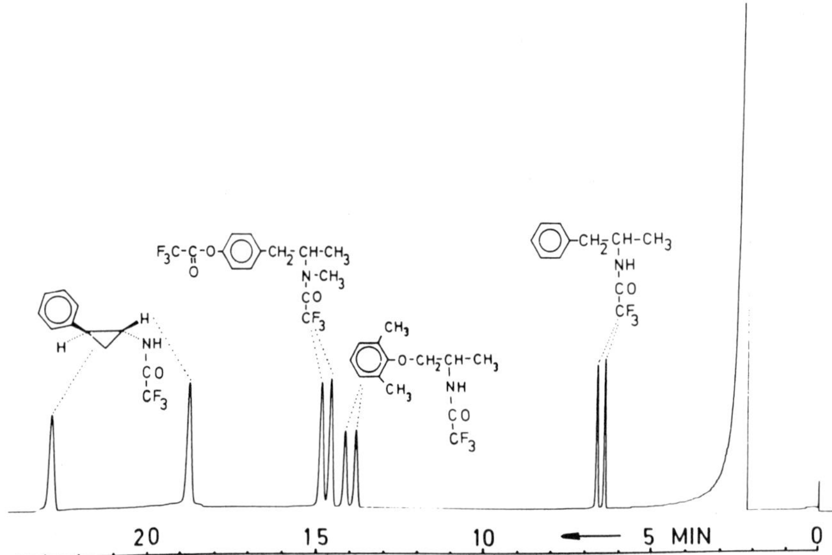

FIGURE 5 Enantiomer separation of trifluoroacetylated amphetamine, mexiletin, pholedrine and tranylcypromine. 45-m Pyrex glass capillary with 3-ac-2,6-pe-β-CD at 175°C.

C. Amino Acid-Type Drugs

Although enantiomer separation of amino acids has always been a domain of chiral diamide phases such as Chirasil-val [1] or XE-60-L-val-(*S*)-α-phenylethylamide [2], the development of hydrophobic cyclodextrin derivatives has affected even this field of applications of enantioselective gas chromatography. Octakis(3-*O*-butyryl-2,6-di-*O*-pentyl)-γ-cyclodextrin [Lipodex E®, (32) (see footnote on p. 115)] is highly selective for amino acid enantiomers. Since hydrogen bonding is not necessary for a determination of the enantiomers, some unusual amino acids can be separated that could not be resolved before, including β-amino acids, α-alkylated, and *N*-alkylated amino acids.

Proof of enantiomeric purity is necessary in the case of L-3,4-dihydroxyphenylalanine (L-DOPA), a drug used to treat Parkinson's disease, because of the high toxicity of the D enantiomer. Previously, D,L-DOPA was resolved by Frank, Nicholson, and Bayer using Chirasil-val (41). Figure 6 shows the separation of D,L-DOPA on a column with perpentylated α-cyclodextrin. In the case of α-methyl-DOPA (Aldomet) [4], a drug that is used in its L form for its antihypertensive activity, the enan-

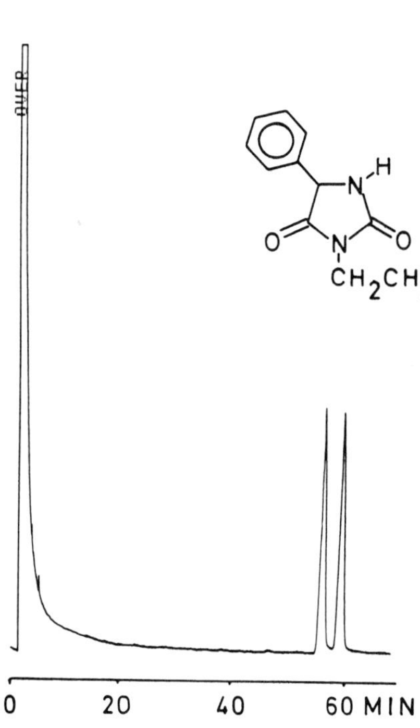

FIGURE 17 Enantiomer separation of ethotoin on a 25-m fused silica capillary with per-pe-α-CD at 170°C.

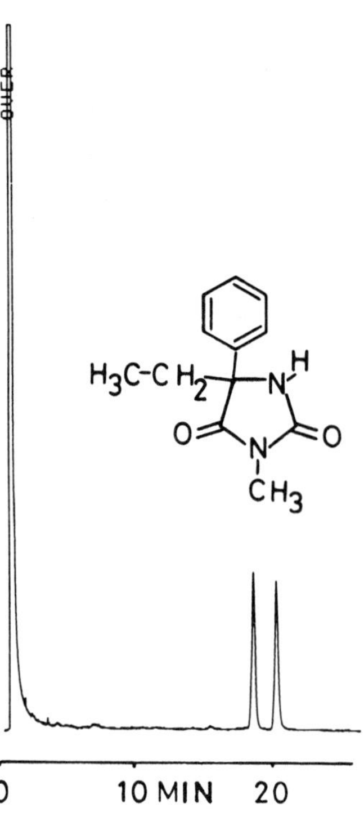

FIGURE 18 Enantiomer separation of mephenytoin on a 10-m fused silica capillary with 3-bu-2,6-pe-γ-CD (60:40, w/w in OV1701) at 160°C.

determined by Knabe and Reischig (55). The enantiomers of methyprylon, also a hypnotic drug, are separated on capillaries with per-*O*-methyl-β-cyclodextrin or on 2-*O*-methyl-3,6-di-*O*-pentyl-β-cyclodextrin, as shown in Fig. 22.

F. Essential Oils

With the advent of hydrophobic cyclodextrin derivatives, many chiral constituents of essential oils became the subject of a detailed investigation of their enantiomeric composition. Several essential oils are used as additives in pharmaceutical formulations, for example, pine needle oil or peppermint oil. They are not only used for their pleasant flavor or fragrance, monoterpene hydrocarbons like α-pinene [8], β-pinene [9], and

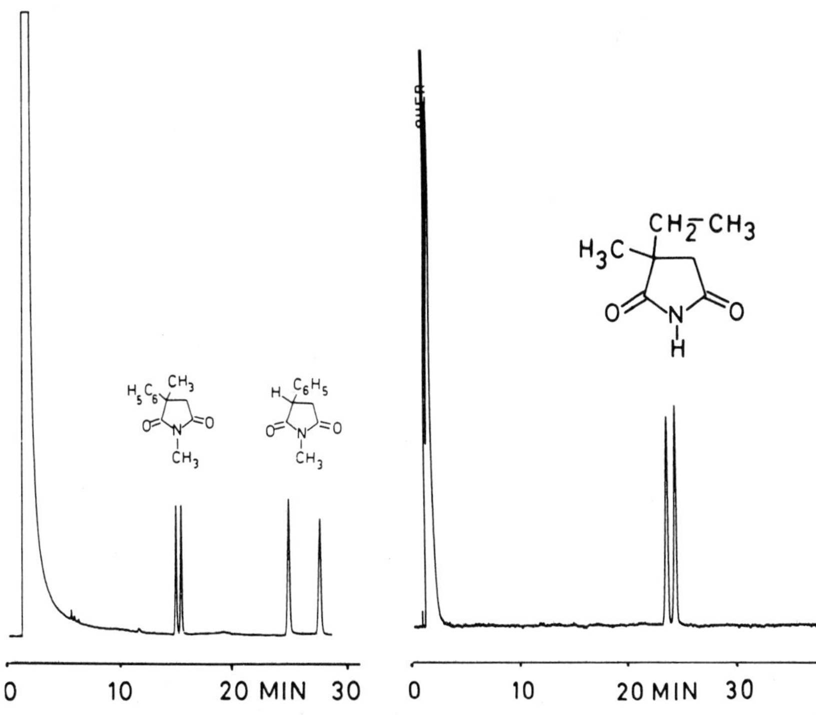

FIGURE 19 Enantiomer separation of mesuximide (left) and phensuximide (right) on a 25-m fused silica capillary with 3-ac-2,6-pe-β-CD at 180°C.

FIGURE 20 Enantiomer separation of ethosuximide on a 25-m fused silica capillary with 6-me-2,3-pe-γ-CD at 115°C.

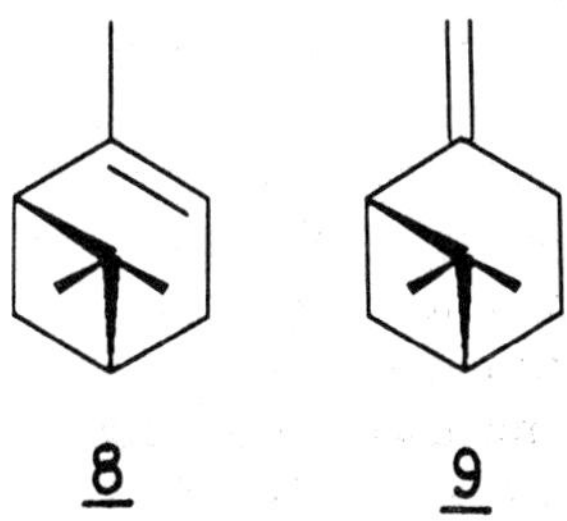

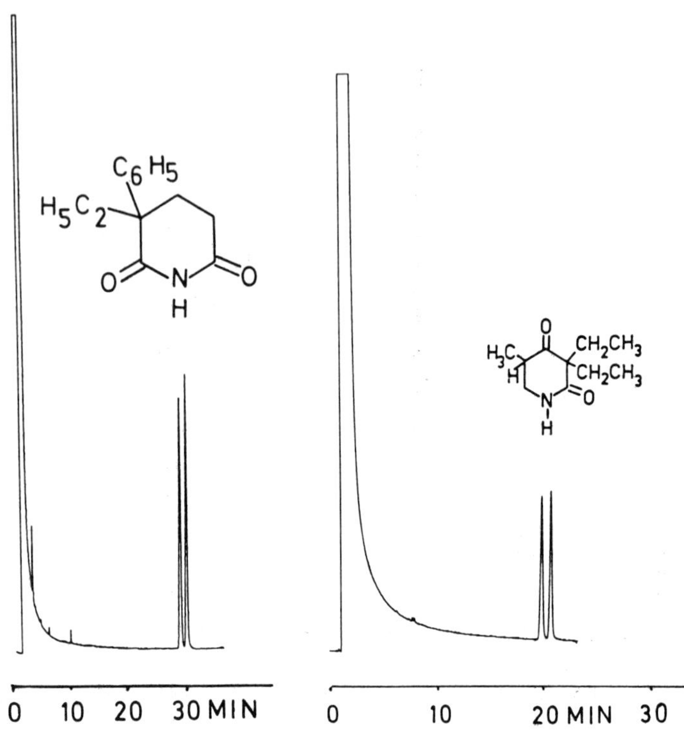

FIGURE 21 Enantiomer separation of gluteth-imide on a 25-m Pyrex glass capillary with 2-me-3,6-pe-β-CD at 168°C.

FIGURE 22 Enantiomer separation of methyprylon on a 25-m Pyrex glass capillary with 2-me-3,6-pe-β-CD at 140°C.

limonene [10], major constituents of all conifer needle oils, are also known for their antibacterial or antifungal activity. Figure 23 shows the separation of the enantiomers of these monoterpenes. Pharmacological activities are also attributed to (−)-menthol [11], the main constituent of peppermint oil, including antiseptic and anesthetic properties. It is an important ingredient of pharmaceutical formulations for the treatment of respiratory infections. Enantioselective gas chromatography allows a differentiation of these constituents with respect to their enantiomeric composition (33, 56, 57) (Fig. 24), thus providing a good means to detect adulterations. Because of the complexity of essential oils, multidimensional gas chromatography may be necessary to obtain unambiguous results (58). (−)-

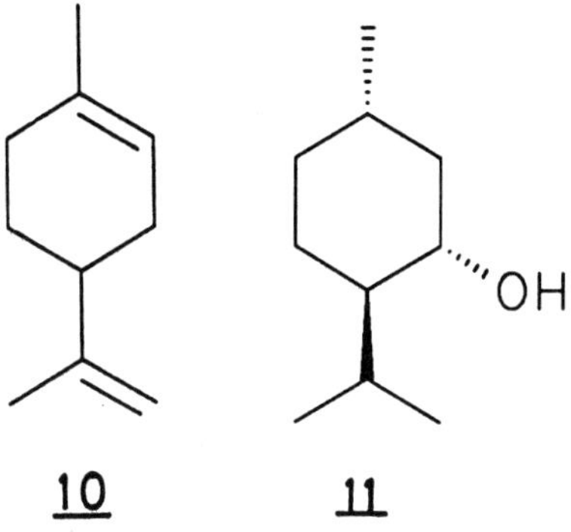

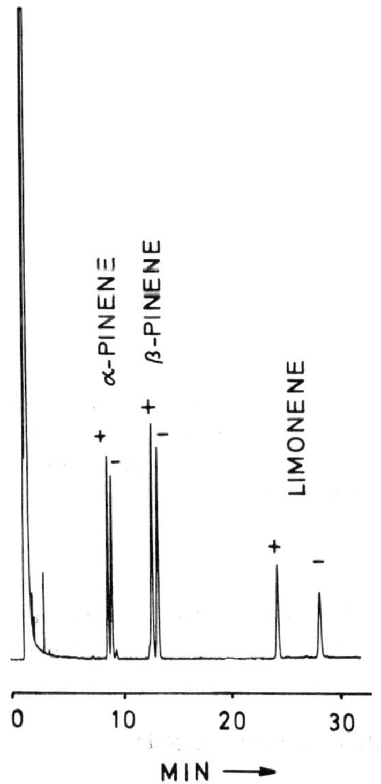

FIGURE 23 Enantiomer separation of α-pinene, β-pinene, and limonene on a 25-m Pyrex glass capillary with 6-me-2,3-pe-γ-CD at 55°C.

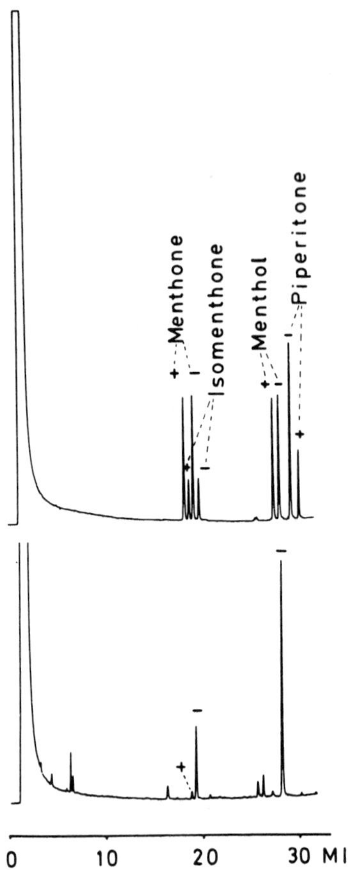

FIGURE 24 Enantiomer separation of menthone, isomenthone, menthol, and piperitone (top) and determination of enantiomeric composition of these compounds in peppermint oil (bottom). 25-m Fused silica capillary with 6-me-2,3-pe-γ-CD at 75°C (15 min), temperature program 2°C/min to 150°C.

Menthyl acetate, also a major component of peppermint oil, should be enantiomerically pure. The presence of the (+) enantiomer is considered an indication of adulteration (59, Fig. 25).

G. Enantioselective Gas Chromatography in Purity Control of Synthetic Building Blocks

A major hurdle on the road to enantiomerically pure drugs is the development of efficient methods of asymmetric synthesis. In many cases, small chiral compounds of synthetic or natural origin are used as starting

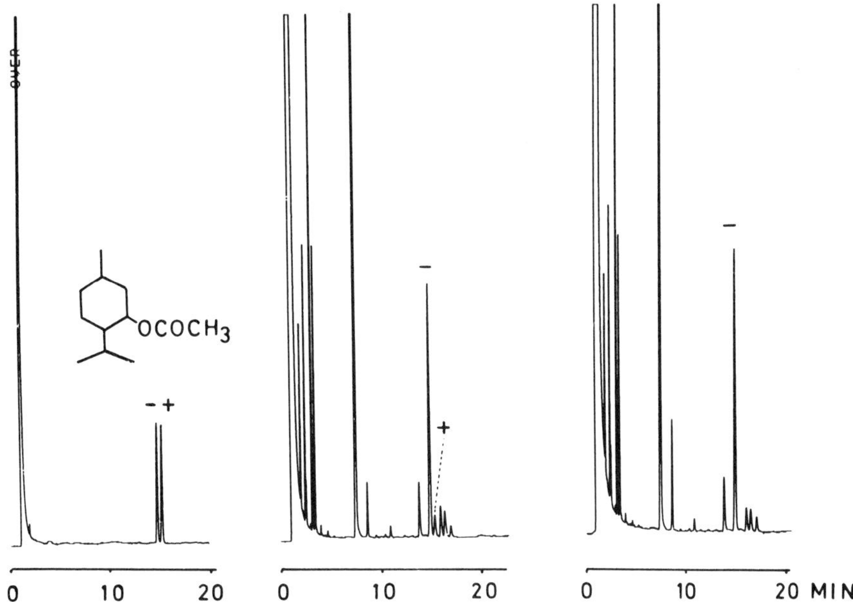

FIGURE 25 Enantiomer separation of menthyl acetate (left) and determination of enantiomeric purity of menthyl acetate in two commercial peppermint oil samples. The presence of (+)-menthyl acetate in one of the samples (middle) may be an indication of adulteration. 22-m Pyrex glass capillary with 6-me-2,3-pe-β-CD at 85°C.

materials or chiral auxiliaries for multistep syntheses. To obtain products of high enantiomeric purity, it is important to have highly pure synthons. A typical example may demonstrate the importance of enantioselective gas chromatography to control the enantiomeric purity of chiral building blocks. Enantioselective epoxidation of allylalcohols according to Katsuki and Sharpless (60) may serve as an important primary step in an asymmetric synthesis. Besides many other applications, this procedure may be used to prepare in only few steps (*S*)-propranolol, an important β-blocker, as shown in Fig. 26 (61). Starting from allylalcohol, Sharpless epoxidation results in glycidol (epoxypropanol) of high enantiomeric purity, which can easily be proved by enantioselective gas chromatography (37). The separation of the enantiomers is demonstrated in Fig. 27.

V. CONCLUSION

It should be emphasized that only a small fraction of separation problems of chiral pharmaceuticals has so far been addressed by enantioselective chromatographic separation of stereoisomeric compounds and, in particular, the spectacular growth in HPLC-CSPs for the separation of enantiomeric molecules. In the past four years, the number of commercially available HPLC-CSPs has grown from 35 to over 50 and will surely increase again by the time this manuscript is published.

The large number of available HPLC-CSPs presents the user with a broad range of possibilities and the problem of deciding which phase to use. These problems can be minimized by grouping the HPLC-CSPs according to the chiral recognition process taking place on the column. The approach described and illustrated below was initially proposed in 1986 (6), revised in subsequent publications up to 1988 (1,7,8), and is now presented in its current form. What is missing from this presentation are the extensive compilations of applications included in the first edition. The use of HPLC-CSPs has grown to such an extent that whole volumes have been dedicated to this topic and the interested reader is directed to some of the recent works in this area (9–11).

II. CLASSIFYING HPLC-CSPS USING THE CHIRAL RECOGNITION MECHANISM

A. The Solute/CSP Complex

HPLC-CSPs are based on molecules of known stereochemical composition immobilized on liquid chromatographic supports. Single enantiomorphs, diastereomers, diastereomeric mixtures, and chiral polymers (such as proteins) have been used as the chiral selector. The chiral recognition mechanisms operating on these phases are the result of the formation of temporary diastereomeric complexes between the enantiomeric solute molecules and immobilized chiral selector. The difference in energy between the resulting diastereomeric solute/CSP complexes determines the magnitude of the observed stereoselectivity, whereas the sum total of the interactions between the solute and CSP, chiral and achiral, determines the observed retention and efficiency.

A simplified version of this process is presented in Fig. 1. In this example, the two solute/CSP complexes, *d*-solute-CSP and *l*-solute-CSP, have different free energies, with the *d*-solute-CSP complex being more stable, that is, the one with the lower free energy. As a result, the *d* isomer will remain on the column longer than the *l* isomer, and the two enantiomers will be resolved.

B. Formation of the Solute CSP Complex

The chiral recognition mechanism can be broken down into parts if one remembers that the parts are interdependent and cannot exist apart from one another. These parts are (1) how the solute/CSP complexes are formed

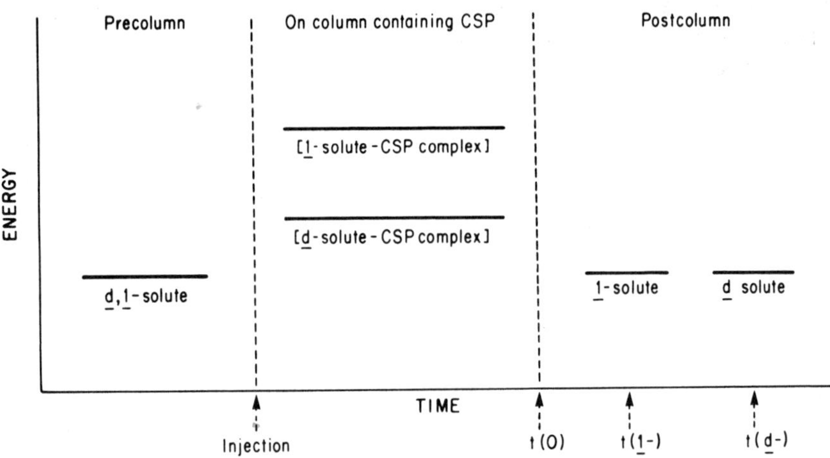

FIGURE 1 A general scheme for the mechanism of chiral resolution on an HPLC chiral stationary phase. Enantioselectivity (α) $= \dfrac{t(d\text{-}) - t(0)/t(0)}{t(l\text{-}) - t(0)/t(0)}$.

and (2) how the stereochemical differences are expressed during and after the formation of these complexes. The *key step* is the formation of the diastereomeric complex, and the current HPLC-CSPs can be broken down into five basic types on the basis of the solute/CSP bonding interactions (8). These are the following:

Type I. When the solute/CSP complexes are formed by attractive interactions, hydrogen bonding, π-π, dipole stacking, etc., between the solute and CSP

Type II. When the primary mechanism for the formation of the solute/CSP complex is through attractive interactions, but inclusion complexes also play an important role

Type III. When the solute enters into chiral cavities within the CSP to form inclusion complexes

Type IV. When the solute is part of a diastereomeric metal complex (chiral ligand-exchange chromatography)

Type V. When the CSP is a protein and the solute/CSP complexes are based on combinations of hydrophobic and polar interactions

C. Chiral Recognition Within Solute/CSP Complexes

Within the solute/CSP complex, chiral recognition is based on the "three-point interaction" model proposed by Dalgliesh (12). According to this mechanism, three interactions occur between the solute and chiral selec-

TABLE 2 Commercially Available Type II CSPs

Chiral stationary phase	Chiral selector
Type I—Cellulose	
CTA	Microcrystalline cellulose triacetate
CTB	Microcrystalline cellulose tribenzoate
Type II—Cellulose	
Chiralcel OA	Cellulose triacetate
Chiralcel OB	Cellulose tribenzoate
Chiralcel OC	Cellulose trisphenylcarbamate
Chiralcel OD	Cellulose tris(3,5-dimethylphenyl carbamate)
Chiralcel OF	Cellulose tris(4-chlorophyl carbamate)
Chiralcel OG	Cellulose tris(4-methylphenyl carbamate)
Chiralcel OJ	Cellulose tris(4-methyl-benzoate)
Chiralcel OK	Cellulose tricinnamate
Type II—Amylose	
Chiralpak AD	Amylose tris(3,5-dimethylphenylcarbamate)
Chiralpak AS	Amylose (*S*)-α-methylbenzylcarbamate

1. *Solute Structure*

Although CTA-I has been used to resolve a large number of compounds, the solutes appear to be limited to molecules that contain a phenyl group. This limitation may be due to the fact that it is the phenyl moiety that enters the chiral cavity to form the solute/CSP complex, as postulated by Hesse and Hagel (40). Francotte and Wolf (41) have recently reported the results of a study on the effect on enantioselectivity of *para*-substituents in the aromatic moieties of a series of compounds. They concluded that there are (41)

> . . . Multiple interaction sites in CTA-I, each exhibiting different chiral discriminations depending upon the local sterical and electronic configuration. In fact, this conclusion is quite conceivable considering the supramolecular structure of CTA-I, which consists of a lamellar arrangement of polysaccharide chains offering a multitude of structurally different cavities available to small intruding molecules.

Enantiomeric cyclic amides and imides, esters, ketones, δ- and γ-lactones, and alkyl-substituted indenes can be resolved on CTA-I without derivatization. Chiral alcohols can also be resolved without derivatization, but the best results are obtained when they are converted to esters. In general, *para*-nitrobenzoic esters are recommended if the hydroxyl group is directly attached to the stereogenic center; benzoate or *para*-bromobenzoate derivatives if the hydroxyl moiety is not at the stereogenic center; and acetate derivatives for benzyl alcohols (8,43). Other recommended derivatives include benzyl amides for amines, *N*-benzyl amide-methyl esters for α-amino acids, and benzyl esters for carboxylic acids (8,43).

Chiral molecules with an axis of dissymmetry (atrop isomers) can also be stereochemically resolved on the CTA-I CSP. The resolution and pharmacological testing of the enantiomers of methaqualone illustrate this type of application (44). Methaqualone (Fig. 3) is a quinazolinone derivative that possesses hypnotic and anticonvulsive activities. The molecule exists in two enantiomeric forms, M(+) and M(−), due to the hindered rotation around the *N*-phenyl bond. Mannschreck et al. (44) were able to partially resolve this compound using the CTA-I CSP to optical purities of 0.75 for the (+) isomer and 0.68 for the (−) isomer. The anticonvulsive activity of the two enantiomers was evaluated by the mouse electroshock test, and the (−) isomer was found to be significantly more potent than the (+) isomer.

2. *Mobile Phases*

An aqueous or predominately aqueous mobile phase cannot be used with the CTA-I CSP. Anhydrous ethanol or ethanol modified with up to 5% water are the most commonly used mobile phases. Other alcohols such as methanol and 2-propanol can be used and dramatic changes can be seen with these solvents (8,43,45,46).

Koller et al. (45) investigated the effect of the structure of the alcohol on the retentions (k′) and enantioselectivities (α) for three enantiomeric solutes on the CTA-I CSP. In the series methanol, ethanol:water (96:4), 2-propanol, the k′ of each enantiomer increased through the series, whereas the effect on the observed α's varied. The authors concluded that the polarity of the eluents may not be the key to their elutropic strength. Instead, the elutropic strength of the solvent may be a function of the steric bulk of the alcohol. By binding to sites near or at the site of inclusion, the alcohol can affect the steric environment of the channels and cavities, thereby changing the fit on the solutes into these cavities. This, in turn, would alter both k′ and α.

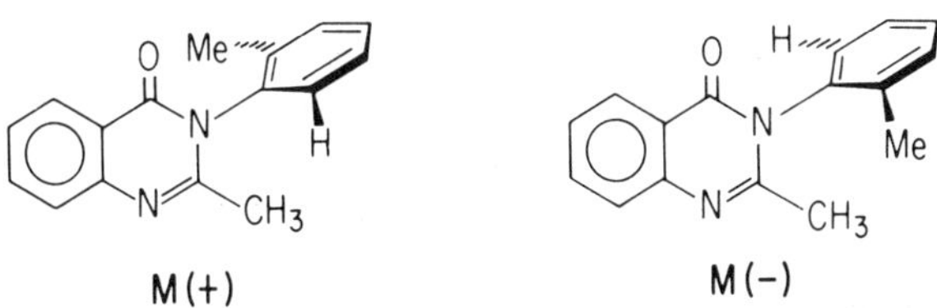

FIGURE 3 The enantiomers of methaqualone.

Nonpolar mobile phases composed of hexane modified with 2-propanol have also been used with the CTA-I CSP (43,45).

B. Microcrystalline Cellulose I: Cellulose Tribenzoate Derivative

Francotte and Wolf (46) have recently reported on the synthesis of a CSP based on tribenzoyl cellulose (TBC). The TBC CSP is composed of beads containing pure TBC and no inorganic carrier such as silica.

1. Solute Structure

This phase can be used in the preparative and analytical stereochemical resolution of a wide variety of compounds, including acetate derivatives of alcohols and diols; benzylic alcohol derivatives; *para*-substituted *d*-phenyl-*d*-valerolactones; 3,4-dihydro-2H-pyran-2-carboxylic acid derivatives; phenylvinylsulfoxide; *trans*-stilbene oxide (46).

2. Mobile Phase

The TBC CSP has been used with a mobile phase composed of hexane:2-propanol (90:10).

C. Microcrystalline Cellulose II: The Cellulosic Phases

Microcrystalline cellulose II, like the cellulose I form, does not possess enough mechanical strength to be used as an HPLC-CSP. Although its triacetate derivative, CTA-II, can be used as an HPLC support, it is not as useful a CSP as CTA-I (47). Ichida et al. (48) were able to improve on the properties of derivatized cellulose II by depolymerization, followed by coating on macroporous silica gel. In this manner, they created a number of new CSPs, the cellulosic CSPs, based on ester and carbamate derivatives of cellulose (see Table 2).

Amylose is another polysaccharide that has been used to create CSPs. Amylose is a D-α-glucose polymer with high helical content. Two amylose-based CSPs are commercially available (Table 2): amylose tris(3,5-dimethylphenylcarbamate), the AD CSP, and amylose (*S*)-α-methylbenzylcarbamate, the AS CSP.

The chiral recognition mechanisms that operate on the cellulosic CSPs have been studied by Wainer and co-workers (49,50). In one study, the chiral recognition of amides on the OB CSP was examined (49). The results indicated that the solute/CSP complexes formed between the OB CSP and the amide solutes were based on attractive hydrogen bonding, π-π, and dipole–dipole interactions. Chiral recognition within the solute/CSP complex was due to the differential inclusion (or fit) of the solute into a chiral cavity or ravine on the CSP. However, studies with aromatic alcohols

suggest that the inclusion of the aromatic moiety of the solute in a chiral cavity or ravine was the key step in the formation of the solute/CSP complex (50). Thus, it appears that the cellulosic CSPs use two different but related mechanisms.

1. *Solute Structure*

The type II CSPs have been used to resolve a wide variety of pharmacologically active compounds, both with and without derivatization (9–11). Some of the compounds that have been resolved without derivatization are hexobarbital (OA-CSP) (51), glutethimide (OB, OC, OK) (51), warfarin (OB, OC) (51), compounds containing a chiral sulfur atom (OB, OC) (52), verapamil (AD) (53), propranolol (OD) (54), and flurbiprofen (AD) (55). The resolution of the latter two compounds, propranolol, an α,β-aminoalcohol, and flurbiprofen, a carboxylic acid, is representative of new applications for the cellulosic CSPs that are primarily used with hexane: alcohol mobile phases. In the case of propranolol, the mobile phase was modified with *N,N*-dimethyloctyl amine, and when flurbiprofen was chromatographed, trifluoroacetic acid was used as the modifier.

Most solutes that can be resolved on the type II CSPs contain one or more aromatic rings or polar π-bonded groups such as carbonyl, sulfinyl, or nitro moieties (9–11). This reflects the two chiral recognition mechanisms discussed above in which hydrophobic moieties are required for inclusion in hydrophobic cavities or ravines and polar groups are required for attractive hydrogen-bonding, dipole, and π-π interactions. Aliphatic mono- or poly-functional alcohols that do not contain these interaction sites must usually be derivatized. In addition, amine and carboxylic acid moieties are often derivatized to improve the chromatographic efficiency of the CSPs. The resolution of the enantiomeric benzylamide derivatives of amphetamine is presented in Fig. 4.

2. *Mobile Phases*

The cellulosic CSPs are most often used with nonpolar mobile phases composed of hexane and a polar modifier such as 2-propanol. These mobile phases are used to maximize the attractive interactions between solute and CSP. Since these interactions contain a polar component, an increase in the polarity of the mobile phase, for example, a lager alcohol content, will reduce the retention and vice versa. These phases have also been used with aqueous mobile phases and a new reversed-phase version of the OD CSP is currrently being marketed (56).

The effect of the structure, type, and concentration of the polar mobile phase modifier (mpm) on retention (k′) and stereoselectivity (α) has been studied (57,58). Wainer et al. (57) chromatographed a series of enantio-

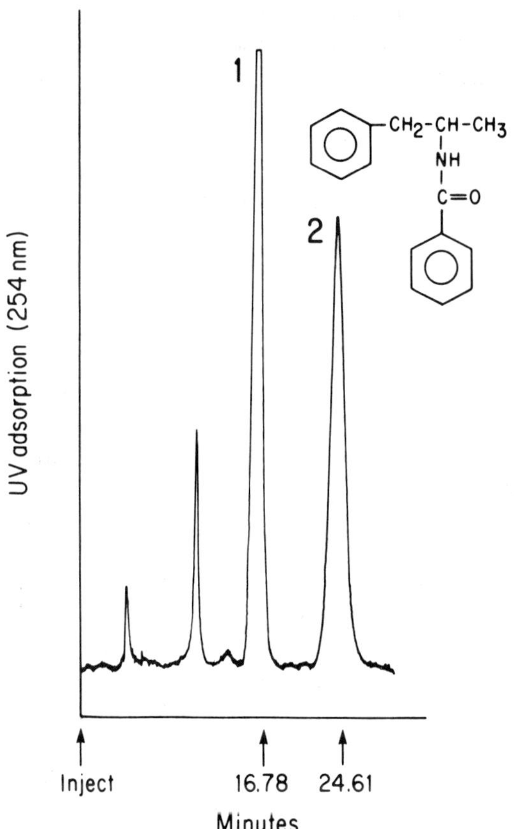

FIGURE 4 The chiral resolution of the benzylamide derivatives of (*R*)- and (*S*)-amphetamine on the OB CSP, where 1 = (*R*)-amphetamine, and 2 = (*S*)-amphetamine.

meric amides on the OB CSP using mobile phases composed of hexane and two homologous series of primary and secondary alcohols. The results of the study indicate that the alcoholic mpms compete with the solutes for achiral and chiral binding sites and that the steric bulk around the hydroxyl moiety of the mpm plays a role in this competition. Increased steric bulk tends to result in increased k′ and α. However, the results also suggest that the effect of the alcoholic mpms on stereoselectivity may also be due to binding to achiral sites near or at the chiral cavities of the CSP that alters the steric environment of these cavities. The latter conclusions are consistent with the observations reported by Koller et al (45) from experiments with the CTA-I CSP.

In a related experiment, Gaffney et al. (58) used the same CSP and mobile phases with three different enantiomeric solutes: phenyl vinyl sulfoxide, glutethimide, and 2-phenylpropanoic acid methyl ester. Reversals in the enantiomeric elution order were observed for the methyl ester solute associated with changes in the chain length of the mpm. The results of the study suggested that at least two chiral recognition mechanisms (attractive interactions and hydrophobic insertion) operate on the OB CSP and that the mpm can change the relative importance of these mechanisms. The results also point out the importance of determining the enantiomeric elution order after each alteration in the composition of the mobile phase.

3. *Pharmaceutical Applications*

The cellulosic CSPs have been used to stereochemically resolve a vast number of compounds (9–11). However, most of the pharmaceutical applications involve bulk substances and only a relatively small number in vivo or in vitro studies. Some examples of the use of cellulosic CSPs in metabolic and pharmacokinetic studies are the following:

1. "Direct enantiomeric separation of betaxolol with applications to the analysis of bulk drug and biological samples" (59). The enantiomers of betaxolol were resolved on the OD CSP using a mobile phase composed of hexane:2-propanol:diethylamine (87:13:0.05, v/v/v). This method was used to determine the concentration of the enantiomers of betaxolol after incubation of a racemic mixture of the compound with rat hepatocytes.

2. "Measurement of underivatized propranolol enantiomers in serum using a cellulose-tris(3,5-dimethylphenylcarbamate) high performance liquid chromatographic (HPLC) chiral stationary phase" (54). A method for the direct measurement of the enantiomers of propranolol in human serum was developed using the OD CSP and a mobile phase composed of hexane:2-propranol:*N*,*N*-dimethyloctylamine (92:8:0.01, v/v/v). The assay was validated for use in pharmacokinetic and metabolic studies and was subsequently used in the investigation of the effect of cimetidine on the metabolism and clearance of propranolol enantiomers (60).

3. "Analytical and preparative high-performance liquid chromatographic separation of the enantiomers of ifosfamide, cyclophosphamide and trofosfamide and their determination in plasma" (61). Three racemic anticancer drugs—ifosfamide (IFF), cyclophosphamide (CTX), and trofosfamide (TFF)—were resolved on the OD CSP and gram quantities of (*R*)-IFF and (*S*)-IFF were prepared for pharmacological testing. An analytical assay was also developed for the determination of the enantiomeric composition of IFF and CTX in plasma using achiral–chiral coupled column

chromatography. A racemic form of another CSP, the D,L- naphthylalanine CSP, was used as the achiral precolumn.

4. "Simultaneous direct determination of the enantiomers of verapamil and norverapamil in plasma using a derivatized amylose HPLC chiral stationary phase" (53). An AD CSP was used to determine the plasma concentrations of the enantiomers of verapamil and its major metabolite norverapamil. After a liquid–liquid extraction, the analytes were resolved on the amylose column using a mobile phase composed of heptane:2-propanol:ethanol, (85:7.5:7.5, v/v/v) modified with 1.0% triethylamine, a flow rate of 1.0 mL/min, and a column temperature of 30°C. The assay was validated for use in human pharmacokinetic studies. A representative chromatogram is presented in Fig. 5.

V. TYPE III CSPs

In type III CSPs, the solute enters into chiral cavities within the CSP to form inclusion complexes and the relative stabilities of the resulting diastereomeric complexes are based on secondary attractive (e.g., hydrogen-bonding) or steric interactions. The driving force for the insertion can be hydrophobic (cyclodextrin and polymethacrylate CSPs) or electrostatic (crown ether CSP). The commercially available CSPs based on these mechanisms are presented in Table 3.

A. Cyclodextrin-Based CSPs

Cyclodextrins (CD) are cyclic oligosaccharides composed of α-D-glucose units linked through the 1,4 position. The molecules are formed by the action of *Bacillus macerous* amylase on starch (62). The three most common forms of this molecule are α-, β-, and γ-cyclodextrin, which contain six, seven, and eight glucose units, respectively. Because of the α-D-glucose units, the cyclodextrin molecule has a stereospecific, toroidal (doughnut-shaped) structure. The interior cavity of the cyclodextrin molecule is relatively hydrophobic, and a variety of water-soluble and insoluble compounds can fit into it to form inclusion complexes (63). All three forms of the molecule have been attached to silica and successfully used as CSPs for the resolution of a variety of enantiomeric compounds as well as geometrical isomers such as *cis*- and *trans*-stilbene (64), structural isomers such as *m*- and *p*-nitroaniline (64), and sugar anomers (65).

The chromatographic utility of the CD CSPs has been expanded through the development of derivatized CD supports. The currently available derivatized cyclodextrin-based CSPs are listed in Table 3 and include acetylated (in α-, β-, and γ-CD forms); (*S*)- and (*R*,*S*)-hydroxypropyl ether (β-CD); (*S*)- and (*R*)-naphthylethyl carbamate (β-CD); 3,5-

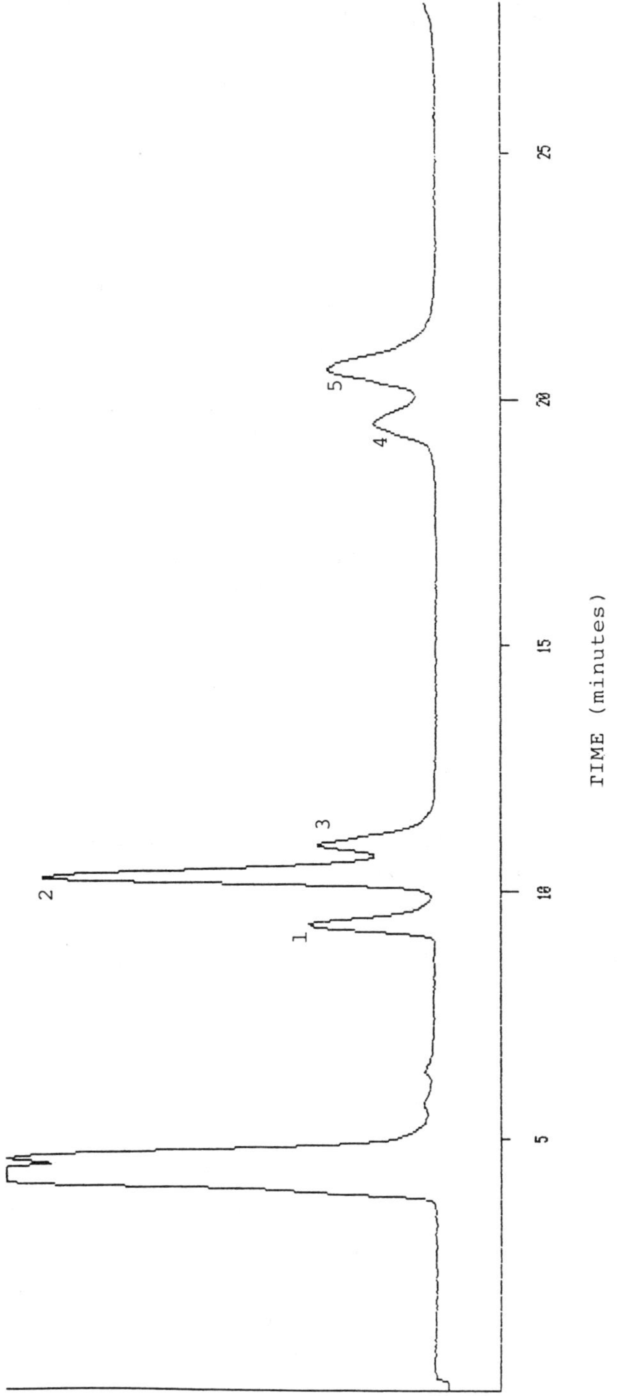

FIGURE 5 Chromatogram of plasma from a healthy volunteer 4 hr after the administration of 240 mg of a sustained-release formulation of racemic verapamil, where 1 = (*S*)-verapamil, 2 = (*R*)-verapamil, 3 = internal standard, 4 = (*S*)-norverapamil, 5 = (*R*)-norverapamil. [From Shibukawa and Wainer (53).]

TABLE 3 Commercially Available Type III CSPs

Chiral stationary phase	Chiral selector
Cyclodextrin	
Cyclobond I	β-Cyclodextrin
Cyclobond II	γ-Cyclodextrin
Cyclobond III	α-Cyclodextrin
Cyclobond I Ac	β-Cyclodextrin acetylate
Cyclobond I SP	β-Cyclodextrin (*S*)-hydroxypropyl ether
Cyclobond I RSP	β-Cyclodextrin (*R,S*)-hydroxypropyl ether
Cyclobond I SN	β-Cyclodextrin (*S*)-naphthylethyl carbamate
Cyclobond I RN	β-Cyclodextrin (*R*)-naphthylethyl carbamate
Cyclobond I DMP	β-Cyclodextrin 3,5-dimethylphenyl carbamate
Cyclobond I PT	β-Cyclodextrin *para*-toluoyl ester
Cyclobond II Ac	γ-Cyclodextrin acetylate
Cyclobond III Ac	α-Cyclodextrin acetylate
Synthetic polymers	
Chiralpak OT (+)	Poly(triphenylmethyl methacrylate)
Chiralpak OP (+)	Poly(2-pyridyldiphenylmethyl methacrylate)
ChiraSpher	Poly-*N*-acryloyl-(*S*)-phenylalanine ethylester
Crown ether	
CR (+)	Chiral 18-crown-6 ether, structure not reported
CR (−)	Chiral 18-crown-6 ether, structure not reported

dimethylphenyl carbamate (β-CD); and *para*-toluoyl ester (β-CD). Unlike the standard CD CSPs, the derivatized CD CSPs can be used with both reverse and normal phases and can operate with a variety of chiral recognition mechanisms. In fact, the (*S*)- and (*R*)-naphthylethyl carbamate-CD CSPs have been described as mixed-mode CSPs and will be discussed in Sec. VI.

1. Solute Structure

To achieve chiral separations on CD CSPs, a part or all of the solute molecules must enter the cyclodextrin cavity. In most cases, the solutes that are successfully resolved contain an aromatic moiety at or adjacent to the stereogenic center, and it is the aromatic portion of the molecule that inserts itself into the chiral cavity of the cyclodextrin molecule to form the inclusion complex (66,67). The size of the aromatic moiety and cyclodextrin cavity determine which CD CSP will form the best inclusion complex, and single aromatic rings fit best in the α-CD, naphthylrings in the β-CD and aromatic system larger than naphthyl in the γ-CD (68). In addition to the aromatic system, the solutes successfully resolved on the CD CSPs also contain either a hydrogen-bonding group or an additional π system (carbonyl moiety or aromatic ring) at the stereogenic center (68). The structural requirements for the derivatized CD CSPs often differ and are discussed below in Sec. IV.

2. *Mobile Phases*

The most common mobile phases used with the CD CSPs are composed of water modified with methanol, ethanol, or acetonitrile. Buffers such as 0.5–1% triethylammonium acetate and phosphate can also be used and tend to improve efficiency and reduce retention of anionic and cationic solutes. As a general rule, the greater the aqueous content of the mobile phase, the greater the retention and stereoselectivity.

Normal phases composed of hexane modified with 2-propanol have been successfully used with the 3,5-dimethylphenyl carbamate and *para*-toluoyl ester forms of the CD CSP. The β-CD CSP has been used in subcritical fluid chromatography with a mobile phase composed of carbon dioxide modified with methanol, ethanol, or 2-propanol (69). Under these conditions, racemic amides and phosphine oxides were enantiomerically resolved.

3. *Pharmaceutical Applications*

The CD CSPs have been used to stereochemically resolve a vast number of compounds (9–11). However, most of the pharmaceutical applications involve bulk substances and only a relatively small number in vivo or in vitro studies. Some examples of the use of CD CSPs in metabolic and pharmacokinetic studies are the following:

1. "The stereoselective metabolism of phenytoin in children" (70). McClanahan and Maguire have reported on a study of the metabolism of phenytoin that is widely prescribed for the treatment of epilepsy, 4-hydroxyphenyl metabolites of phenytoin (Fig. 6). Although phenytoin itself

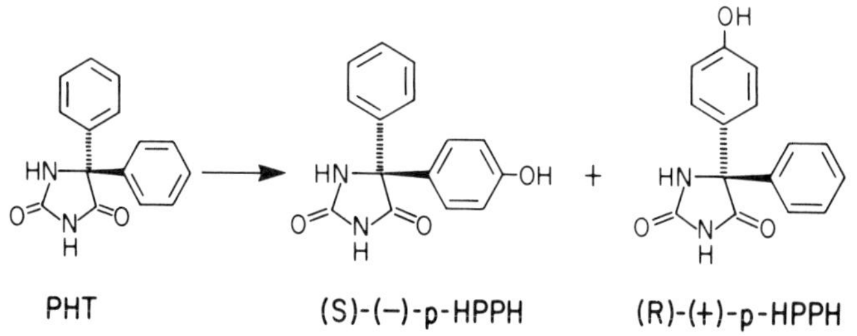

PHT (S)-(−)-p-HPPH (R)-(+)-p-HPPH

FIGURE 6 The metabolism of phenytoin to 5*S*-(−)- and 5*R*-(+)-5-(4-hydroxyphenyl)-5-phenylhydantoin, e.e., (*S*)-(−)-*p*-HPPH and (*R*)-(+)-*p*-HPPH, respectively.

amines have been resolved after conversion to benzamide derivatives; enantiomeric 1,2- and 1,3-dicarboxylic acids have been resolved as benzyl esters or benzamides; and racemic alkyl alcohols, 1,2- and 1,3-diols, and some sugars have been resolved as benzoate esters (9–11).

Solutes containing asymmetric sulfur atomes or asymmetric phosphorous atoms can be resolved with or without an aromatic moiety in the molecule. For example, the anticancer drug ifosfamide, an oxazaphosphorane not containing an aromatic moiety, has been resolved on this type of CSP (73). In addition, these are excellent CSPs for the resolution of enantiomeric molecules with an axis of dissymmetry (atrop isomers) (9–11).

2. *Mobile Phases and Temperature*

The most common mobile phases used with these CSPs are methanol modified with 2-propanol or mixtures of toluene and dioxane. The CSPs derived from triphenyl methacrylate and diphenyl-2-pyridyl-methyl methacrylate operate best at a temperature of 15°C or less.

C. Chiral Crown Ethers

Crown ethers are synthetic macrocyclic polyethers that can form selective complexes with various cations. An 18-crown-6 ether is illustrated in Fig. 8, where 18 indicates the total number of atoms in the polyether ring and 6 the number of oxygen atoms. Chiral crown ethers have been synthesized by

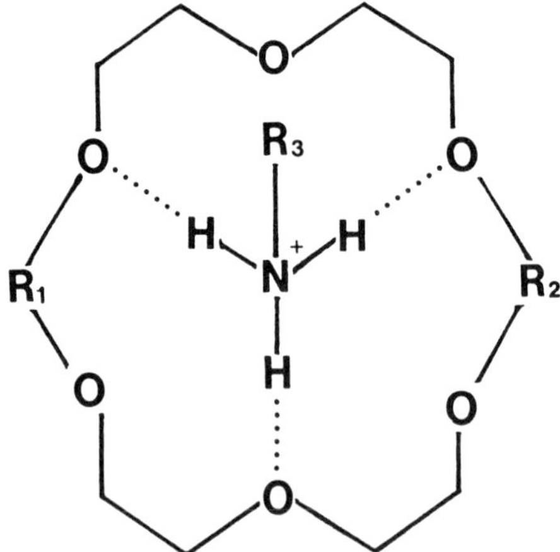

FIGURE 8 The structure of an 18-crown-6 ether ammonium complex, where R_1 and R_2 = the chiral elements of the crown ether and R_3 = rest of the included molecule.

adding substituents to the ether ring that hinder its rotation about an axis of dissymmetry, that is, producing atrop isomers, and have been immobilized on chromatographic supports, yielding CR CSPs (75).

1. Solute Structure

Compounds based on the 18-crown-6 ring have the ability to form stable complexes with potassium and ammonium ions (Fig. 8). This is a function of the cavity size and the electrostatic interactions between the charged ion and free electrons on the oxygen atoms. Thus, the solutes that can be successfully resolved on the CR CSP must contain a primary amine moiety at or near the stereogenic center. Steric bulk, especially aromatic moieties, that are close to the stereogenic center also enhance chiral recognition. Given these requirements, the CR CSP is ideally suited for the stereochemical resolution of free and *O*-derivatized amino acids, small peptides, and many pharmaceutically interesting compounds containing primary amine moieties.

2. Mobile Phase

The mobile phases used on the CR CSP are limited to perchloric acid (usually around 10 *mM*) modified with small amounts of organic modifiers such as alcohols or acetonitrile. Although the mobile phases are limited, retention (k') and stereoselectivity (α) can be altered by varying the pH between 1 and 3 and adjusting the temperature. Udvarhelyi and Watkins (75) have investigated these parameters on the chromatography of a series of racemic phenylglycines. The results of this study indicated that when the pH was raised from 1–3, the k's decreased dramatically, as did most of the observed α's. However, the magnitude of the change in α, as well as the direction, was a function of solute structure. The effect of temperature was consistent throughout the series and an increase from 5–25°C resulted in a decrease in k' and α.

3. Applications

At the present time, the reported applications of the CR CSP have been limited to the separation of amino acids and some dipeptides as bulk substances. One example of the use of the CR CSP in a complex matrix was the direct stereochemical resolution of aspartame stereoisomers and their degradation products in coffee and diet soft drinks (76). Aspartame (*N*-DL-α-aspartyl-DL-phenylalanine methyl ester) is a dipeptide whose L,L-isomer is a low-calorie sweetener sold under the name NutraSweet. The structure of aspartame and its major degradation products are presented in Fig. 9 and the stereochemical separation of these compounds on the CR CSP in Fig. 10A. The resolutions were accomplished using a mobile phase

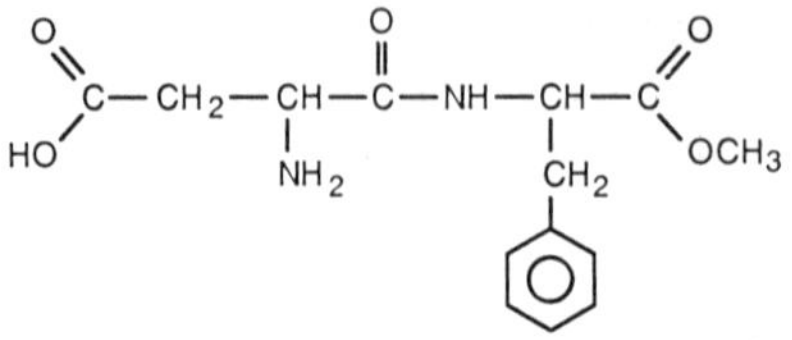

Aspartame (APM)

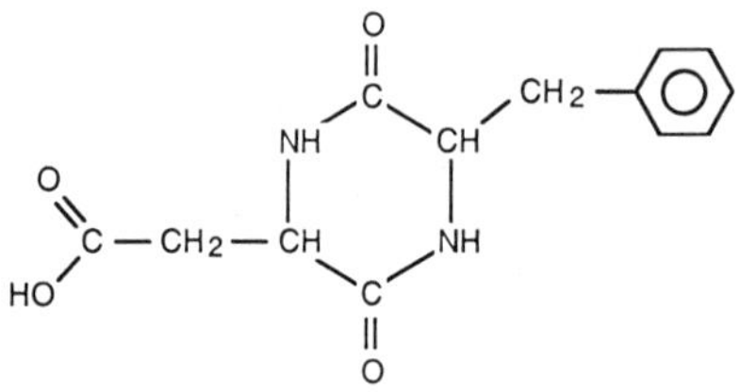

Diketopiperazine (DKP)

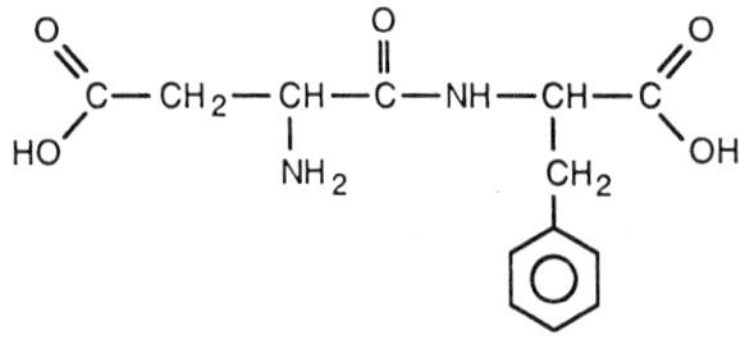

Aspartyl-phenylalanine (Asp-Phe)

FIGURE 9 The structure of aspartame and its major degradation products.

composed of aqueous perchloric acid (pH 2.8) modified with 1.5% 2-propanol, a flow rate of 0.6 mL/min, and a temperature gradient (10–40°C). The application of this method to the determination of L,L-aspartame and its major decomposition product, L,L-diketopiperazine, in a diet cola is presented in Fig. 10B.

VI. MIXED-MODE CSPs

A mixed-mode, or multimodal, CSP is one that has been designed to operate with two or more different types of chiral recognition mechanisms. Armstrong et al. (19,78) have developed such a CSP by reacting β-cyclodextrin with (*R*)- or (*S*)-naphthyl-ethylisocyanate. The resulting naphthylethylcarbamate derivative of β-cyclodextrin, the NEC-β-CD CSP, can function as either a type I CSP or type III CSP.

When the NEC-β-CD CSP is used with a nonpolar mobile phase such as hexane modified with 2-propanol, the solute/CSP complexes are formed

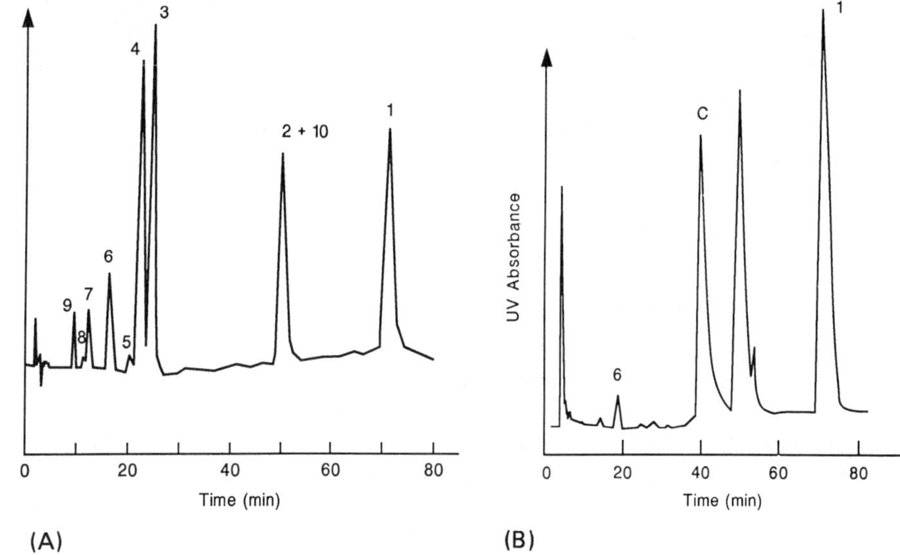

FIGURE 10 (A) Chromatogram of a 1-day-old synthetic mixture of the four stereoisomers of APM (0.1 mM of each isomer), where 1 = L,L-APM; 2 = L,D-APM; 3 = D,D,-APM; 4 = D,L-APM; 5 = L-Asp-D-Phe; 6 = D,D-DKP + L,L,-DKP; 7 = D,L-DKP + L,D-DKP; 8 = D-Asp-D-Phe; 9 = D-Asp-L-Phe; 10 = L-Asp-L-Phe. Column: Crownpack CR(+), Daicel, 150 × 4 mm i.d.; injection loop: 20 μl; mobile phase: aqueous $HClO_4$ pH = 2.8/2-PrOH, (98.5:1.5), (v:v); flow rate: 0.6 ml/min.; temperature gradient: 0 to 26 min, T° = 10°C; 26 to 70 min, linear gradient from T° = 10°C to T° = 40°C; detector: = 210 nm, RS = 0.02 AUFS, ATT – 2^5. (B) The determination of L,L-aspartame and its decomposition products in a diet cola, where 1 = L,L-aspartame; 6 = L,L-diketopiperazine; C = caffeine. [From Motellier and Wainer (76).]

by attractive interactions, hydrogen bonding, π-π, dipole stacking, etc., between the solute and CSP, type I interactions. These interactions take place between the naphthylethylcarbamate moiety and complimentary functionalities on the solute. As with the other type I CSPs, if the solutes do not contain the appropriate interaction sites, they must be added through derivatization. Thus, nonaromatic amines and alcohols, amino acids, and carboxylic acids should be converted to their dinitrophenyl derivatives before chromatography on the NEC-β-CD CSP.

For solutes meeting the requirements for use with a standard or derivatized β-CD CSP, the NEC-β-CD CSP can be used with aqueous mobile phases modified with acetonitrile or another organic modifier. Since chiral recognition is a function of the interaction of the solute with the chirality of the CD as a whole, the configuration about the stereogenic center of the naphthylethyl moiety can affect the observed stereochemical resolutions. Thus, the (*R*)-NEC-β-CD and (*S*)-NEC-β-CD CSP can display different enantioselectivities.

A number of solutes have been stereochemically resolved on this multimodal CSP, using both chiral recognition mechanisms (19), and a flowchart has been developed to help in the application of the NEC-β-CD CSP (19). This chart is presented in Fig. 11.

VII. TYPE IV CSPs

Chiral ligand-exchange chromatography is based on the formation of diastereomeric ternary complexes that involve a transition metal ion (M), usually copper II; a single enantiomer of a chiral molecule (L), usually an amino acid; and the enantiomers of the racemic solute (*R* and *S*). The diastereomeric mixed chelate complexes formed in this system are represented by the formulas L-M-*R* and L-M-*S*. When these complexes have different stabilities, the less stable complex is eluted first, and the enantiomeric solutes are separated.

Chiral ligand-exchange chromatography can be accomplished using an achiral chromatographic support and a mobile phase containing M and L. The efficiency and reproducibility of the chromatography can be improved through the immobilization of the chiral ligand (L) on a chromatographic support that produces the type IV CSPs. The commercially available type IV CSPs are presented in Table 4.

A. Solute Structure

The solutes resolved by the Type IV CSPs must be able to form coordination complexes with transition metal ions. This limits the classes of compounds to α-amino acids and mono- and dicarboxylic acids containing an α-hydroxyl moiety. The most common solutes resolved on these CSPs are underivatized amino acids, although a number of derivatized amino acids can be resolved, including *N*-acetyl, amide, *N*-carbamoyl, *N*-carbobenzoxy, and hydantoin derivatives, as well as complex amino acids such as the biochemical modulator BSO (79).

B. Mobile Phases

The Type IV CSPs are used with aqueous mobile phases that contain copper(II) sulfate (CS). The usual starting concentration of CS is 0.25 *mM* and retention times can be manipulated by increasing or decreasing the CS concentration; an increase in CS concentration usually shortens retention, whereas a decrease has the opposite effect. Concentrations as low as 0.05 *mM* and as high as 20.0 *mM* have been used.

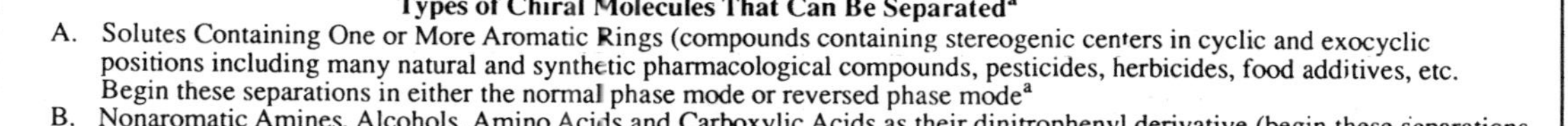

Types of Chiral Molecules That Can Be Separated[a]

A. Solutes Containing One or More Aromatic Rings (compounds containing stereogenic centers in cyclic and exocyclic positions including many natural and synthetic pharmacological compounds, pesticides, herbicides, food additives, etc. Begin these separations in either the normal phase mode or reversed phase mode[a]

B. Nonaromatic Amines, Alcohols, Amino Acids and Carboxylic Acids as their dinitrophenyl derivative (begin these separations in the normal phase mode).

Normal Phase Mode[a]

Beginning Mobile Phase Composition 90:10, hexane:isopropanol

peak(s) elute

no elution

Try 50:50, hexane: ipa

no separation

separation

peak(s) elute

no elution

Alter mobile phase ratio until k' is between 3 and 10. Try different modifiers (neutral, acid & base). Sometimes a different solvent helps (chloroform, methylene chloride, toluene, etc.). If no separation is evident go to a reversed phase approach.

Optimize separtaion by varying % ipa. Sometimes altering the modifier (e.g., butanol, *sec*-butanol, pentanol, etc.) helps. Occasionally adding a small amount of HOAc or organic amine helps.

Use mobile phase of neat ethanol, acetonitrile or occasionally methanol. Also, add 0.1 to 1.0% of HOAc or diethylamine if your racemate is a carboxylate or amine respectively.

separation

no separation

Optimize separation by varying % HOAc or amine. Sometimes using a binary mobile phase helps (i.e., 80:20 AcN:EtOH). Sometimes using a different organic acid or amine modifier improves the separation.[c]

Try the opposite configuration NEC-β-CD column

separation

no separation

Try reversed phase LC[d]

Reversed Phase Mode[a]

Beginning Mobile Phase Composition 35:65, acetonitrile:buffer (pH = 7)[b]

peak(s) elute

no elution

no separation

separation

Try 75:25, acetonitrile:buffer

Change pH of buffer to 3.5 or 4.0 and repeat 35:65, AcN:buffer. Change organic modifier.

separation

no separation

no separation

separation

Adjust concentration of organic modifier to obtain optimum k'. Optimize by varying pH, buffer strength, buffer type and sometimes the nature of the organic modifier.[c]

Change pH of buffer to 3.5 or 4.0 and repeat 35:65 AcN:buffer. Change organic modifier.

Try opposite configuration NEC-β-CD column

separation

separation

no separation

no separation

Try normal phase LC[d]

Try opposite configuration NEC-β-CD column

separation

no separation

Try normal phase LC[d]

FIGURE 11 Basic flowchart for the use and optimization of the NEC-β-CD multimodal CSP. [a]Before choosing whether to begin the separation in the normal phase or reversed phase mode, one must know the solubility and stability of the racemate in these mobile phases. For example, there is no sense in trying a normal phase separation if the compound is insoluble in that mobile phase. [b]Recommended buffers include ammonium nitrate and triethylammonium acetate. [c]Two additional things can be done to optimize the separation. One is to decrease the flow rate. The other is to lower the temperature. Selectivity factors (α's) usually increase at lower temperatures. [d]If the separation does not occur in any mode, another CSP must be used. If a racemate is slightly volatile (or a volatile derivative can be made), gas chromatography is an excellent option. [From Armstrong et al. (19).]

TABLE 4 Commercially Available Type IV CSPs

Chiral stationary phase	Chiral selector
WE	*N*-Carboxymethyl-(1*R*,2*S*)-diphenylamino ethanol
WH	L-Proline
WM	L-*tert*-Leucine
MCI gel	Amino acid, structure not reported

Chromatographic retention on a type IV CSP can be manipulated by altering the pH of the mobile phase, adding mobile-phase modifiers, or changing the temperature. Examples of these effects are the following: (1) The pH of the mobile phase can be varied between 3 and 7, and the lower the pH, the shorter the chromatographic retention. (2) The addition of methanol and acetonitrile usually increases retention, whereas sodium chloride has the opposite effect. (3) The chromatography is usually carried out at 50°C and a reduction in temperature increases the retention time.

VIII. TYPE V CSPs

Proteins are complex polymers of high molecular weight that are composed of chiral subunits (L-amino acids). These biopolymers play a variety of different roles in a biological system, including the uptake and transport of drug substances. One of the key aspects of the uptake and transport mechanisms is the reversible binding of small molecules to the protein and this process is often stereospecific.

An example of this stereospecificity is the in vitro binding of propranolol to α_1-acid glycoprotein (AGP) and human serum albumin (HSA) (80,81). When human AGP is the binding protein, (*S*)-propranolol is bound to a greater extent than the *R*-isomer (80), whereas the reverse is found with HSA (81). Other serum albumins, including bovine serum albumin (BSA), have also displayed stereo-specific binding (82,83).

The ability of proteins to stereoselectively bind small molecules has been used to develop a series of commercially available protein-based CSPs (the type V HPLC CSPs), including phases that contain immobilized AGP (84), HSA (85), BSA (86), and ovomucoid (OVM) (87) (see Table 5). All these CSPs are useful in the HPLC resolution of enantiomeric compounds and appear to have an extremely wide range of applications, and the AGP CSP seems to have the broadest utility of any of the current CSPs (9–11). However, although the type V CSPs display high enantioselectivities, they also have low capacities due to the relatively small amounts of the chiral selector that can be immobilized per g silica. Thus, these CSPs are useful

TABLE 5 Commercially Available Type V CSPs

Chiral stationary phase	Chiral selector
AGP	α_1-Acid glycoprotein
OVM	Ovomucoid
HSA	Human serum albumin
BSA	Bovine serum albumin

analytical tools, but not applicable to the large-scale preparation of pure enantiomers.

A. General Mechanism

Type V CSPs are normally used with an aqueous mobile phase containing a phosphate buffer at pH 7.0. Under these conditions, hydrogen bonding and hydrophobic interactions take place between the protein part of the CSP and the solute. In addition, electrostatic interactions can occur. For example, the isoelectric points of AGP and BSA are 2.7 and 4.7, respectively, and both proteins have a net negative charge at neutral pH. Cationic and anionic solutes are also changed at this pH; this situation suggests the possibility of ion-exchange and/or ion-pairing types of interactions between the solute and CSP. Studies by Schill et al. (88) have shown that, under certain conditions, charged compounds are retained on the AGP CSP as ion pairs.

The binding interactions between the solute and protein usually involves stereospecific and nonstereospecific mechanisms. These mechanisms make the type V CSPs sensitive to the composition of the mobile phase, temperature, flow rate, and pH. These parameters can be adjusted to improve the chromatography and stereoselectivity of specific solutes on the AGP CSP (88,89), OVM CSP (90,91), BSA CSP (92,93), and HSA CSP (94).

B. The AGP CSP

1. Solute Structure

The AGP CSP can be used with a wide range of cationic and anionic compounds. The cationic compounds include relatively simple molecules such as methylphenidate, tetrahydroxoline, and terbutaline; α,β-amino alcohols such as ephedrine, labetalol, and nadolol; and complex molecules such as atropine, methorphan, and verapamil (9–11). Anionic molecules such as the α-methylarylacetic acid antiinflammatory agents ibuprofen, fenoprofen, and naproxen can also be resolved on the AGP CSP (9–11). All these compounds can be resolved without precolumn derivatization.

2. Mobile Phases

The basic mobile phase used with the AGP CSP is composed of phosphate buffer and one or more modifiers. The type and concentration of the mobile phase modifier are extremely important and both k′ and α can be altered by changing the mobile phase (9–11).

1. *Mobile-phase modifiers.* The modifiers that have been used with the AGP CSP encompass uncharged and charged compounds and include 2-propanol, *N*,*N*-dimethyloctylamine, tetrapropyl- and tetrabutyl-ammonium bromide, and octanoic acid (see Fig. 12). The addition of these modifiers to the mobile phase usually results in a decrease in the retention of solutes, with an accompanying decrease in enantioselectivity (α). However, in some instances, the addition of a modifier results in an increase in α.

The effect of various mobile-phase modifiers on α is the result of a highly complex relationship involving the structure of the solute, as well as the structure and type of modifier, and may also involve changes in the conformation of the AGP (95). The complexity of this relationship is illustrated by the effect of various modifiers on the chiral resolution of two closely related compounds, atropine and *N*-methylhomatropine (Fig. 13).

For *N*-methylhomatropine, the use of the cationic modifier dimethyloctylamine reduces the value of α relative to the value obtained when the uncharged modifier 2-propanol is used. The same result occurs when the anionic modifier octanoic acid is used. However, the addition of the tetrabutylammonium bromide almost doubles the enantioselectivity relative to the value of α obtained with 2-propanol. When atropine is the solute, poor enantioselectivity is observed with neutral and cationic modifiers, whereas the addition of an anionic modifier, octanoic acid, results in an α of 1.6.

2. *The effect of pH.* In addition to the use of a modifier, adjusting the pH of the mobile phase can change the retention and enantioselectivity. The magnitude of the effect depends on the nature of the solute and composition of the mobile phase. The resolution of cyclopentolate illustrates the effect of pH on α (89). When the mobile phase contains sodium chloride (0.1 *M*) and 2-propanol (0.33 *M*), a decrease in the pH from 7.5–6.5 produces a decrease in α from 1.96–1.79. However, when tetrabutylammonium bromide (0.003 *M*) is the modifier, a corresponding decrease in the pH produces an increase in α from 1.70–2.09.

3. Pharmaceutical Applications

Because of its broad enantioselectivity, the AGP CSP has been used in a large number of pharmacokinetic studies. A representative selection is presented below:

UNCHARGED MODIFIERS	
monovalent alcohols	1-propanol
	2-propanol
	ethanol
diols	ethylene glycol
	propylene glycol
	1,2-butanediol
amino acids	6-aminohexanoic acid
	ß-alanine
	leucine
CATIONIC MODIFIERS	
tertiary amines	N,N-dimethyloctylamine
	N,N-dimethylethylamine
quaternary amines	tetrapropylammonium bromide
	tetrabutylammonium bromide
amino acid	1,2-diaminobutyric acid
ANIONIC MODIFIERS	
carboxylic acids	octanoic acid
	butyric acid
	decanoic acid
amino acid	aspartic acid
sulfamic acids	cyclohexylsulfamic acid

FIGURE 12 Mobile-phase modifiers used with the AGP CSP.

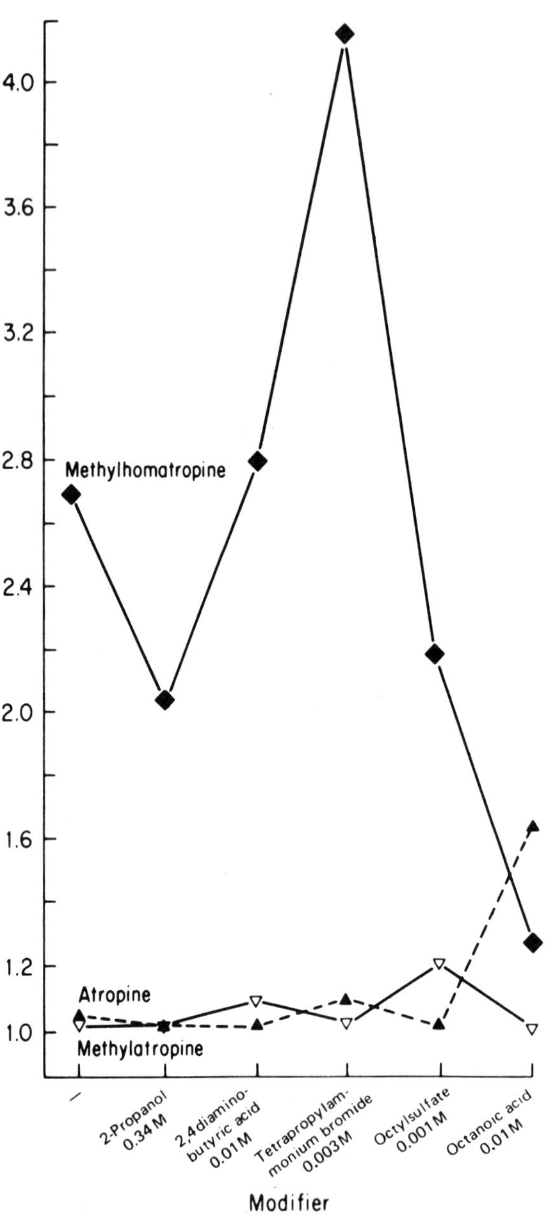

FIGURE 13 The effect of mobile-phase modifiers on the enantioselective resolution of atropine, methylatropine, and *N*-methylhomatropine.

1. "Separation and quantitation of (*R*)- and (*S*)-atenolol in human plasma and urine using an α_1-AGP column" (96). (*R*)- and (*S*)-atenolol were extracted from human serum and urine using dichloromethane containing 3% heptafluorobutanol; acetylated using acetic anhydride; separated on an AGP CSP using a mobile phase of phosphate buffer (0.01 *M*, pH 7.1) modified with 0.25% (v/v) acetonitrile. Quantitation was performed using fluorescence detection and the limit of detection was less than 6 ng/mL. The acetylation was required to eliminate a chromatographic overlap with endogenous materials and metabolites. The plasma concentration-time curves for (*R*)- and (*S*)-atenolol did not differ significantly in two subjects who received a single oral dose of 50 mg rac-atentolol and the *R*/*S* ratio of atenolol in the urine was approximately 1.

2. "Improved performance of the second generation α_1-AGP columns: Applications to the routine assay of plasma levels of alfuzosin hydrochloride" (97). The concentrations of (*R*)- and (*S*)-alfuzosin in rat plasma were determined by (1) addition of 1 mL 0.1 *N* sodium hydroxide to 1 mL rat plasma; (2) extraction with 7 mL of dichloromethane:ethyl ether (3:4, v/v); (3) evaporation of the organic phase and redissolution in 80 μL of the mobile phase; (4) chromatography on the AGP CSP using a mobile phase composed of phosphate buffer (0.05 *M*, pH 7.4; containing 0.025 *M* tetrabutyl ammonium bromide):acetonitrile (96:4, v/v); (5) fluorimetric detection. The chromatography took less than 10 min and plasma concentrations of 5 ng/mL could be quantitated. Initial pharmacokinetic studies demonstrated no significant difference in the plasma concentration-time curves of the two enantiomers.

C. The OVM CSP

Miwa and co-workers (87) have recently reported on the synthesis of a CSP based on ovomucoid, OVM CSP, an acid glycoprotein found in chicken egg white. The primary, secondary, and most of the tertiary structure of the protein have been elucidated (98). The molecule consists of a single 186 amino acid chain divided into three tandem homologous domains by 9 disulfide bonds, carbohydrate moieties [4–5 glycosylated asparagine residues (99)], and sialic acid moieties composing 0.5–1.0% of the total weight of the protein (100).

As with the other protein-based CSPs, initial studies indicate a relationship between the structure of the protein and chromatographic properties of the OVM CSP (101). When the sialic acid residues were enzymatically removed from the protein, the capacity factors (k′) of the enantiomers of an acidic solute (ketoprofen) were reduced, whereas the k′s of the enantiomers of a basic solute (chlorpheniramine) were unaffected (101). The observed enantioselectivities for both solutes remained unchanged. However, when the carbohydrate moieties were chemically removed, the k's of the enantiomers of both solutes were reduced and the enantioselectivities lost.

1. *Solute Structure*

Since ovomucoid is a glycoprotein the solutes that can be stereochemically resolved on the OVM CSP are similar to the basic, neutral, and acidic compounds separated on the AGP CSP. The major differences between the two CSPs appears to be quantitative (i.e., the magnitude of the chiral separation), rather than qualitative. The two CSPs have been recently compared and a representative series of successfully resolved solutes is contained in this article (102).

2. *Mobile Phases*

The basic mobile phase used with the OVM CSP is composed of phosphate buffer and one or more modifiers. The type and concentration of the mobile-phase modifier as well as the pH are extremely important and both k′ and α can be altered by changing the mobile phase (102).

1. *The effect of pH.* Okamoto and Nakazawa (90) have observed an unexpected decrease in the retention of abscisic acid enantiomers when the pH of the mobile phase was lowered from 4.0–3.5. This was attributed to a change in the properties of the protein. Ovomucoid has been shown to undergo a low-pH transition, during which the protein adopts a more folded and ordered configuration (99,101).

2. *The effect of organic modifiers.* The effect of organic mobile-phase modifiers on retention and stereoselectivity of abscisic acid enantiomers has also been examined using methanol, ethanol, and 2-propanol (91). The results of this study indicate that hydrophobic interactions are involved in the retention, but not the stereochemical resolution of the solutes. Miwa and co-workers (91) have also postulated that hydrophobic and hydrogen-bonding interactions are important in the retention of basic compounds.

The effects of alcoholic modifiers and pH on the chromatographic properties of the OVM CSP have been investigated using acidic, basic, and neutral solutes (103). A series of primary, secondary, and tertiary alcohols and pH's ranging from 3.5–6.0 were used in this study. The results indicate that both the shape and hydrophobicity of the alcoholic modifier affect retention (k′) and enantioselectivity (α). In general, an increase in the hydrophobicity of the modifier results in a decrease in k's and α's. However, this is not the case when *t*-butanol is the modifier, suggesting that the size of the alkyl moiety attached to the carbinol carbon also contributes to the chromatographic results. The pH studies indicated that

Coulombic interactions play a role in the retention of the acidic and basic solutes. The results also suggest that in addition to ethanol and 1-propanol, *t*-butanol should be considered during optimization and that maximum efficiencies may be obtained at pH 5.0

3. *An approach to optimization.* A systematic approach to the determination of the optimum chromatographic conditions has recently been published (104) and is presented in Fig. 14.

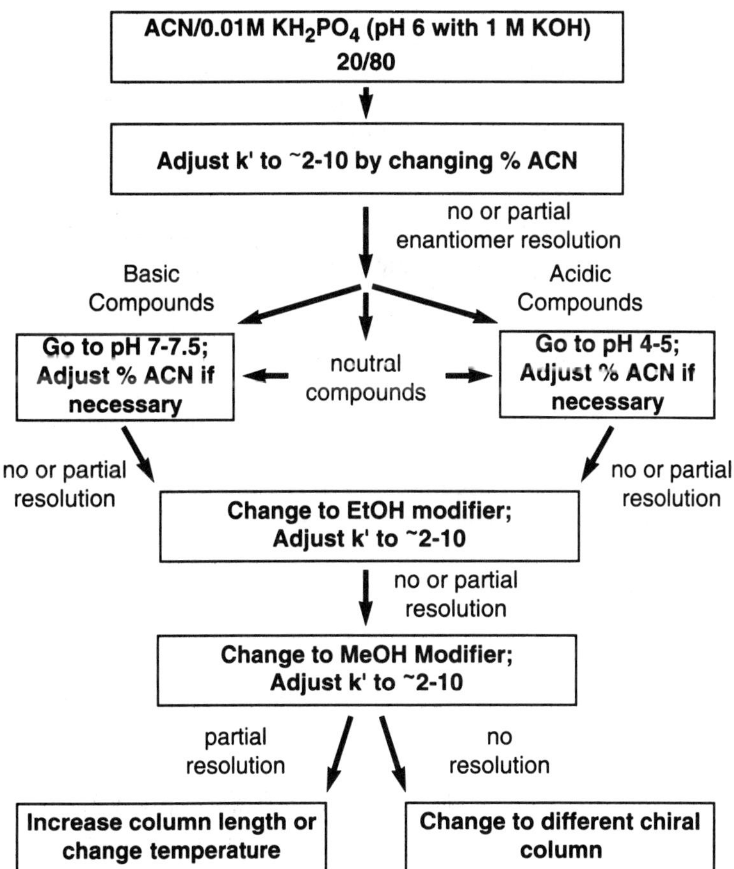

FIGURE 14 Basic flowchart for the use and optimization of the OVM CSP. [From Kirkland et al. (104).]

3. *Pharmaceutical Applications*

Because the OVM CSP is a relatively new CSP, only one pharmacokinetic application has been reported. Because of the utility of this phase, it is evident that such a situation will quickly change.

1. "The determination of the enantiomers of halofantrine and monodesbutylhalofantrine in plasma and whole blood using sequential achiral/chiral high performance liquid chromatography" (105). An HPLC method employing sequential achiral–chiral chromatography was developed for the determination of enantiomers of halofantrine (HF) and its monodesbutylated metabolite (HFM) in plasma and whole blood. After extraction from the biological matrix, HF and HFM were separated and quantitated on an achiral RP2 HPLC column. The eluent fractions containing HF and HFM were collected and reinjected onto the chiral HPLC system that was composed of an OVM CSP. The enantiomers of HF and HFM were stereochemically resolved on the OVM CSP and the enantiomeric ratios determined. The assay has been validated and applied to a pilot study of the pharmacokinetics of halofantrine in humans.

D. The HSA CSP and BSA CSP

Human serum albumin and bovine serum albumin are closely related proteins and, consequently, the chromatographic properties of the CSPs based on these proteins are similar. The only difference between the two phases appears to be due to inherent differences in stereoselectivity between HSA and BSA. For example, on the HSA-CSP, (*S*)-warfarin elutes before (*R*)-warfarin, whereas on the BSA CSP, the opposite elution order is observed (85). This is consistent with the enantioselectivities of the native proteins (106). However, even though there are differences between the CSPs, the selectivity, mobile-phase effects, and chromatographic properties of the HSA CSP and BSA CSP are so similar that the two phases will be discussed together.

1. *Solute Structure*

A wide variety of anionic and neutral compounds can be resolved on serum albumin (SA) CSPs, but not cationic compounds and, in general, the solute should contain aromatic and polar moieties (92). Some representative classes of compounds are the following:

1. *Amino acids.* Some aromatic amino acids such as kynurenin can be resolved on the BSA CSP without derivatization (92). However, most amino acids require precolumn derivatization of the amine moiety. The *N*-derivatives that have been used include acetyl, benzenesulphonyl, phthalimido, DANSYL, 2,4-dinitrophenyl, and 2,4,6-trinitrophenyl (85,92,107,108).

2. *Sulfur-containing compounds.* Molecules containing asymmetric sulphoxide or sulphoximine moieties and an aromatic group near or at the stereogenic center have been resolved (92,107).

3. *Coumarin derivatives.* Phenprocoumon and warfarin have been resolved (85,92).

4. *Benzodiazepine derivatives.* A wide variety of chiral benzodiazepines including oxazepam, lorazepam, and temazepam as well as their hemisuccinate and acetate derivatives have been resolved (85,92).

5. *2-Aryl propionic acid nonsteroidal antiinflammatory drugs.* The compounds from this class of agents that have been stereochemically resolved include ibuprofen, flurbiprofen, naproxen, benoxaprofen, pirprofen, suprofen, fenoprofen, indoprofen, and ketoprofen (94).

6. *Reduced folates.* Leucovorin and 5-methyltetrahydrofolate have been stereochemically resolved on SA CSPs (85,109).

7. *Neutral compounds.* Some neutral compounds including benzoin have been separated on SA CSPs (92).

2. Mobile Phases

The basic mobile phase used with the SA CSP is composed of phosphate buffer and one or more modifiers. The buffer concentration can range from 0.01–0.20 *M* and the pH from 4.5–8.0. These parameters can affect both retention (k′) and enantioselectivity (α). In addition, both k′ and α can be manipulated using organic and inorganic mobile-phase modifiers.

1. *The effect of pH.* In general, for *N*-derivatized amino acids, an increase in pH will result in a decrease in retention and stereochemical resolution (107,108). This effect is believed to be due to the fact that a decrease in the pH will result in a corresponding decrease in the net negative charge of the BSA. This means that the Coulomb interaction between the BSA and *N*-derivatized amino acids or other uncharged carboxylic acids will increase, resulting in an increase in k′ (108). The same effect was found for the enantiomers of oxazepam hemisuccinate (85).

A similar effect of pH of k′ has been found for the enantiomers of warfarin, but the effect on α was opposite that found with the *N*-derivatized amino acids (110). An increase in the pH of the mobile phase from 5.0–8.0 resulted in a decrease in k′ for (*S*)-warfarin from 33.6–3.4 (a 90% drop) and a decrease in k′ for (*S*)-warfarin from 38.6–4.6 (an 88% drop), the net result being an increase in α from 1.15 (pH 5.0) to 1.35 (pH 8.0).

2. *The effect of buffer concentration.* For solutes containing a carboxylic acid moiety, the general rule appears to be that the lower the buffer

3. Pharmaceutical Applications

The SA CSPs have been used in a variety of pharmacokinetic and pharmacodynamic studies. Because of the nature of these columns, they are most often used in coupled achiral–chiral systems. Two examples of this approach are the following:

1. "The measurement of warfarin enantiomers in serum using coupled achiral/chiral high-performance liquid chromatography" (110). An assay for the serum concentrations of (*R*)-warfarin and (*S*)-warfarin was developed using the BSA CSP coupled to a Pinkerton internal-surface reverse-phase (ISRP) achiral column. The ISRP column was used to separate (*R,S*)-warfarin from the serum components and warfarin metabolites and to quantitate the total warfarin concentration. The eluent containing the (*R,S*)-warfarin was then selectively transferred to the BSA CSP, where the enantiomers were enantioselectively resolved ($\alpha = 1.19$) and the enantiomeric composition determined.

2. "Determination of low levels of the stereoisomers of leucovorin and 5-methyltetrahydrofolate in plasma using a coupled chiral-achiral high performance liquid chromatographic system with post-chiral column peak compression" (114). A method for the determination of low levels of the stereoisomers of leucovorin (LV) and 5-methyltetrahydrofolate (5-MTH) was developed and validated for plasma levels of both compounds ranging from 15–500 ng/mL. The assay involved initial chromatography on a BSA CSP, followed by postcolumn peak compression and elution on two C_{18} columns. The BSA CSP separated LV and 5-MTH from interfering plasma components and from each other and achieved the stereochemical resolution of the diastereomeric (6*S*)- and (6*R*)-LV. The eluent containing (6*S*)-LV was directed onto one C_{18} column and the eluent containing (6*R*)-LV and 5-MTH was directed onto the other. This was followed by sequential rapid gradient elution of the target compounds from the respective C_{18} columns.

REFERENCES

1. I. W. Wainer, in *Drug Stereochemistry: Analytical Methods and Pharmacology* (I. W. Wainer and D. E. Drayer, eds.), Marcel Dekker, New York, 1988, p. 164.
2. W. H. Decamp, *Chirality, 1*:2 (1989).
3. H. Shindo and J. Caldwell, *Chirality, 3*:91 (1991).
4. A. J. Hutt, *Chirality, 3*:161 (1991).
5. M. N. Cayen, *Chirality, 3*:94 (1991).
6. I. W. Wainer, *Trends Anal. Chem., 6*:125 (1987).
7. I. W. Wainer, R. M. Stiffin, and Y.-Q. Chu, in *Chiral Separations* (D. Stevenson and I. D. Wilson, eds.), Plenum Press, New York, 1988, pp. 11–22.

8. I. W. Wainer, *A Practical Guide to the Selection and Use of HPLC Chiral Stationary Phases*, J. T. Baker, Phillipsburg, N.J., 1988.
9. A. M. Krstulovic, *Chiral Separations by HPLC: Applications to Pharmaceutical Compounds*, Wiley, New York, 1989.
10. W. J. Lough, *Chiral Liquid Chromatography*, Blackie, Glasgow, 1989.
11. D. Stevenson and I. D. Wilson, *Recent Advances in Chiral Separations*, Plenum Press, New York, 1990.
12. C. E. Dalgliesh, *J. Chem. Soc.*, *137*:3940 (1952).
13. W. H. Pirkle, M. H. Hyung, and B. Bank, *J. Chromatogr.*, *316*:585 (1984).
14. W. H. Pirkle and M. H. Hyung, *J. Chromatogr.*, *322*:287 (1985).
15. W. H. Pirkle and J. E. McCune, *J. Chromatogr.*, *469*:67 (1989).
16. I. W. Wainer and T. D. Doyle, *J. Chromatogr.*, 284:117 (1984).
17. I. W. Wainer an dM. C. Alembik, *J. Chromatogr.*, *367*, 59 (1986).
18. D. M. McDaniel and G. B. Snider, *J. Chromatogr.*, *404*:123 (1987).
19. D. W. Armstrong, M. Hilton, and L. Coofin, *LC•GC*, *9*:646 (1991).
20. I. W. Wainer and T. D. Doyle, *L. C. Mag.*, *2*:88 (1984).
21. I. W. Wainer and T. D. Doyle, *J. Chromatogr.*, *259*:465 (1983).
22. I. W. Wainer and T. D. Doyle, *J. Chromatogr.*, *284*:117 (1984).
23. N. Oi and H. Kitahara, *J. Chromatogr.*, *265*:117 (1983).
24. I. W. Wainer, T. D. Doyle, K. H. Donn, and J. R. Powell, *J. Chromatogr.*, *306*:405 (1984).
25. I. W. Wainer, T. D. Doyle, Z. Hamidzadeh, and M. Aldreidge, *J. Chromatogr.*, *261*:123 (1983).
26. Q. Yang, Z. P. Sun, and D. K. Ling, *J. Chromatogr.*, *477*:208 (1988).
27. T. D. Doyle, W. M. Adams, F. S. Fry, Jr., and I. W. Wainer, *J. Liq. Chromatogr.*, *9*:455 (1986).
28. M. Ziet, L. J. Crane, and J. Horvath, *J. Liq. Chromatogr.*, *7*:709 (1984).
29. P. Pescher, A. Tambute, M. Caude, and R. Rosset, *J. Chromatogr.*, *371*:159 (1986).
30. P. Macaudiere, A. Tambute, M. Caude, R. H. Rosset, M. A. Alembik, and I. W. Wainer, *J. Chromatogr.*, *371*:177 (1986).
31. D. J. Mazzo, C. J. Lindemann, and G. S. Brenner, *Anal. Chem.*, *58*:636 (1986).
32. W. H. Pirkle, J.-P. Chang, and J. A. Burke, III, *J. Chromatogr.*, *479*:377 (1989).
33. W. H. Pirkle and J. A. Burke, *J. Chromatogr.*, (in press).
34. S. V. Kakodkar and M. Zief, *Chirality*, 2:124 (1990).
35. I. W. Wainer, M. C. Alembik, and L. J. Fisher, *J. Pharmac. Biomed. Anal.*, *5*:735 (1987).
36. C. O. Meese, P. Thalheimer, and M. Eichelbaum, *J. Chromatogr.*, *423*:344 (1987).
37. P. Delatour, E. Benoit, M. Caude, and A. Tambute, *Chirality*, *2*:156 (1990).
38. F. Gimenez, R. Farinotti, A. Thuillier, G. Hazebroucq, and I. W. Wainer, *J. Chromatogr.*, *529*:339 (1990).
39. J. Blackwell, in *Cellulose and Other Natural Polymer Systems* (R. M. Brown, Jr., ed.), Plenum Publishers, New York, 1982, pp. 403–428.
40. G. Hess and R. Hagel, *Chromatographia*, *6*:277 (1973).

41. E. Francotte and R. M. Wolf, *Chirality*, *2*:16 (1990).
42. G. Blaschke, *J. Liq. Chromatogr.*, *9*:341 (1986).
43. A. Mannschreck, H. Koller, and R. Wernicke, *Kontakte*, *1*:16 (1985).
44. A. Mannschreck, H. Koller, G. Stühler, M. A. Davies, and J. Traber, *Eur. J. Med. Chem.—Chim. Ther.*, *19*:381 (1984).
45. H. Koller, K. H. Rimbock, and A. Mannschreck, *J. Chromatogr.*, *282*:80 (1983).
46. E. Francotte and R. M. Wolf, *Chirality*, *3*:43 (1991).
47. E. Francotte, R. M. Wolf, D. Lohmann, and R. Mueller, *J. Chromatogr.*, *347*:25 (1985).
48. A. Ichida, T. Shibata, I. Okamoto, Y. Yuki, H. Namikoshi, and Y. Toga, *Chromatographia*, *19*:280 (1984).
49. I. W. Wainer and M. C. Alembik, *J. Chromatogr.*, *385*:85 (1986).
50. I. W. Wainer, R. M. Stiffin, and T. Shibata, *J. Chromatogr.*, *411*:139 (1988).
51. *Bulletin on Chiracel Columns*, Daicel Chemicals Ltd., New York, 1985.
52. I. W. Wainer, M. C. Alembic, and C. R. Johnson, *J. Chromatogr.*, *361*:374 (1986).
53. A. Shibukawa and I. W. Wainer, *J. Chromatogr.* (in press).
54. Y. Okamoto, M. Kawashima, R. Aburatani, K. Hatada, T. Nishiyama, and M. Masuda, *Chem. Lett.*:1237 (1986).
55. Y. Okamoto, R. Aburatani, Y. Kaida, K. Hatada, N. Inotsune, and M. Nakano, *Chirality*, *1*:219 (1989).
56. K. Ikeda, T. Hamasaki, H. Kohno, T. Ogawa, T. Matsumoto, and J. Sakai, *Chem. Lett.*:1089 (1989).
57. I. W. Wainer, M. C. Alembik, and E. Smith, *J. Chromatogr.*, *388*:65 (1987).
58. M. H. Gaffney, R. M. Stiffin, and I. W. Wainer, *Chromatographia*, *27*:15 (1989).
59. A. M. Krstulovic, M. H. Fouchet, J. T. Burke, G. Gillet, and A. Durand, *J. Chromatogr.*, *452*:477 (1988).
60. R. J. Straka, R. L. Lalonde, and I. W. Wainer, *Pharm. Res.*, *5*:187 (1988).
61. D. Masurel and I. W. Wainer, *J. Chromatogr.*, *490*:133 (1989).
62. D. W. Armstrong, *J. Liq. Chromatogr.*, *7*:353 (1984).
63. D. W. Armstrong and W. DeMonde, *J. Chromatogr. Sci.*, *22*:411 (1984).
64. J. Debowski, J. Jurczak, and D. Sybilska, *J. Chromatogr.*, *282*:83 (1983).
65. D. W. Armstrong and H. L. Jin, *Chirality*, *1*:27 (1989).
66. D. W. Armstrong, T. J. Ward, R. D. Armstrong, and T. E. Beesley, *Science*, *232*:1132 (1986).
67. K. B. Lipkowitz, S. Raghothama, and J. Yang, *J. Am. Chem. Soc.* (in press).
68. L. Coventry, in *Chiral Liquid Chromatography* (W. J. Lough, ed.), Blackie, Glasgow, 1989, pp. 148–165.
69. P. Macaudiere, A. Tambute, M. Caude, and R. H. Rosset, *J. Chromatogr.*, *405*:135 (1987).
70. J. S. McClanahan and J. H. Maguire, *J. Chromatogr.*, *381*:438 (1986).
71. A. M. Krstulovic, J. M. Gianviti, J. T. Burke, and B. Mompon, *J. Chromatogr.*, *426*:417 (1988).
72. L.-E. Edholm, C. Lindberg, J. Paulson, and A. Walhegen, *J. Chromatogr.*, *424*:61 (1988).

73. Y. Okamoto and K. Hatada, in *Chiral Separations by HPLC: Applications to Pharmaceutical Compounds* (A. M. Krstulovic, ed.), Ellis Horwood Ltd., Chichester, U.K., 1989, pp. 316–335.
74. T. Shinbo, T. Yamaguchi, K. Nishimura, and M. Sugiura, *J. Chromatogr.*, *405*: 145 (1987).
75. P. M. Udvarhelyi and J. C. Watkins, *Chirality*, *2*:200 (1990).
76. S. Motellier and I. W. Wainer, *J. Chromatogr.*, *516*:365 (1990).
77. D. W. Armstrong, C. D. Chang, and S. H. Lee, *J. Chromatogr.*, *593*:83 (1991).
78. D. W. Armstrong, A. N. Stalcup, M. L. Hilton, J. D. Duncan, J. R. Faulkner, Jr., and S. C. Chang, *Anal. Chem.*, *62*:1610 (1990).
79. I. W. Wainer (unpublished results).
80. U. K. Walle, T. Walle, S. A. Bal, and L. S. Olanoff, *Clin. Pharmacol.*, *18*:718 (1984).
81. F. Albani, R. Riva, M. Contin, and A. Baruzzi, *Br. J. Clin. Pharmacol.*, *18*:244 (1984).
82. T. Albec-Kolbac, S. Rendic, Z. Fuks, V. Sunkic, and F. Kajez, *Acta Pharm. Jugosl.*, *29*:53 (1979).
83. S. Allenmark, *Chem. Scr.*, *20*:5 (1982).
84. J. Hermansson, *J. Chromatogr.*, *269*:71 (1983).
85. E. Domenici, C. Bertucci, P. Salvadori, G. Felix, I. Cahagne, S. Motellier, and I. W. Wainer, *Chromatographia*, *29*:170 (1990).
86. S. Allenmark, B. Bomgren, and H. Boren, *J. Chromatogr.*, *264*:63 (1983).
87. T. Miwa, T. Miyakawa, M. Kayano, and Y. Miyake, *J. Chromatogr.*, *408*:316 (1987).
88. G. Schill, I. W. Wainer, and S. A. Barkan, *J. Liq. Chromatogr.*, *9*:641 (1986).
89. G. Schill, I. W. Wainer, and S. A. Barkan, *J. Chromatogr.*, *365*:73 (1986).
90. M. Okamoto and H. Nakazawa, *J. Chromatogr.*, *504*:445 (1990).
91. T. Miwa, T. Miyakawa, M. Kayano, and Y. Miyake, *J. Chromatogr.*, *408*:316 (1987).
92. S. Allenmark, *J. Liq. Chromatogr.*, *9*:425 (1986).
93. I. W. Wainer and Y.-Q. Chu, *J. Chromatogr.*, *455*:316 (1988).
94. T. A. G. Noctor, G. Felix, and I. W. Wainer, *Chromatographia*, *31*:55 (1991).
95. A.-F. Aubry, F. Gimenez, R. Farinotti, and I. W. Wainer, *Chirality*, *4*:30 (1992).
96. M. Enquist and J. Hermansson, *Chirality*, *1*:209 (1989).
97. A. M. Krstulovic and J. L. Vende, *Chirality*, *1*:243 (1989).
98. I. Kato, J. Schrode, W. J. Kohr, and M. Laskowski, *Biochemistry*, *26*:193 (1987).
99. J. G. Beeley, *Biochem. J.*, *159*:335 (1976).
100. M. D. Melamed, in *Glycoproteins: Their Composition, Structure and Function* (A. Gottschalk, ed.), Elsevier, New York, 1972, pp. 317–334.
101. T. Miwa, H. Kuroda, and S. Sakashita, *J. Chromatogr.*, *511*:89 (1990).
102. K. M. Kirkland, K. L. Nielson, and D. A. McCombs, *J. Chromatogr.*, *545*:43 (1991).
103. J. Iredale, A.-F. Aubry, and I. W. Wainer, *Chromatographia*, *31*:329 (1991).
104. K. M. Kirkland, K. L. Neilson, and D. A. McCombs, *LC•GC* (in press).
105. F. Gimenez, A.-F. Aubry, R. Farinotti, K. Kirkland, and I. W. Wainer, *J. Biomed. Pharm. Anal.* (in press).

leads, ultimately, to a compound (starting material) that is commercially available or whose synthesis is known.

Usually, several routes to the target will be generated in this fashion. The choice of which one is to be attempted is often subjective, based on the prejudices of the chemist involved. On a more logical basis, the factors leading to the synthesis choice can involve the cost and availability of the starting material, the length of the synthesis, the overall probability of success, and the options available should one reaction not occur as predicted.

The preparation of an enantiomerically homogeneous chiral molecule adds another element of difficulty to the retrosynthetic analysis. Either the analysis must include a specific step for obtaining one enantiomer, or it must lead ultimately to an enantiomeric starting material. These possibilities are examined more fully below.

B. Introduction of Chirality

The practicing organic chemist now has available a variety of synthetic tools for preparing enantiomerically pure compounds (2). These methods all derive, ultimately, from a naturally occurring chiral molecule. The means by which this natural chirality is applied to preparing other chiral molecules varies widely in concept and execution. These concepts fall, however, into three general areas: resolution, asymmetric synthesis, and the use of the chiral carbon pool. Comprehensive reviews of these methods exist (3–5), and thus only a brief outline of each will be presented here.

1. Resolution

Classical Resolution and Variants. Resolution is the process by which a chiral recemic molecule is combined with a second chiral, but enantiomerically homogeneous, molecule. The resultant mixture of diastereomers is separated and the appropriate diastereomer is then cleaved to recover the resolving agent and the desired enantiomer. As opposed to enantiomers, diastereomers have different physical properties, for example, melting points and solubilities, thus allowing for separation.

The most classical of resolutions is exemplified by the separation, by crystallization, of the diastereomeric salts formed by treatment of a racemic acid with one enantiomer of a chiral base, typically an alkaloid such as quinine. Unfortunately, despite significant recent advances (3,6), the relative solubilities of two diastereomers, and thus the probability for success of a classical resolution, are difficult to predict. It thus remains, for most chemists, a largely empirical method. On the other hand, a successful resolution often provides both enantiomers, even when both enantiomers of the resolving agent are not at hand, by recovery from the enriched mother liquors. A careful study of the pharmacological and toxicological properties of the individual enantiomers can determine whether, in fact, the cost of separation is necessary or justified.

The development of newer separation techniques, in particular preparative gas and liquid chromatography, has broadened the scope of resolution in recent years. An alternative to the acid–base salt separation by crystallization, for example, would be formation of the covalent amide linkage, chromatographic separation of the diastereomers, and then chemical hydrolysis.

Resolution, by its very nature, is an inefficient process. The maximum obtainable yield is 50%; in practice, inefficient separation requiring more than one crystallization or chromatography and/or mechanical losses during processing often make the actual yield significantly lower. As a practical matter of synthetic strategy then, it is important to carry out the resolution as early in the synthesis as possible, when the material to be lost carries the minimum value. For economic viability, a drug synthesis involving a resolution usually must contain an efficient recycle of the wrong enantiomer. In most cases, this recycle is effected by racemization and reresolution. There are examples, however, where clever synthetic design allows the carrying forward of both enantiomers of a chiral intermediate; such syntheses have been termed chirally economic (7).

Second-Order Asymmetric Transformations. A modification of the classical resolution occurs in the specific case where equilibration of the chiral center can be achieved during the resolution. By judicious choice of reaction conditions, one diastereomeric salt can be induced to crystallize under the equilibration conditions. As this material precipitates, solution equilibrium is reestablished by racemization of the now-major isomer remaining. In the best cases, over 90% of a single diastereomeric salt can be obtained.

Examples of second-order asymmetric transformations are relatively rare. By far the best known case (Fig. 1) is the preparation of methyl

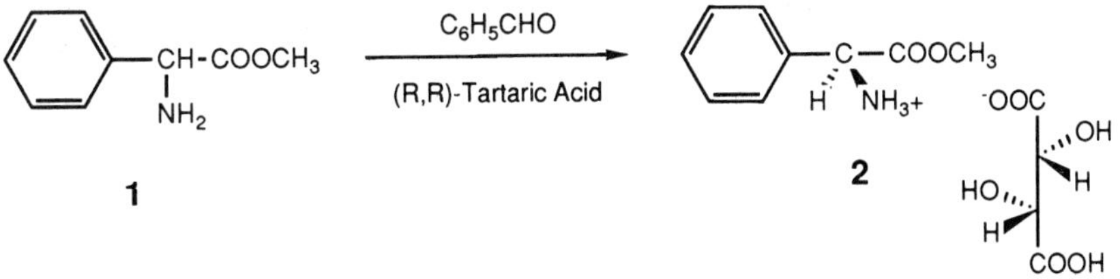

FIGURE 1 Second-order asymmetric transformation.

R-phenylglycinate-*R*,*R*-hydrogen tartrate [2], a key building block for the β-lactam antibiotic, ampicillin. The addition of one mole each of benzaldehyde and *R*,*R*-tartaric acid to a 10% solution of racemic methyl phenylglycinate [1] in ethanol results in precipitation, after 24 hr, of the desired salt [2] in 85% yield. Reuse of the salt mother liquors as feed in subsequent runs results in, ultimately, an overall 95% conversion to the desired material (8). The presence of benzaldehyde greatly facilitates the racemization process by forming, reversibly, a Schiff base.

The finding of a second-order asymmetric transformation involves not only the empiricism of the classical resolution but also the finding of resolution conditions that simultaneously allow the diastereomeric interconversion. It is not, then, surprising that these rigid criteria have kept the number of demonstrated examples small.

Kinetic Resolution. The selective reaction of one member of a racemic pair with a chiral reagent is the basis for a kinetic resolution. This reaction provides recovered starting material in one enantiomeric series with a product in the opposite series.

The reagents giving a kinetic resolution can be either chemical or enzymatic. The most generally useful of such reagents, to date, have been enzymes (9). Perhaps the best known example is the acylase-mediated hydrolysis of, for example, racemic *N*-acetylphenylalanine [3] (Fig. 2). The process gives *S*-amino acid [5] of, usually, very high enantiomeric purity, as well as recovered *R*-*N*-acetyl amino acid [4]. As in classical resolution, the obtainable yield is 50%, and recycle of the unwanted enantiomer is required for maximum efficiency. Fortunately, there are several simple methods available for racemization of *N*-acyl amino acids and, thus, by recycling, an excellent yield of *S*-amino acid is often achieved. This methodology is now practiced industrially, principally in Japan, yielding many tons annually of synthetic amino acids (10). The industrial applications are particularly elegant in that often an immobilized enzyme is used. The kinetic resolution is effected by simply

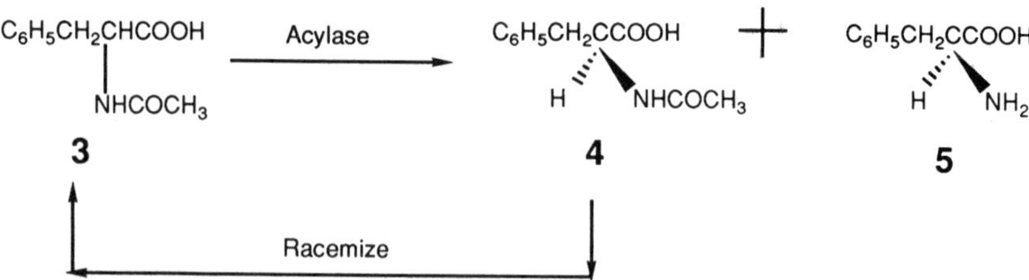

FIGURE 2 Enzymatic kinetic resolution of amino acids.

passing a solution of the racemate through a column containing the immobilized enzyme.

Other enzymic kinetic resolutions are known. Of particular value to the synthetic chemist are the lipase and/or esterase-mediated hydrolyses of esters of chiral racemic alcohols (11) or acids (12). The resultant product alcohols or acids and recovered esters are often of high enantiomeric purity.

Methods for chemical kinetic resolution to give products of high enantiomeric purity are less well known. Perhaps the most successful, and one complementary in terms of the products obtained with the enzymic methods, is the epoxidation of a racemic secondary allylic alcohol (13). When this epoxidation is carried out using *t*-butylhydroperoxide as oxidant in the presence of a titanium catalyst that is chirally modified by an ester of tartaric acid, the selectivity for one enantiomer of the starting alcohol is often virtually complete.

Thus, a chiral secondary alcohol, extremely useful as an intermediate for many synthetic targets, can be prepared by either a chemical or enzymatic kinetic resolution. The choice depends on the particular molecule sought and the prejudices of the chemist involved. Recycling of the unwanted enantiomer in these cases is simple, involving oxidation, then reduction to the racemate.

2. Asymmetric Synthesis

Asymmetric synthesis is the chemical or biochemical conversion of a prochiral substrate to a chiral product. In general, this involves reaction at an unsaturated site having prochiral faces (C=C, C=N, C=O, etc.) to give one product enantiomer in excess over another. The reagents effecting the asymmetric synthesis are used either catalytically or stoichiometrically. Clearly, the former is to be preferred, for economic reasons, when applicable. The reagents can be either chemical or enzymatic.

Asymmetric synthesis is, in itself, a very active and exciting field for scientific exploration, with major discoveries being reported continually. The reader is referred to the five-volume treatise by Morrison (4) for a comprehensive review and an assessment of recent developments.

The methodologies for asymmetric synthesis have now matured to the extent that they form the basis for commercial syntheses of several chiral compounds (14). Two such examples involve the preparation of pharmaceuticals. Shown in Fig. 3 are the key chirality-introducing steps in the synthesis of L-dopa [8] and cilastatin [11].

L-dopa, used in the treatment of Parkinson's disease, is best prepared by asymmetric catalytic hydrogenation (15) of the enamide [6]. The hydrogenation, performed with a soluble rhodium catalyst modified with the

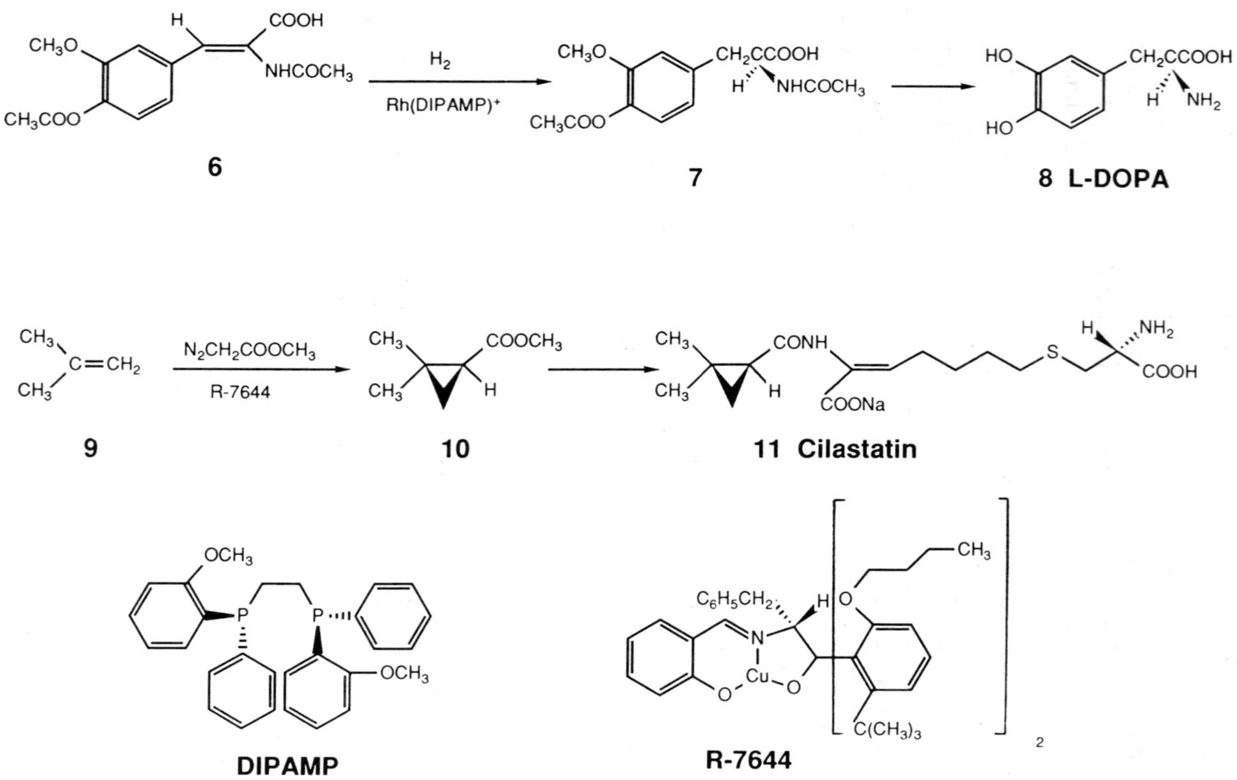

6 → 7 → 8 L-DOPA

H_2 / Rh(DIPAMP)$^+$

9 → 10 → 11 Cilastatin

$N_2CH_2COOCH_3$ / R-7644

DIPAMP

R-7644

FIGURE 3 Industrial asymmetric syntheses of pharmaceuticals.

chiral bisphosphine DIPAMP, gives the protected amino acid [7] in 94% enantiomeric excess (e.e.). Enantiomeric enrichment and removal of the protecting groups then provide the desired amino acid. It was this industrial preparation of L-dopa that firmly established asymmetric synthesis as a viable synthetic tool, rather than an exotic curiosity, in the minds of most organic chemists.

Thienamycin and its derivatives are exciting new antibiotics. Their clinical use is limited, however, by their susceptibility to the kidney enzyme dehydropeptidase I. Reversible inhibition of this enzyme is provided by cilastatin [11]. The preparation of the *S*-cyclopropane portion [10] of cilastatin is achieved (16) by decomposition of ethyl diazoacetate in isobutylene [9] in the presence of the chiral copper catalyst R-7644. The product [10] is obtained in 92% e.e. and then further processed to cilastatin. Cilastatin is now marketed in combination with the thienamycin derivative imipenem as a very-broad-spectrum antibiotic.

Asymmetric synthesis, when applicable, is a very valuable tool for chiral drug synthesis. Although the number of examples giving high e.e.'s is growing, it is still limited, and the method will not be applicable in all cases. Of particular concern in any asymmetric synthesis is the fact that no such reported reaction yet gives absolute (i.e., 100%) introduction of chirality, and thus asymmetric synthesis must be paired with an enantiomeric enrichment step. A reaction giving a 95% e.e. may be of little use in drug synthesis if a method for reaching enantiomeric homogeneity cannot be found.

3. *Chiral Carbon Pool*

The third major source of chiral pharmaceuticals involves synthesis using naturally occurring chiral molecules as starting materials (5,17). Those compounds most generally used are carbohydrates, amino acids, terpenes, and smaller, microbiologically derived compounds such as lactic acid or tartaric acid. In addition, the synthetic chemist now has in his or her repertoire a variety of rather standard building blocks derived by manipulation of the natural substances; a list of such compounds has been compiled (5).

A retrosynthetic analysis may well lead to a molecule recognizably derived from the chiral carbon pool. Presumably, the resulting synthesis will then be subject only to the vagaries encountered in the preparation of any target molecule, chiral or not. Unfortunately, the actual situation is not always that simple. If the target molecule contains more than one chiral center, the introduction of the later centers must be highly stereoselective to avoid diastereomer formation. As noted above, though, diastereomers usually are separated fairly readily and the loss of a small amount of material as a diastereomer usually can be tolerated. Synthetic operations offering the possibility of racemization are, of course, to be avoided if at all possible.

Of most concern in using the chiral carbon pool, however, is the enantiomeric homogeneity of the natural products themselves. Although it is generally accepted that most carbohydrates and amino acids are enantiomerically pure, it is known that many terpenes are not. The small molecules may or not be enantiomerically pure. The only sure method of avoiding a nasty surprise during the projected synthesis is to use a starting material, the enantiomeric composition of which is known with certainty.

A further limitation of the chiral pool approach may be the availability of only one member of an enantiomeric pair. Strategies that circumvent this problem are available in certain cases, however (5).

II. EXEMPLIFICATION

The prostaglandins are extremely bioactive substances. Their availability in only very small amounts from natural sources, as well as their potential use in pharmacology in their native or altered form, has made them the subject of intense synthetic interest in recent years. These syntheses amply illustrate, as a coherent whole, the methods outlined above for obtaining chiral molecules.

It is by no means possible here to describe all synthetic work on prostaglandins. The reader is referred to a leading review (18) for that purpose. The examples chosen were those best illustrating the ingenuity of the synthetic chemist who needed to prepare a complex and relatively unstable chiral molecule. Emphasis in the discussion and figures is placed on the means used for introduction of chirality.

A. Corey Lactone

A by now classic retrosynthesis of prostaglandins $PGF_{2\alpha}$ and PGE_2 (Fig. 4) leads to the bicyclic lactone [12], five-carbon phosphonium salt [13], and phosphonate [14] (19). These compounds contain all the carbon atoms of the prostaglandins and, in [12], all but one of the chiral centers. Lactone [12] has come to be known generically as the Corey lactone, and its synthesis in one enantiomeric form has been the subject of numerous complementary investigations.

Several of the seminal routes to the lactone, as devised by Corey, are summarized in Fig. 5. Diels–Alder reaction of (methoxymethyl)cyclopentadiene [15] with chloroacrylonitrile and then basic hydrolysis gave the bicyclic ketone [16] (20). Ring expansion in a selective Baeyer–Villiger reaction led to lactone [17] that was then hydrolyzed to hydroxy acid [18].

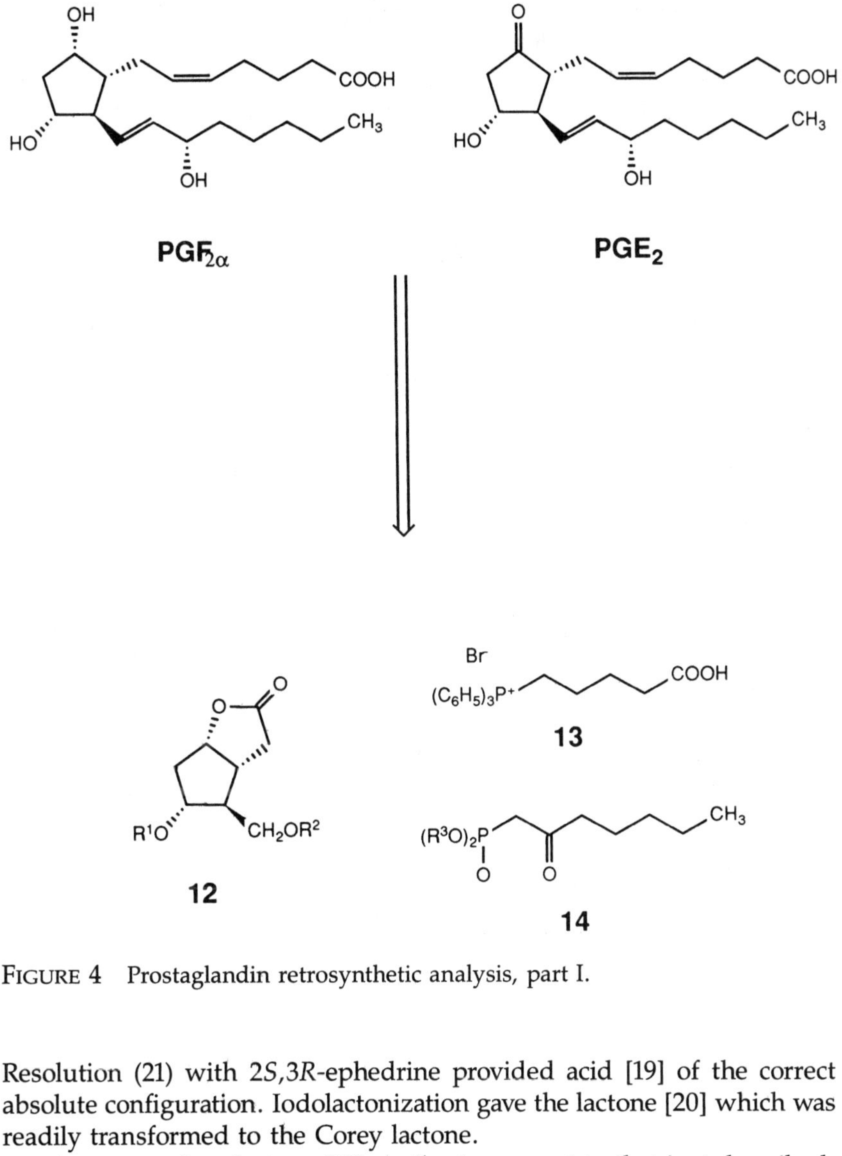

FIGURE 4 Prostaglandin retrosynthetic analysis, part I.

Resolution (21) with 2*S*,3*R*-ephedrine provided acid [19] of the correct absolute configuration. Iodolactonization gave the lactone [20] which was readily transformed to the Corey lactone.

An approach to lactone [12] similar in concept to that just described, but not requiring a resolution, involved asymmetric Diels–Alder reaction of (benzyloxymethyl)cyclopentadiene [21] with the chiral ester of acrylic acid and 8-phenylmenthol (22). The adduct [22] was obtained in undetermined but apparently quite high e.e. Oxidation of the ester enolate of [22], followed by lithium aluminum hydride reduction, gave diol [23] as an

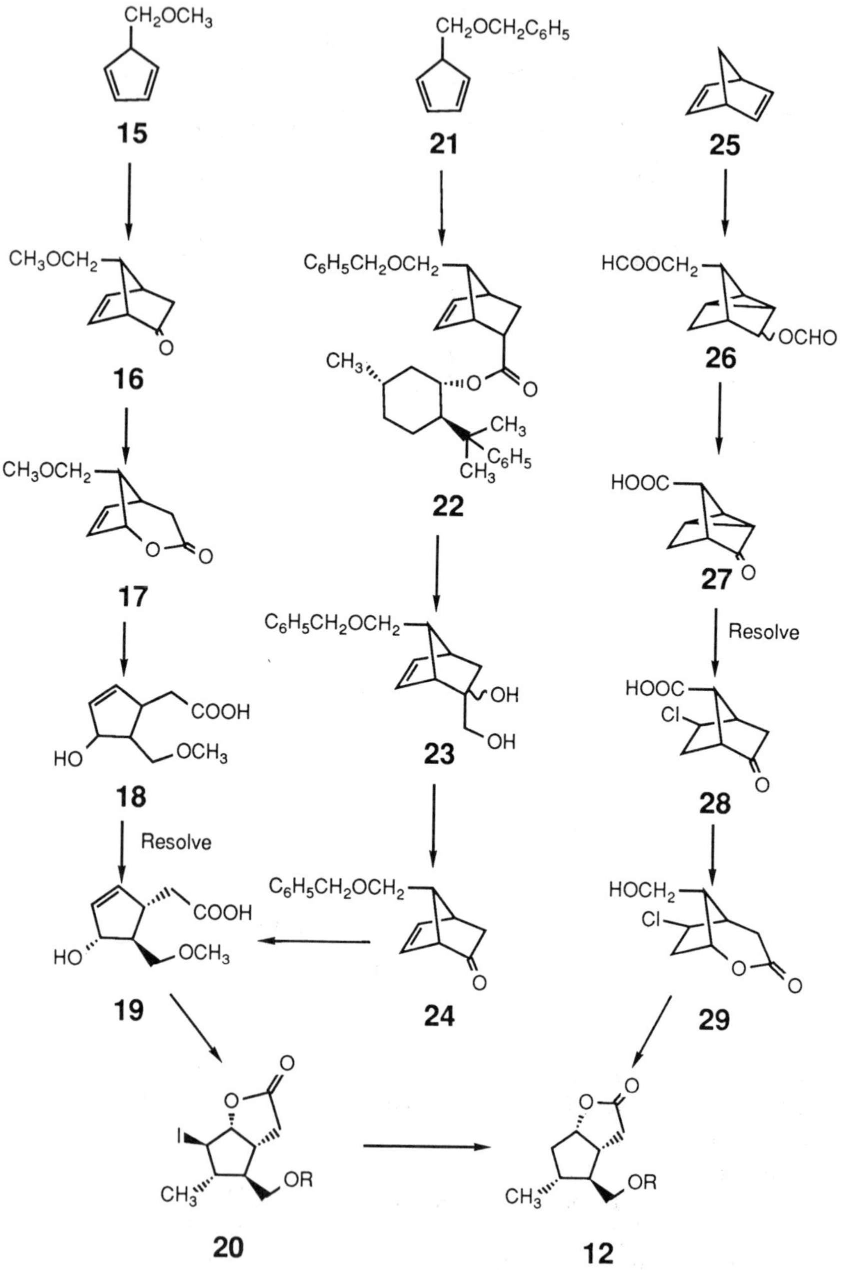

FIGURE 5 Corey approaches to lactone [12].

endo/exo mixture. As a by-product of this reaction, the 8-phenylmenthol could be efficiently recovered for reuse. Oxidative removal of the excess carbon atom gave ketone [24]. This synthetic equivalent to the resolved form of [16] was oxidized with basic hydrogen peroxide to hydroxy acid [19], from which the desired Corey lactone is readily obtained. Crystallization of this compound gave enantiomerically pure material; the enantiomer and any diastereomers, if present at all, were lost in this operation.

A third Corey approach involving bicyclic compounds started with the reaction of norbornadiene [25] with paraformaldehyde and formic acid, catalyzed by sulfuric acid (23). The mixture of formates [26] so produced could be directly oxidized with Jones reagent to keto acid [27]. Classical resolution, requiring two to three crystallizations, was effected with *S*-α-methylbenzylamine. One conversion of the resolved acid [27] to the Corey lactone involved cleavage with hydrochloric acid to chloro ketone [28]. Baeyer-Villiger oxidation followed by selective reduction of the acid functionality gave lactone [29]. After protection of the alcohol as its tetrahydropyranyl ether, base-catalyzed ring opening and relactonization with expulsion of chloride gave the desired Corey lactone.

A route to the Corey lactone that was devised by a Hoffmann-La Roche group (24) also involved bicyclic intermediates and a resolution (Fig. 6). However, use of the "meso trick" made introduction of the necessary chirality an efficient process. Thus, treatment of the symmetrical and hence achiral diol [30] with phosgene and then isobornylamine gave the mixture of diastereomers [31] and [32]. These urethanes were separable by fractional crystallization. Although the isolated yield of the desired diastereomer [32] was only 25%, the mother liquors, enriched in [31], were recycled by hydrolysis to the starting material, diol [30]. A continued resolution/recycle led to a quite efficient overall conversion of [30] to [32]. Elaboration of alcohol [32] to the one-carbon-homologated nitrile, followed by hydrolysis, gave lactone [33] and recovered isobornylamine. A several-step series of reactions involving ring opening and amide formation with pyrrolidine, oxidation to the aldehyde, epimerization, reduction, and ether formation led to amide [34]. Ozonolysis and oxidation then generated the diacetate [35] possessing the desired stereochemistry and oxidation level. Conversion of [35] to [12] was straightforward.

An application of the meso trick that does not, in principle, require recycling, has been provided (Fig. 7) by a Japanese group (25). Reaction of *cis*-2-cyclopenten-1,4-diol [36] with *N*-mesyl-*S*-phenylalanyl chloride gave, in addition to diester and recovered starting material, the diastereomeric esters [37] and [38]. Separation was effected by either chromatography or crystallization. Conversion of the free alcohol of [37] to its tetrahydropyranyl ether and saponification gave alcohol [39]. Transfer of the

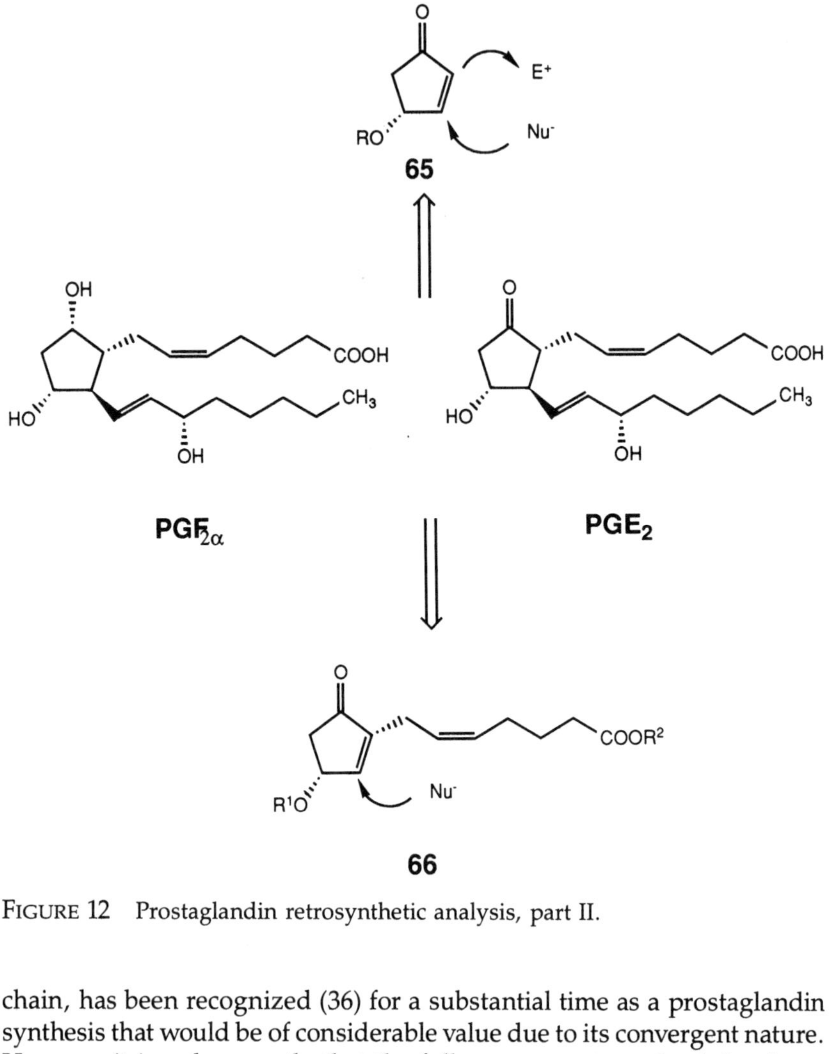

FIGURE 12 Prostaglandin retrosynthetic analysis, part II.

chain, has been recognized (36) for a substantial time as a prostaglandin synthesis that would be of considerable value due to its convergent nature. However, it is only recently that the fully convergent synthesis has been achieved.

In earlier work, a less convergent variant was developed in which a nucleophile was added in the 1,4 fashion to enone [66] containing the preformed upper chain. As practiced by the Sih group (37,38), the enone [66] (Fig. 13) was prepared starting from ethyl acetoacetate [67]. A nine-step chain-lengthening provided keto ester [68]. Reaction of this material with diethyl oxalate, followed by acidic hydrolysis and reesterification, gave the trione [69]. The key, chirality-inducing step involved reduction

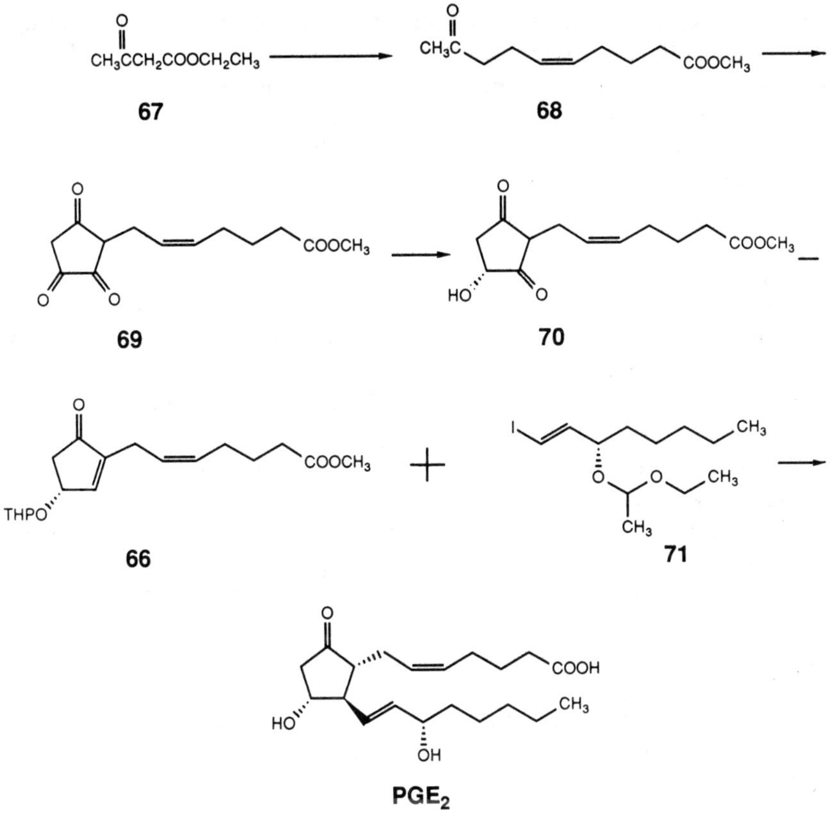

FIGURE 13 Sih synthesis of PGE_2.

of [69] with *Dipodascus uninucleatus* to the 4*R*-alcohol [70]. Apparently, the reaction was enantiospecific, although no evidence was presented to support this claim. Conversion of [70] to enone [66] involved reduction of the derived enol mesitylenesulfonate with sodium *bis*(2-methoxyethoxy)-aluminum hydride, acidic rearrangement and elimination, and THP ether formation. Addition of the cuprate prepared from the enantiomeric iodide [71] gave, after removal of the ether protecting groups with acid and microbiological ester hydrolysis, PGE_2.

An alternative preparation of the unprotected alcohol corresponding to structure [66] involved elaboration of the enantiomer of the bicyclic lactone [41] (Figs. 7 and 10) (39).

For the synthesis shown in Fig. 13 to be of value, a source of the enantiomeric vinyl iodide [71] or its equivalent must be available. A number of solutions to the problem have been devised. The initial approaches involved resolution. The hydrogen phthalate of racemic *E*-3-hydroxy-1-iodo-1-octene was resolved with *S*-α-methylbenzylamine (40). Alternatively, resolution in a similar manner of 1-octyn-3-ol (41) and then conversion of the acetylenic unit to the *E*-1-iodoalkene gave the same result (40). In either case, the overall efficiency of the resolution, due to the necessary chemical manipulations, was quite low.

Sih (38) has described the reduction of *E*-1-iodo-1-octen-3-one with *Penicillium decumbens* to give the desired *S*-alcohol. Based on optical rotation, the e.e. was about 80%. An asymmetric chemical reduction of this same ketone, using lithium aluminum hydride that had been partially decomposed by one mole each of *S*-2,2′-dihydroxy-1,1′-binaphthol and ethanol (42), gave the desired alcohol in 97% e.e. This reagent also reduced 1-octyn-3-one in 84% e.e. to the corresponding alcohol (43). A 92% e.e. could be obtained with B-3-pinanyl-9-borabicyclo[3.3.1]nonane as the reducing agent (44).

The more convergent prostaglandin synthesis in which the two side chains are added to enone [65] in a one-pot operation has been difficult to achieve (36) because the second step (reaction of the ketone enolate with an electrophile) initially failed. A variety of solutions (45) to the problem with varying degrees of sophistication have been developed. One memorable step along the way was Stork's synthesis (46,47) of $PGF_{2\alpha}$ (Fig. 14). The racemic 4-cumyloxy-2-cyclopentenone [72], upon reaction with the organocuprate derived from iodide [73] and subsequent trapping with formaldehyde, a very powerful electrophile, gave a 1:3 mixture of diastereomeric keto alcohols [74] and [75]. Mesylation and base-induced elimination gave [76]. This enone successfully underwent a second conjugate addition, this time with the cuprate from the *Z*-iodide [77]. Manipulation of the protecting groups and oxidation level of the resulting adduct [78] gave $PGF_{2\alpha}$ methyl ester [79]. At this point, the offending diastereomer was removed by chromatography.

The ultimate in the three-component coupling approach to prostaglandins has now been achieved by Noyori (48). As illustrated in Fig. 15, the cuprate derived from iodide [82] was added to enone [80] in the usual fashion. Then, after addition of hexamethylphosphoramide, triphenyltin chloride was used to effect enolate interchange. As opposed to lithium (or copper) enolates, the tin enolate is cleanly alkylated with allylic iodide [81]. The protected PGE_2 [83] was obtained in 78% yield. Two-step deprotection to PGE_2 was straightforward.

For the cyclopentenone conjugate addition approach to prostaglandins to be useful, good syntheses of the chiral lower chain and cyclopentenones must be available. Some preparations of the former have already

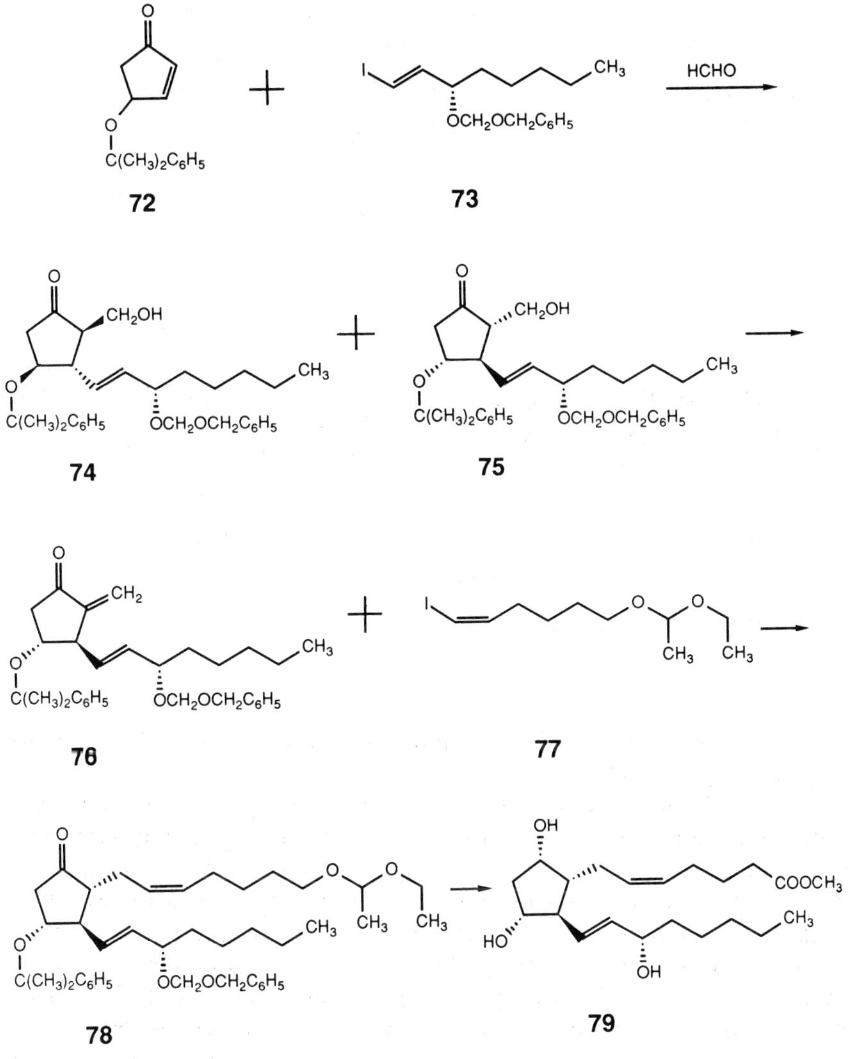

FIGURE 14 Stork synthesis of $PGF_{2\alpha}$ part I. Although diastereomer separation was postponed to the $PGF_{2\alpha}$ stage, only the desired isomers of compounds [76] and [78] are shown, for convenience.

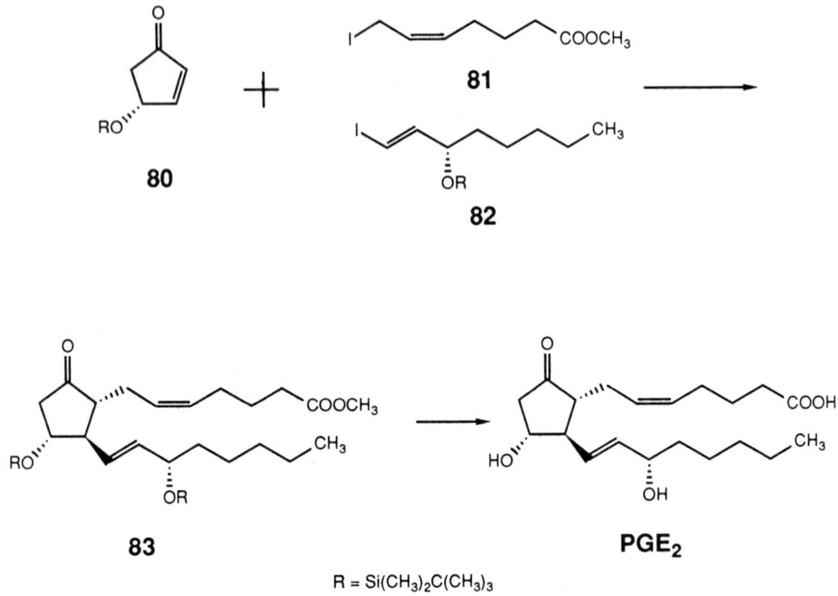

80 81 82 83 PGE2

R = $Si(CH_3)_2C(CH_3)_3$

FIGURE 15 Noyori synthesis of PGE_2.

been discussed. A few of the more interesting approaches to the latter are shown in Figs. 16–18.

One synthesis of cyclopentenone [80], requiring a resolution, involved initial ring contraction of phenol when treated with alkaline hypochlorite (49). Resolution of the resulting *cis* acid [85] was effected with brucine. The desired enantiomer [86] formed the more soluble brucine salt and was thus obtained from the mother liquors of the initial resolution. Oxidative decarboxylation with lead tetracetate, partial dechlorination with chromous chloride, and alcohol protection gave chloro enone [87]. Zinc-silver couple (50) dechlorinated [87] to the desired cyclopentenone [80].

Use of the chiral carbon pool for cyclopentenone preparation is also known. The fungal metabolite terrein [88] was selectively monoacetylated and then reduced with chromous chloride to enone [89]. Acetylation and olefin cleavage with ruthenium tetroxide and sodium periodate led to aldehyde [90], which was readily decarbonylated to [65] (51). An alternative route (52) began with the less common *S,S*-tartaric acid [91], converted in four steps to diiodide [92]. Dialkylation of methyl methylthiomethyl sulfoxide with [92] gave the cyclopentane derivative [93]. Treatment of [93]

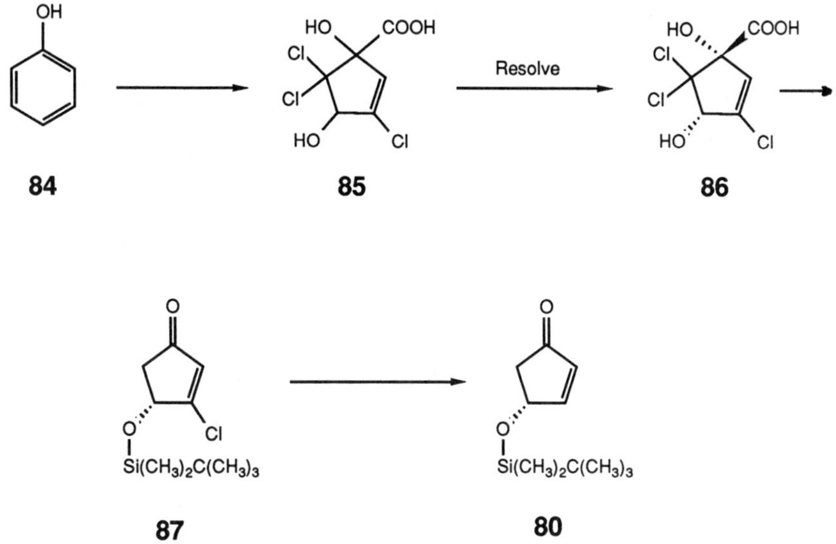

84 → 85 → Resolve → 86 →

87 → 80

FIGURE 16 Cyclopentenone synthesis involving resolution.

with sulfuric acid in ether liberated the masked carbonyl and caused elimination and deprotection of the C—3 alcohol to give directly [65]. The material thus obtained had an e.e. of about 85%, as estimated by nuclear magnetic resonance.

An intriguing synthesis of chiral cyclopentenone [100] from D-glucose has recently been described (53). The readily available diacetone glucose [94] was benzylated, selectively deprotected, and oxidatively cleaved to the aldehyde, which was condensed with nitromethane to adduct [95]. Acidic hydrolysis of the product gave hemiacetal [96], cleaved with periodate in methanol to aldehyde [97]. Aldol-type cyclization was effected with triethylamine; subsequent dehydration to [98] was induced by mesylation. The nitro olefin [98], upon treatment with activated lead in an acidic media, was converted to ketone [99]. Mesylation in the presence of triethylamine then led directly to cyclopentenone [100].

Two asymmetric synthesis approaches to chiral cyclopentenone derivatives can be envisaged. The first, reduced to practice by Noyori (43), involved reduction of cyclopentene-1,4-dione with lithium aluminum hydride chirally modified with binaphthol to give *R*-4-hydroxycyclopent-2-en-1-one in 94% e.e. Alternatively, manganese dioxide oxidation of allylic alcohol [40] (Fig. 7), in analogy to the *cis* isomer (54), would be expected to give the same enone.

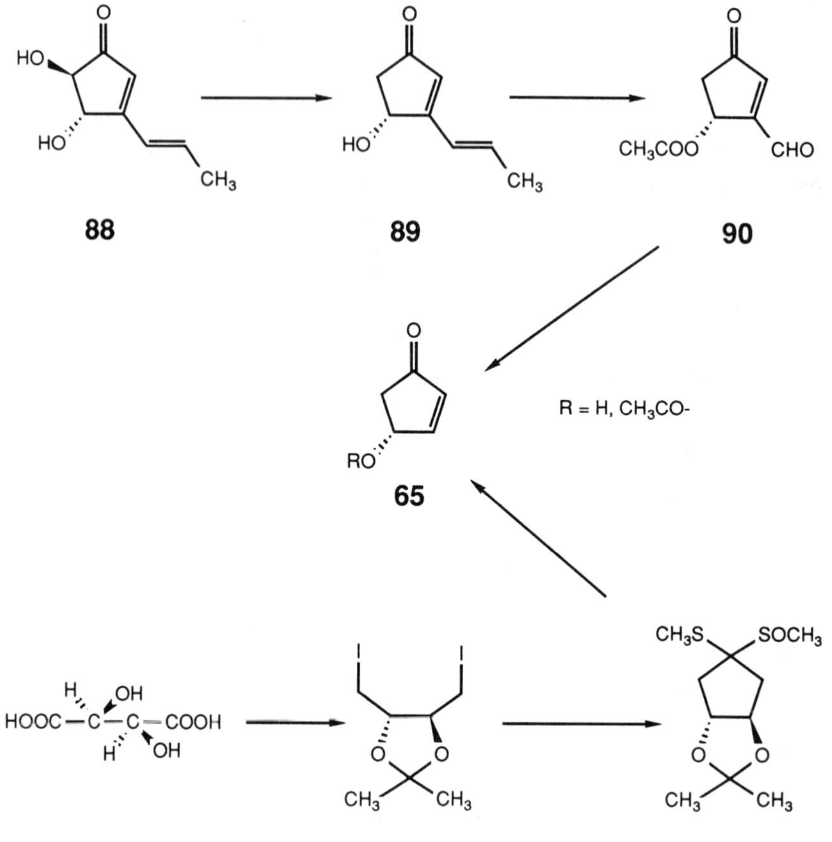

FIGURE 17 Chiral carbon pool approaches to cyclopentenones, part I.

C. Stork Synthesis

There is one synthesis of $PGF_{2\alpha}$ that cannot be classified with any others. It is, however, such an elegant route to $PGF_{2\alpha}$ both in concept and execution, that its inclusion in this discussion of prostaglandin syntheses is mandatory. Stork (55) initiated the synthesis with lactone [101] (Fig. 19), a commercially available material obtained from D-glucose [52] by homologation with cyanide, followed by hydrolysis. Lactone partial reduction with sodium borohydride and *bis*(isopropylidenation) gave [102]. Further reduction with borohydride, selective acetylation of the primary alcohol, and elimination of the vicinal hydroxyl groups by heating with dimethyl-

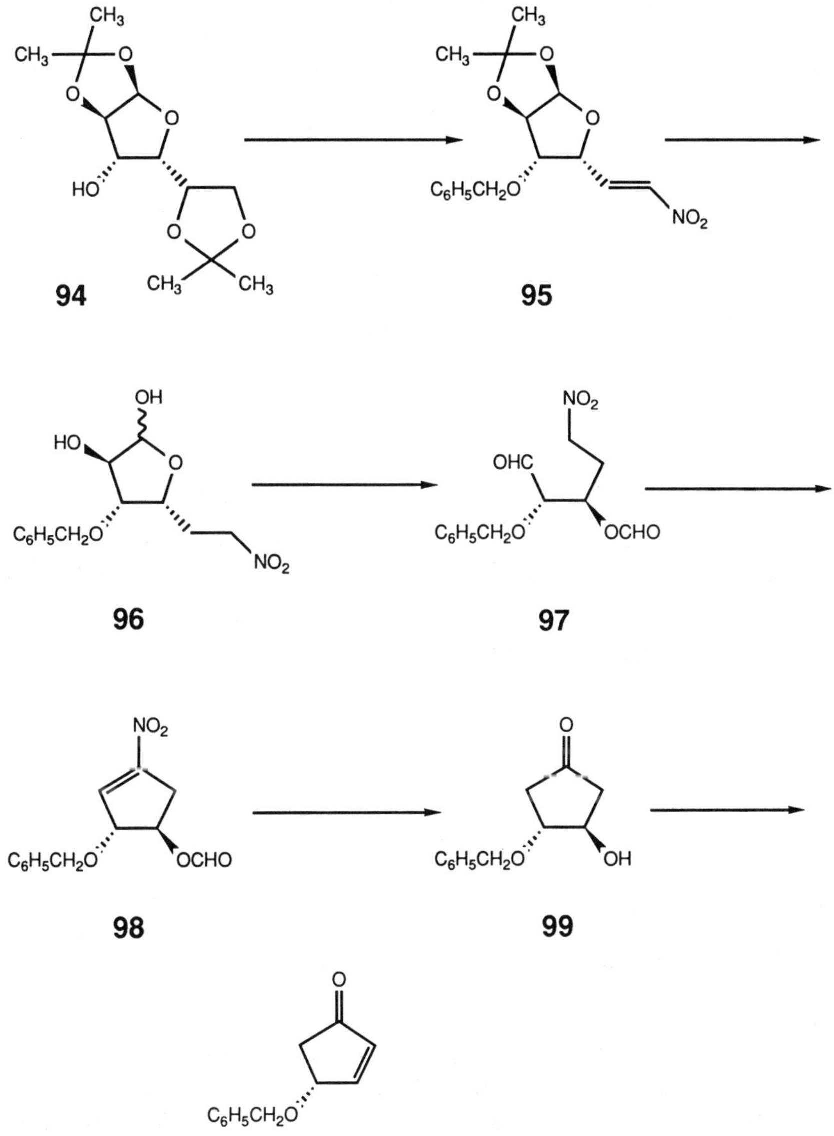

94

95

96

97

98

99

100

FIGURE 18 Chiral carbon pool approaches to cyclopentenones, part II.

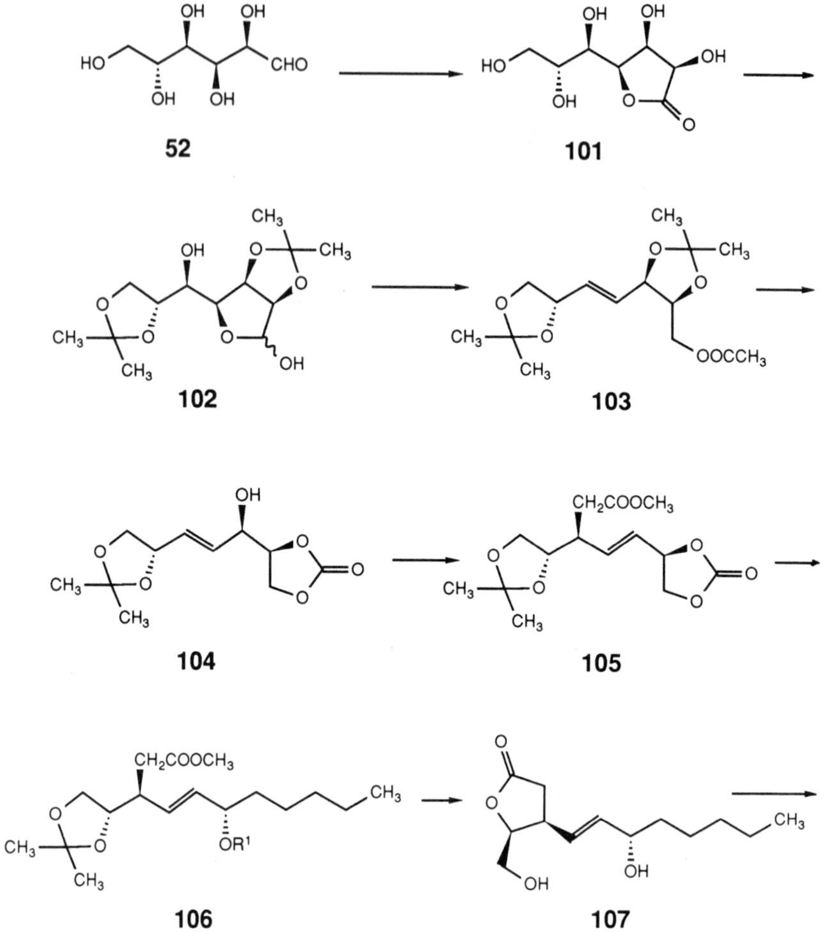

FIGURE 19 Stork synthesis of $PGF_{2\alpha}$, part II.

formamide dimethyl acetal led to olefin [103]. The allylic alcohol [104], needed for a projected orthoester Claisen rearrangement, was obtained from [103] by ester hydrolysis, treatment with methyl chloroformate to give the mixed carbonate, isopropylidene group hydrolysis (with concomitant cyclic carbonate formation), and reprotection of the terminal glycol. The Claisen rearrangement proceeded as anticipated, with complete transfer of asymmetry, to give ester [105].

The lower side chain of $PGF_{2\alpha}$ was elaborated by selective cleavage of

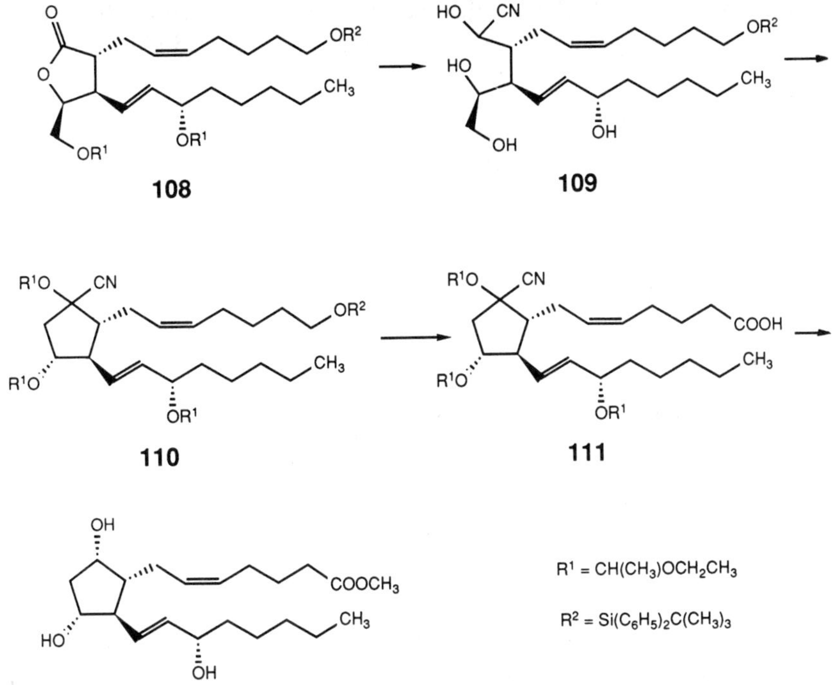

108

109

110

111

$R^1 = CH(CH_3)OCH_2CH_3$

$R^2 = Si(C_6H_5)_2C(CH_3)_3$

79

FIGURE 19 Continued

the carbonate, protection of the primary and secondary hydroxyl groups as a tosylate and ethoxyethyl ether, respectively, and reaction with lithium dibutylcuprate. Treatment of [106] with acid led to deprotection and lactonization to [107]. Protection of the alcohol groups and alkylation of the lactone enolate with the diphenyl-*t*-butylsilyl either of *Z*-7-bromo-5-hepten-1-ol introduced the top side chain.

At this point, formation of the cyclopentanone ring was undertaken. Lactone to lactol reduction of [108] was followed by cyanohydrin formation, giving monoprotected pentaol [109]. The primary alcohol was selectively tosylated and the remaining three alcohols converted to their ethoxyethyl ethers. Base-induced cyclization gave protected cyclopentanone [110]. Selective cleavage of the silyl protecting group and oxidation gave acid [111]. Ether hydrolysis (with cyanohydrin reversal) and selective ketone reduction completed the synthesis of $PGF_{2\alpha}$, which was characterized as its methyl ester [79].

It is worthy or note that, of the five chiral centers in lactone [101], two (C—2 and C—6) appear in the final product as C—11 and C—15.

III. CONCLUSION

The preparation of a chiral pharmaceutical in enantiomerically homogeneous form is clearly a viable proposition. The tools—resolution, asymmetric synthesis, and the chiral carbon pool—are available. As exemplified by the prostaglandins, the manner in which these tools are used is limited only by the imagination and inventiveness of the chemist.

REFERENCES

1. D. A. Pensak and E. J. Corey, in *Computer-Assisted Organic Synthesis* (W. T. Wipke and W. J. Howe, eds.), ACS, Washington DC, 1977, p. 1.
2. J. D. Morrison, in *Asymmetric Synthesis* (J. D. Morrison, ed.), Vol. 1, Academic Press, New York, 1983, p. 1.
3. J. Jacques, A. Collet, and S. H. Wilen, *Enantiomers, Racemates and Resolutions*, Wiley-Interscience, New York 1981.
4. *Asymmetric Synthesis* (J. D. Morrison, ed.), Vols. 1–5, Academic Press, New York, 1983–1985.
5. J. W. Scott, in *Asymmetric Synthesis* (J. D. Morrison and J. W. Scott, eds.), Vol. 4, Academic Press, New York, 1984, p. 1.
6. E. Fogassy, F. Faigl, and M. Ács, *Tetrahedron, 41*:2837 (1985).
7. A. Fischli, *Chimia, 30*:4 (1976).
8. J. C. Clark, G. H. Phillipps, and M. R. Steer, *J. C. S. Perkin I*:475 (1976).
9. J. B. Jones, in *Asymmetric Synthesis* (J. D. Morrison, ed.), Vol. 5, Academic Press, New York, 1985, p. 309.
10. T. Kaneko, Y. Izumi, I. Chibata, and T. Itoh, *Synthetic Production and Utilization of Amino Acids*, Kodansha, Tokyo, 1974.
11. W. E. Ladner and G. M. Whitesides, *J. Am. Chem. Soc., 106*:7250 (1984).
12. M. Schneider, N. Engel, and H. Boensmann, *Angew. Chem. Int. Ed. Engl., 23*: 64 (1984).
13. V. S. Martin, S. S. Woodard, T. Katsuki, Y. Yamada, M. Ikeda, and K. B. Sharpless, *J. Am. Chem. Soc., 103*:6237 (1981).
14. J. W. Scott, *Industr. Chem. News, 7*:32 (1986).
15. W. S. Knowles, *Acc. Chem. Res., 16*:106 (1983).
16. T. Aretani, *Pure Appl. Chem., 57*:1839 (1985).
17. S. Hanessian, *Total Synthesis of Natural Products: The "Chiron" Approach*, Pergamon Press, Oxford, 1983.
18. S. M. Roberts and F. Scheinmann, *New Synthetic Routes to Prostaglandins and Thromboxanes*, Academic Press, New York, 1982.
19. E. J. Corey, *Ann. N.Y. Acad. Sci., 180*:24 (1971).
20. E. J. Corey, N. M. Weinshenker, T. K. Schaaf, and W. Huber, *J. Am. Chem. Soc., 91*:5675 (1969).

21. E. J. Corey, T. K. Schaaf, W. Huber, V. Koelliker, and N. M. Weinshenker, *J. Am. Chem. Soc.*, *92*:397 (1970).
22. E. J. Corey and H. E. Ensley, *J. Am. Chem. Soc.*, *97*:6908 (1975).
23. J. S. Bindra, A. Grodski, T. K. Schaaf, and E. J. Corey, *J. Am. Chem. Soc.*, *95*: 7522 (1973).
24. A. Fischli, M. Klaus, H. Mayer, P. Schönholzer, and R. Rüegg, *Helv. Chim. Acta*, *58*:564 (1975).
25. M. Nara, S. Terashima, and S. Yamada, *Tetrahedron*, *36*:3161 (1980).
26. I. Tömösközi, L. Gruber, G. Kovács, I. Székely, and V. Simonidesz, *Tetrahedron Lett.*:4639 (1976).
27. E. J. Corey and J. Mann, *J. Am. Chem. Soc.*, *95*:6832 (1973).
28. M. Asami, *Tetrahedron Lett.*, *26*:5803 (1985).
29. Y.-F. Wang, C.-S. Chen, G. Girdaukas, and C. J. Sih, *J. Am. Chem. Soc.*, *106*: 3695 (1984).
30. F. Johnson, K. G. Paul, D. Favara, R. Ciabatti, and U. Guzzi, *J. Am. Chem. Soc.*, *104*:2190 (1982).
31. R. J. Ferrier and P. Prasit, *J. C. S. Chem. Comm.*:983 (1981).
32. E. J. Corey, K. C. Nicolaou, and D. J. Beames, *Tetrahedron Lett.*:2439 (1974).
33. J. J. Partridge, N. K. Chadha, and M. R. Uskoković, *J. Am. Chem. Soc.*, *95*:7171 (1973).
34. J. J. Partridge, N. K. Chadha, and M. R. Uskoković, *Org. Syn.*, *63*:44 (1984).
35. R. F. Newton, J. Paton, D. P. Reynolds, S. Young, and S. M. Roberts, *J. C. S. Chem. Comm.*:908 (1979).
36. R. Noyori and M. Suzuki, *Angew. Chem. Int. Ed. Engl.*, *23*:847 (1984).
37. J. B. Heather, R. Sood, P. Price, G. P. Peruzzotti, S. S. Lee, L. F. H. Lee, and C. J. Sih, *Tetrahedron Lett.*:2313 (1973).
38. C. J. Sih, J. B. Heather, R. Sood, P. Price, G. Peruzzotti, L. F. H. Lee, and S. S. Lee, *J. Am. Chem. Soc.*, *97*:865 (1975).
39. L. Gruber, I. Tömözközi, E. Major, and G. Kovács, *Tetrahedron Lett.*:3729 (1974).
40. A. F. Kluge, K. G. Untch, and J. H. Fried, *J. Am. Chem. Soc.*, *94*:7827 (1972).
41. J. Fried, C. H. Lin, M. M. Mehra, and P. Dalven, *Ann. N.Y. Acad. Sci.*, *180*:38 (1971).
42. R. Noyori, I. Tomino, and M. Nishizawa, *J. Am. Chem. Soc.*, *101*:5843 (1979).
43. R. Noyori, *Pure Appl. Chem.*, *53*:2315 (1981).
44. M. M. Midland, D. C. McDowell, R. L. Hatch, and A. Tramontano, *J. Am. Chem. Soc.*, *102*:869 (1980).
45. T. Tanaka, N. Okamura, K. Bannai, A. Hazato, S. Sugiura, K. Manabe, and S. Kurozumi, *Tetrahedron Lett.*, *26*:5575 (1985).
46. G. Stork and M. Isobe, *J. Am. Chem. Soc.*, *97*:6260 (1975).
47. G. Stork and M. Isobe, *J. Am. Chem. Soc.*, *97*:4745 (1975).
48. M. Suzuki, A. Yanagisawa, and R. Noyori, *J. Am. Chem. Soc.*, *107*:3348 (1985).
49. M. Gill and R. W. Rickards, *J. C. S. Chem. Comm.*:121 (1979).
50. M. Gill and R. W. Rickards, *Tetrahedron Lett.*:1539 (1979).
51. L. A. Mitscher, G. W. Clark, III, and P. B. Hudson, *Tetrahedron Lett.*:2553 (1978).

52. K. Ogura, M. Yamashita, and G. Tsuchihashi, *Tetrahedron Lett.*:759 (1976).
53. S. Torii, T. Inokuchi, R. Oi, K. Kondo, and T. Kobayashi, *J. Org. Chem.*, *51*:254 (1986).
54. T. Tanaka, S. Kurozumi, T. Toru, S. Miura, M. Kobayashi, and S. Ishimoto, *Tetrahedron*, *32*:1713 (1976).
55. G. Stork, T. Takahashi, I. Kawamoto, and T. Suzuki, *J. Am. Chem. Soc.*, *100*:8272 (1978).

8

Enzymatic Synthesis and Resolution of Enantiomerically Pure Compounds

David I. Stirling *Celgene Corporation, Warren, New Jersey*

I. INTRODUCTION

Biocatalysis is one of a number of forms of chemical catalysis (Fig. 1) that can be utilized to synthesize a variety of organic chemicals. Over 60% of the 135 MM tons of organic chemicals produced in the United States involve a catalytic step somewhere in their manufacture (1,2). In recent years many reports and reviews extolling the virtues of biocatalysis for the production of chemicals have been released (e.g., 3–9). However, there have still been very few examples of commercial chemical processes introduced in the last few years that utilize a biocatalyst, for example, the acrylamide process (10–12). There has been small but growing concern as to the validity of the expectations placed on bioconversion-based chemical process (13).

The two major advantages that a biocatalyst may offer over a chemical counterpart can be summarized as mild reaction conditions and catalytic specificity. Enzyme/substrate interactions can significantly lower the activation energy of a chemical reaction. For this reason, enzymes exhibit high catalytic activities under mild reaction conditions, low temperatures and pressures, primarily around neutral pH. These egregious reaction conditions also result in the production of less toxic/undesirable by-products, that is, waste minimization, a growing advantage to biology-based processes.

The reaction specificity associated with enzymes reinforces the benefit of waste minimization due to their inherent ability to produce a homoge-

1. HOMOGENEOUS

2. HETEROGENEOUS
—nonuniform, e.g. Co, Rh, Pd on supports
—uniform, e.g. zeolites
—biocatalysis

FIGURE 1 Types of chemical catalysis.

neous product. Enzymes can also exhibit remarkable regiospecificity toward their substrate. They can differentiate between equivalent position or groups of similar reactivity. Finally, there is one form of enzymatic specificity that most observers, as well as practitioners, agree is perhaps the single attribute of greatest potential: stereoselectivity (13–15).

II. METHODS OF PRODUCTION OF CHIRAL COMPOUNDS

Process options for the production of homochiral compounds are summarized in Fig. 2. The three basic routes are separation of racemic mixture, synthesis using a naturally occurring chiral synthon, and asymmetric synthesis using a prochiral intermediate. Historically, the efficiency of asymmetric synthesis has been capricious in terms of chemical and optical yield. Hence, from a practical, commercial process perspective, resolution via diastereomer crystallization has remained important for many commercial scale processes, for example, diltiazem.

Asymmetric synthesis has advanced significantly in recent years with the advent of optically active reagents, auxiliaries, and catalysts. Both chemical and biological systems have been developed during this time. For example, Fig. 3 illustrates a recently published synthetic route to diltiazem, an important calcium channel blocker (16). The process starts with the chiral auxiliary (1*R*,2*S*)-2-phenylcyclohexanol, and proceeds through a chiral (aryl)oxirane, with two chiral centers formed by induction from the original auxiliary. The chiral epoxide is then opened with 2-aminobenzenethiol and the original chiral auxiliary recovered (Step 4). Finally, the molecule is cyclized, aminoalkylated, and acylated, resulting in the desired product.

Similarly, an example of a biological asymmetric synthesis is shown in Fig. 4. In this case, α-methylbenzylamine is synthesized in high chiral

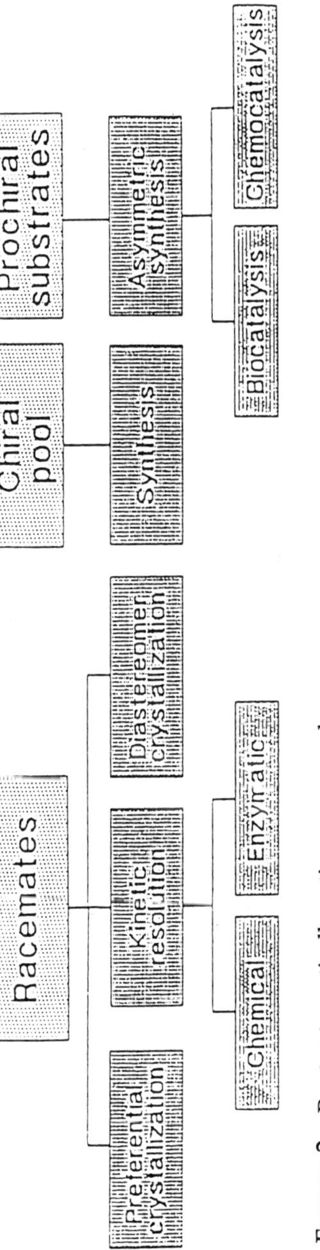

FIGURE 2 Routes to optically active compounds.

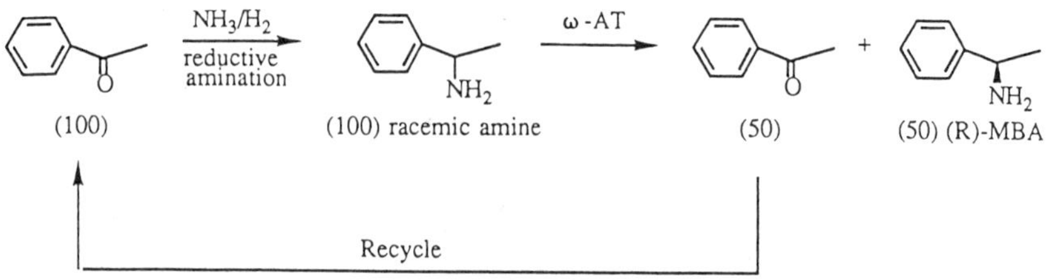

FIGURE 9 Kinetic resolution of α-methylbenzylamine using an ω-aminotransferase.

TABLE 1 Enzyme Types, Function, and Potential Products

Enzyme group	Reaction types	Potential chiral products
Oxidoreductase	Oxidation	Alcohol Epoxide Sulfoxide Amino acid Lactone
	Reduction	Alcohol Lactone
Transferase	Hydroxymethyl transfer	Hydroxyamino acid
	Amino group transfer	Amino acid Amine
Hydrolase	Ester hydrolysis	Alcohol/carboxylic acid/carboxylic ester
	(*Trans*) esterification	Alcohol/carboxylic acid/carboxylic ester
	Nitrile/amide hydrolysis	Carboxylic acid
	Hydantoin hydrolysis	Amino acid
	Alkylhalide hydrolysis	Haloalkanoic acid Alcohol/epoxide
Lyase	C-C formation	Amino acid Acyloin Cyanohydrin
	C-O formation	Alcohol Amino acid
	C-N formation	Amino acid
Isomerase	Lactone formation	Lactone

the various groups. The following potpourri of enzyme-catalyzed chiral resolutions and syntheses is not intended to be comprehensive in nature, but rather to exemplify the diverse yet prodigious selectivity exhibited by these biocatalysts.

A. Oxidoreductase: Oxidation

The insertion of an oxygen atom(s) into a molecule in a regioselective/stereoselective manner can be catalyzed by a broad group of enzymes. The target molecules are diverse in nature as illustrated by the following examples. The source of oxygen can be dioxygen or water.

1. Alcohol

Oxygenases can insert one or both atoms of dioxygen into many types of compounds and in some cases creating chiral centers (27). Toluene dioxygenase has been shown to oxidize toluene (Fig. 10) to produce *cis*-dihydrodiol (28,29). The overall oxidation involves the insertion of two atoms of oxygen, producing two chiral alcohol centers. Interestingly, the same enzyme can catalyze the stereoselective mono-hydroxylation of certain substrates. Figure 11 shows the products obtained from the oxidation of deuterated indene using toluene dioxygenase (30).

2. Epoxide

Many mono-oxygenases have been shown to epoxidize a variety of olefinic substrates (27). In a few cases, this fortuitous enzyme reaction has been demonstrated to be stereoselective in nature. The ω hydroxylase from *Pseudomonas oleovorans* was reported to oxidize 1,7-octadiene, producing the *R* enantiomer of the epoxide in high enantiomeric excess (31). Workers at Shell Research Limited showed that a similar *Pseudomonas oleovorans* mono-oxygenase system could produce a key epoxide intermediate used for the production of the β-adrenergic receptor-blocking drug (*S*)-atenolol (Fig. 12) (32).

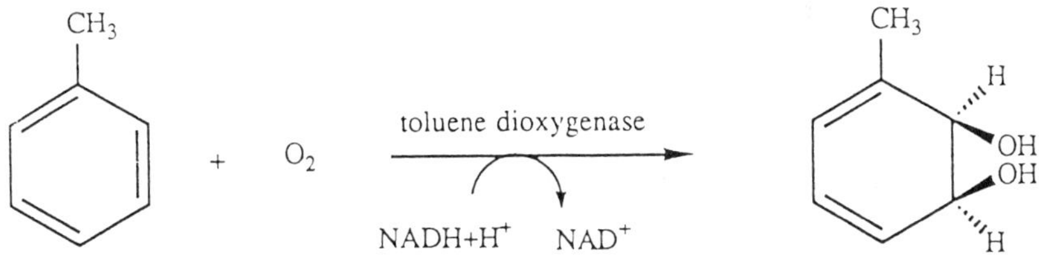

FIGURE 10 Oxidation of toluene.

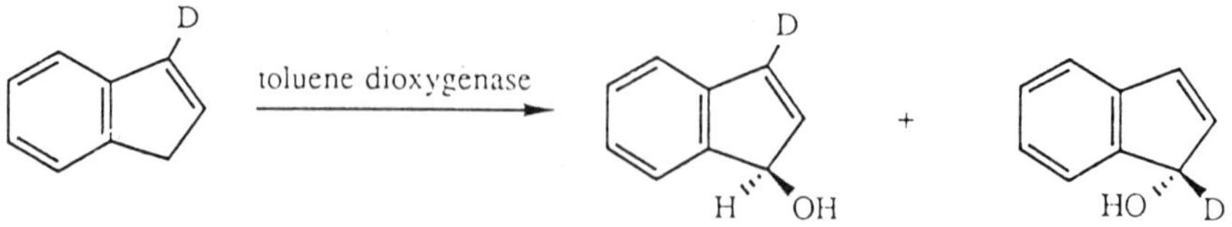

FIGURE 11 Benzylic mono-oxygenation of indene.

3. *Sulfoxide*

Reports on the ability of microorganisms to selectively oxidize various types of sulfides to the corresponding chiral sulfoxides have grown significantly in the last few years (33). Chiral sulfoxides are evident in a variety of pharmaceutical and agricultural chemical compounds (33). The ω hydroxylase from *Pseudomonas oleovorans* has been shown to produce a variety of chiral aliphatic sulfoxides from corresponding sulfides (34). Vinyl sulfox-

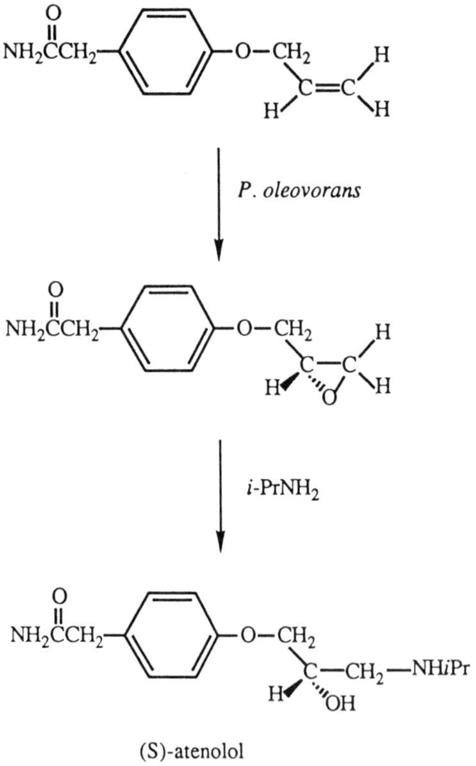

FIGURE 12 Microbial epoxidation route to (*S*)-atenolol.

ides can be synthesized using the fungus *Mortierella isabellina* (33). Figure 13 shows the enantioselective sulfoxidation of the substrate vinyl sulfide, which gave a 45% yield and an enantiomeric excess of 95% (33).

4. Amino Acid

Microbial enzymes catalyzing the oxidative deamination of primary amines are classified into two groups: flavoprotein amine oxidases and copper-containing amine oxidases. Some of the copper-dependent enzymes have recently been shown to contain the prosthetic group pyrroloquinoline quinone (PQQ) (35,36). Amino acid oxidases show a stereoselective preference for either the D or L isomers of amino acids. The enzymes catalyze the dioxygen-dependent oxidation of the amino acid to the corresponding α-keto acid, ammonia, and the concomitant reduction of molecular oxygen to hydrogen peroxide.

5. Lactone

Chiral lactones can be formed from ketones via the Baeyer–Villiger reaction. Such lactones are potentially useful synthons for a number of natural products (37). Many of the examples of enantioselective lactone formation have been demonstrated using cyclohexanone oxygenase isolated from various *Acinetobacter* species (37,38). Figure 14 shows the enzymatic lactonization of methylcyclohexanone, which gave an 80% yield with an enantiomeric excess greater than 98%.

B. Oxidoreductase: Reduction

1. Alcohol

Many microorganisms possessing alcohol dehydrogenases that are capable of reducing ketones and diketones have been demonstrated to produce chiral alcohols. Examples of such enantioselective reductions have been reviewed on many occasions (3–8). The main advantage of a secondary alcohol dehydrogenase for the production of chiral alcohols

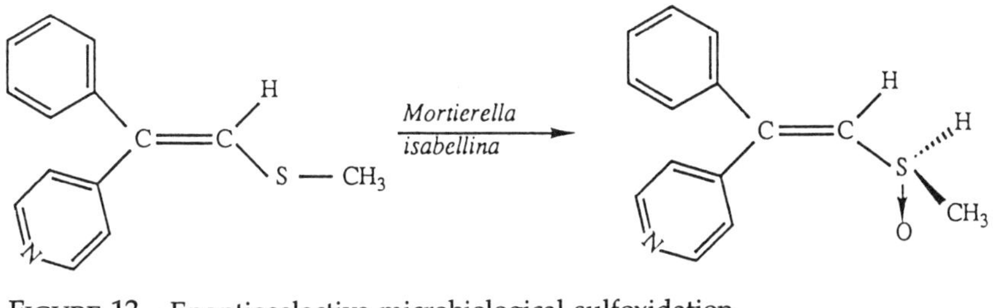

FIGURE 13 Enantioselective microbiological sulfoxidation.

1. *Alcohol/Ester*

A process utilizing a stereospecific ester cleavage yielding the product *R*-glycidylbutyrate in high optical purity has recently been scaled to commercial levels (63). The process catalyzed by porcine pancreatic lipase is shown in Fig. 19. *R*-glycidylbutyrate is a useful chiral synthon for a variety of commercial products, for example, β-blockers.

2. *Carboxylic Acid*

Two groups of chiral aromatic carboxylic acids are important commercial intermediates for the agricultural and pharmaceutical industries. The *R* enantiomer of α-phenoxypropionic acids confers biological activity for a number of herbicides. The *S* enantiomer of a variety of arylpropionic acids is the biologically active form of the nonsteroidal antiinflammatory products labeled profens. Racemic mixtures of the alkyl esters of these propionic acid derivatives have been effectively resolved to yield the desired optically active carboxylic acids (64–66). Figure 20 shows examples of the resolution of aromatic propionic acid esters.

F. Hydrolase: (*Trans*)Esterification

It is generally held that an aqueous environment is desirable, although not a prerequisite, for the optimal activity of enzymes. It is assumed that in most cases, placing an enzyme in an organic medium will cause deleterious effects to the tertiary structure of the protein. However, if this was not the case, there would be potential advantages for enzyme catalysis in an organic medium. It would clearly facilitate conversion of water-insoluble substrates and aid in product recovery. The thermodynamic equilibria of certain reactions are unfavorable in aqueous systems. Enzyme stereoselectivity has been shown to improve in some nonaqueous enzyme transformations (67,68).

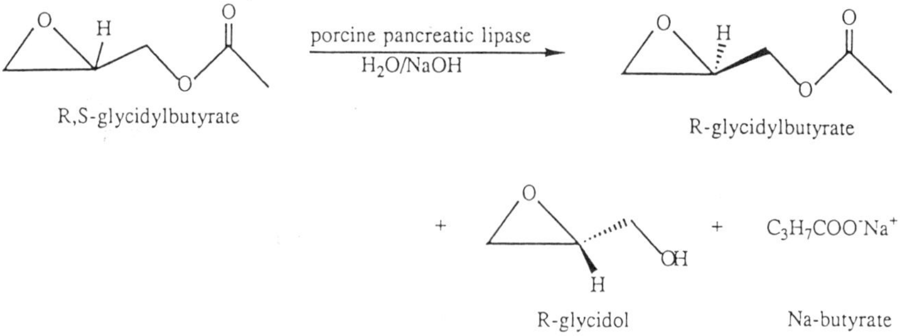

R,S-glycidylbutyrate → (porcine pancreatic lipase, H_2O/NaOH) R-glycidylbutyrate + R-glycidol + $C_3H_7COO^-Na^+$ (Na-butyrate)

FIGURE 19 Resolution process for the production of *R*-glycidylbutyrate.

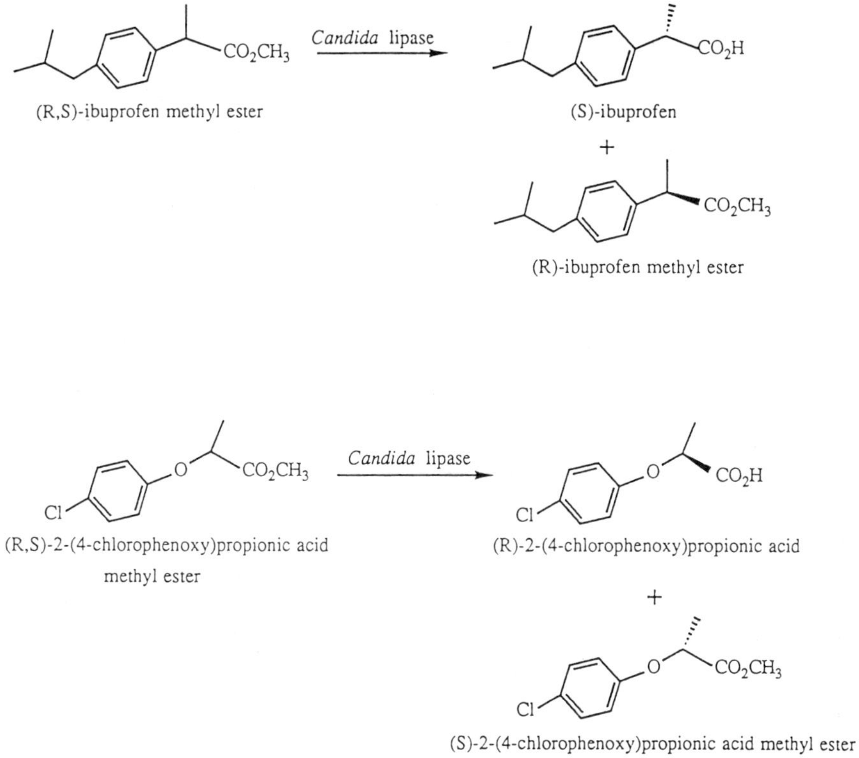

(R,S)-ibuprofen methyl ester — *Candida* lipase → (S)-ibuprofen + (R)-ibuprofen methyl ester

(R,S)-2-(4-chlorophenoxy)propionic acid methyl ester — *Candida* lipase → (R)-2-(4-chlorophenoxy)propionic acid + (S)-2-(4-chlorophenoxy)propionic acid methyl ester

FIGURE 20 Lipase solution of (*R*,*S*)-arylpropionic acid and (*R*,*S*)-α-phenoxypropionic acid esters.

The types of solvent systems employed range from water-miscible cosolvents, for example, acetone, dimethylformamide, through biphasic water/water-immiscible solvent systems to anhydrous organic media with less than 1% water present. Guidelines are beginning to emerge as to the choice of organic solvent plus desirable water content of the medium for a particular enzyme and reaction (9,67). The vast majority of enzymes used for preparative purposes in organic media have been hydrolases.

1. *Alcohol/Carboxylic Acid/Ester*

Racemic mixtures of carboxylic acids can be resolved by esterification using lipases in organic media. Figure 21 illustrates such a process when racemic α-bromopropionic acid is stereoselectively esterified with *n*-butanol using a lipase from the yeast *Candida cylindraceae* (68). This reaction was carried out in virtually anhydrous hexane.

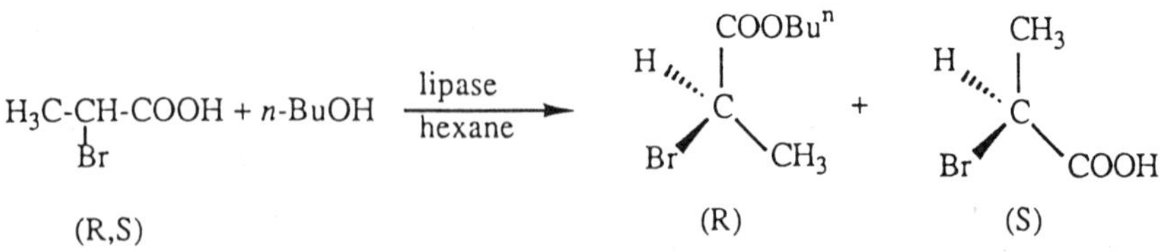

$$H_3C\text{-}\underset{\displaystyle Br}{CH}\text{-}COOH + n\text{-}BuOH \xrightarrow[\text{hexane}]{\text{lipase}} \text{(R)-}CH_3CHBrCOOBu^n + \text{(S)-}CH_3CHBrCOOH$$

(R,S) (R) (S)

FIGURE 21 Stereoselective esterification of α-bromopropionic acid.

Similarly, enantiomeric mixtures of carboxylic acid esters can be separated using transesterification (Fig. 22). An *n*-butanol/water biphasic medium aryloxypropionic acid methyl ester was transesterified stereoselectively, yielding the butylester of the *R* enantiomer (69).

G. Hydrolase: Nitrile Hydrolysis

The enzymatic hydrolysis of nitriles to yield either the corresponding amides or carboxylic acids has been studied in some detail over the last 10 yr (70,71). The enzymatic hydrolytic cleavage of nitriles can be achieved by two types of hydrolase: nitrile hydratase or nitrilase (Fig. 23). Nitrile hydratase has been commercially exploited for the production of various amides, the most notable being acrylamide (10–12,71,72).

1. Acid

Various optically active carboxylic acids can be produced via the activities of nitrile-metabolizing enzymes, for example, from the corresponding cyanohydrins, α-hydroxy acids, α-amino acids, and arylalkanoic acids. Figure 6 shows two types of products derived from the enantioselective action of nitrilase. (*S*)-ibuprofen, the nonsteroidal antiinflammatory agent, can be prepared in high enantiomeric excess by resolving the corresponding racemic nitrile using a nitrilase system from an *Acinetobacter* sp. (73). (*S*)-nitrile is hydrolyzed in one step to (*S*)-ibuprofen. Similarly, (*S*)-aminopropionitrile can be selectively hydrosized by a nitrilase in a single step to L-alanine (74).

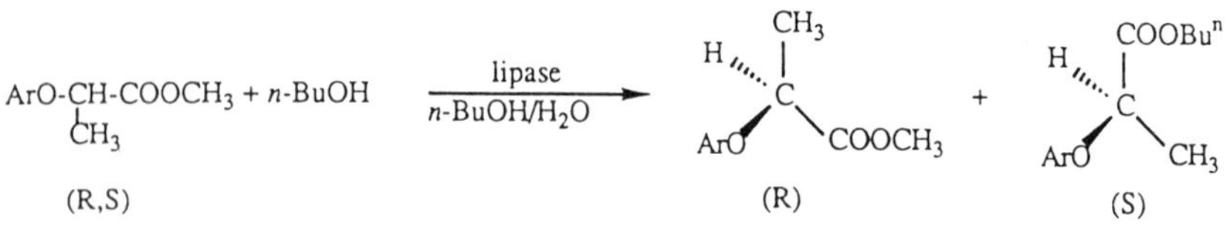

$$ArO\text{-}\underset{\displaystyle CH_3}{CH}\text{-}COOCH_3 + n\text{-}BuOH \xrightarrow[n\text{-}BuOH/H_2O]{\text{lipase}} \text{(R)-}ArOCH(CH_3)COOCH_3 + \text{(S)-}ArOCH(CH_3)COOBu^n$$

(R,S) (R) (S)

FIGURE 22 Stereoselective *trans*-esterification of aryloxypropionic acid methyl ester.

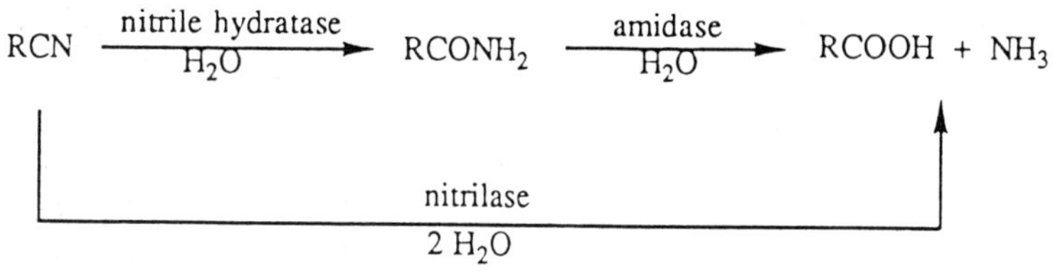

$$RCN \xrightarrow[H_2O]{\text{nitrile hydratase}} RCONH_2 \xrightarrow[H_2O]{\text{amidase}} RCOOH + NH_3$$

$$RCN \xrightarrow[2\,H_2O]{\text{nitrilase}} RCOOH + NH_3$$

FIGURE 23 Microbial metabolism of nitriles.

Chiral carboxylic acids can also be obtained via a two-enzyme step process involving nitrile hydratase and amidase. Figure 24 shows the scheme for producing L-amino acids using a nonenantiospecific nitrile hydratase and L-enantiospecific amidase in combination (75). An amino peptidase from a *Pseudomonas putida* strain was shown to stereoselectively hydrolyze amides of L-amino acids, yielding the corresponding amino acids (76). In this particular process, the racemic amides were prepared chemically from the nitriles under alkaline conditions in the presence of a catalytic amount of acetone.

H. Hydrolase: Hydantoin Hydrolysis

1. Amino Acid

Dihydropyrimidinase- (hydantoinase-) based processes have been successfully employed for the production of D-amino acids, particularly D-*p*-hydroxyphenylglycine (7,25,77,78). D,L-hydantoins are chemically synthesized from the corresponding aldehydes using the Bucherer–Berg

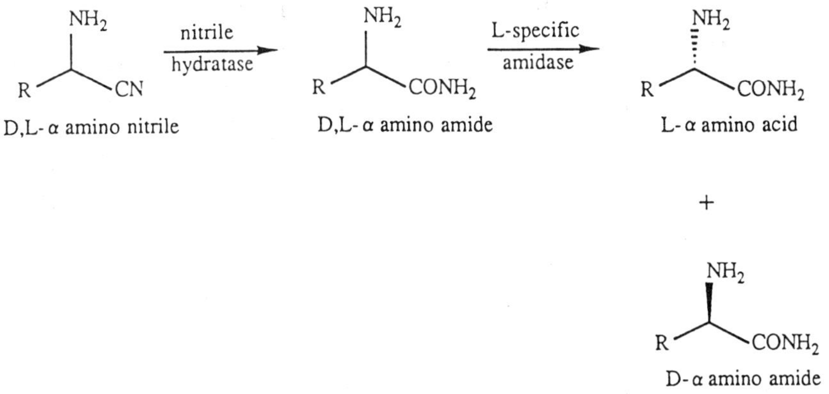

$$\underset{\text{D,L-}\alpha\text{ amino nitrile}}{R\text{–}CH(NH_2)\text{–}CN} \xrightarrow[\text{hydratase}]{\text{nitrile}} \underset{\text{D,L-}\alpha\text{ amino amide}}{R\text{–}CH(NH_2)\text{–}CONH_2} \xrightarrow[\text{amidase}]{\text{L-specific}} \underset{\text{L-}\alpha\text{ amino acid}}{R\text{–}CH(NH_2)\text{–}CONH_2} + \underset{\text{D-}\alpha\text{ amino amide}}{R\text{–}CH(NH_2)\text{–}CONH_2}$$

FIGURE 24 Production of L-amino acids using a nonselective nitrile hydratase and a L-stereospecific amidase.

reaction. A D-specific hydantoinase hydrolyzes the D-hydantoin to the D-*N*-carbamyl amino acid. The free D-amino acid is recovered either by treatment with nitrous acid (7,25) or enzymatically using an amidohydrolase (77,78).

The advantage of a hydantoin-based approach for the production of D-amino acids is that the remaining unhydrolyzed *R*-hydantoin is easily racemized under reaction conditions. Therefore in principle, an overall yield approaching 100% is possible. Kanegafuchi has a commercial process based on hydantoinase technology for the production of D-*p*-hydroxyphenylglycine. It produces the racemic hydantoin using an amidoalkylation reaction between phenol, glyoxylic acid, and urea. This circumvents the need to use the relatively expensive *p*-hydroxybenzaldehyde as a raw material for Bucherer–Berg synthesis.

I. Hydrolase: Alkylhalide Hydrolase

Halidohydrolases that catalyze halide hydrolysis are rapidly becoming important commercial catalysts for the synthesis of chiral products and also the bioremediation of industrial waste (79).

1. Haloalkanoic Acid

Stereoselective dehalogenation of 2-haloalkanoic acids has been demonstrated for a number of halidohydrolases (80–82). Figure 77 details the production of an L-haloacid intermediate used in the production of phenoxypropionic acid herbicides. The *R* enantiomer of chloropropionic acid is selectively hydrolyzed to (*S*)-lactic acid due to an inversion of configuration that occurs during the hydrolysis (83). (*S*)-2-chloroproprionic acid is used as a chiral synthon to produce a number of (*R*)-phenoxypropionic acid herbicides, for example, Fusilade 2000 (ICI).

2. Alcohol/Epoxide

Racemic 2,3-dichloropropan-1-ol has been used as a source of (*S*)-2,3-dichloropropan-1-ol, which is subsequently converted to (*R*)-epichlorohydrin, a very useful chiral synthon (84). Figure 25 shows the overall process whereby a *Pseudomonas* sp. containing an (*R*)-2,3-dichloropropan-1-ol-specific halidohydrolase metabolizes the *R* enantiomer of the raw

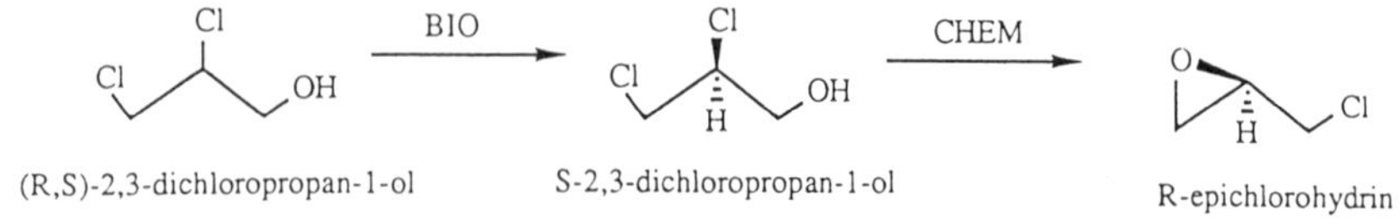

FIGURE 25 Production of (*S*)-2,3-dichloropropan-1-ol.

material, leaving the desired product. The *S*-dichloropropanol can be easily converted chemically to the chiral epichlorohydrin.

J. Lyase: C—C Formation

Lyases are enzymes that cleave C—C, C—O, and C—N bonds by reactions other than hydrolysis or oxidation. They utilize two substrates in one direction but only one in the other. In the latter case, a molecule is eliminated, leaving an unsaturated moiety. When the reverse reaction is the most important, that is, condensation, the enzymes are often designated as synthases.

1. *Amino Acid*

L-threonine aldolase (L-threonine acetaldehyde-lyase) catalyzes the reversible condensation of acetaldehyde and glycine to form L-threonine. The enzyme has been shown to be an activity distinct from serine hydroxymethyltransferase that also catalyzes the above reaction (85,86). The substrate specifically of the adolase has been demonstrated to be flexible with respect to the aldehyde involved. The enzyme has been shown to form phenylserine derivatives from substituted benzaldehydes and glycine (86).

Tyrosine phenol lyase (β-tyrosinase) has been shown to catalyze the efficient synthesis of the L-amino acids L-tyrosine and L-dopa from pyruvate, ammonia and phenol, or catechol, respectively (87–89).

2. *Acyloin*

Pyruvate decarboxylase catalyzes the nonoxidative decarboxylation of pyruvate to acetaldehyde and carbon dioxide. When an aldehyde is present with pyruvate, the enzyme promotes an acyloin condensation reaction. The mechanistic reason for this fortuitous reaction is well understood and involves the aldehyde outcompeting a proton for bond formation with a reactive thiamine pyrophosphate-bound intermediate (90,91). When acetaldehyde is present, the product formed is acetoin. Benzaldehyde results in the production of phenylacetylcarbinol (Fig. 26). Both of these condensations are enantioselective, forming the *R* enantiomer preferentially in both cases.

The ability of pyruvate decarboxylase from many microbial sources to produce phenylacetylcarbinol has been exploited for many years in the synthesis of ephedrine, a natural adrenergic compound (92). The acyloin is reductively aminated to produce the ethanolamine product, (1*R*,2*S*)-ephedrine, with two chiral centers.

3. *Cyanohydrin*

The enzyme mandelonitrile lyase (oxynitrilase) isolated from almond flour has been shown to catalyze the stereospecific cyanohydrination of

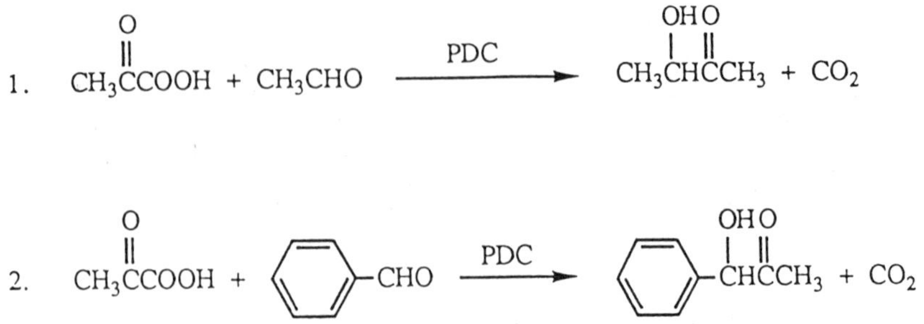

$$1.\quad CH_3\overset{\overset{O}{\|}}{C}COOH + CH_3CHO \xrightarrow{PDC} CH_3\overset{\overset{OH}{|}}{C}H\overset{\overset{O}{\|}}{C}CH_3 + CO_2$$

$$2.\quad CH_3\overset{\overset{O}{\|}}{C}COOH + C_6H_5\text{-}CHO \xrightarrow{PDC} C_6H_5\text{-}\overset{\overset{OH}{|}}{C}H\overset{\overset{O}{\|}}{C}CH_3 + CO_2$$

FIGURE 26 Fortuitous reactions of pyruvate decarboxylase (PDC).

aldehydes, yielding the (*R*)-cyanohydrins (Fig. 27) (93). Due to spontaneous nonselective chemical additions of hydrogen cyanide to benzaldehyde, high optical purity of the cyanohydrin is difficult to obtain. The purity can be improved by the judicious use of certain organic solvents (94).

Recently, it has been demonstrated that the enzymatic synthesis of (*S*)-cyanohydrins was possible using an oxynitrilase isolated from *Sorghum biocolor* (95,96). These optically active cyanohydrins can be subsequently converted chemically to chiral α-hydroxyacids, aminoalcohols, and acyloins.

K. Lyase: C—O Formation

1. *Alcohol*

The stereospecific addition of water to fumaric acid catalyzed by the enzyme fumarase yields optically pure L-malic acid (Fig. 28). A *Brevibacterium flavum* strain with high fumarase activity has been used industrially for the commercial production of L-malic acid (97). The substrate specificity of fumarase is narrow and hence its broader application in organic synthesis has been somewhat limited. However, it has been shown to synthesize L-*threo*-chloromalic acid in very high optical purity (98).

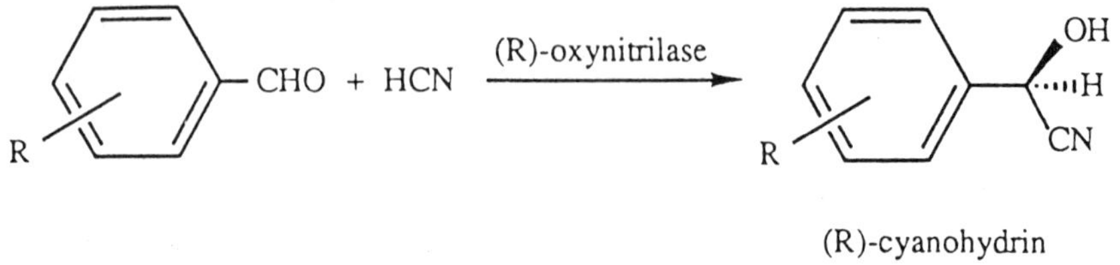

$$R\text{-}C_6H_4\text{-}CHO + HCN \xrightarrow{(R)\text{-oxynitrilase}} R\text{-}C_6H_4\text{-}C(OH)(H)(CN)$$

(R)-cyanohydrin

FIGURE 27 Formation of chiral cyanohydrins.

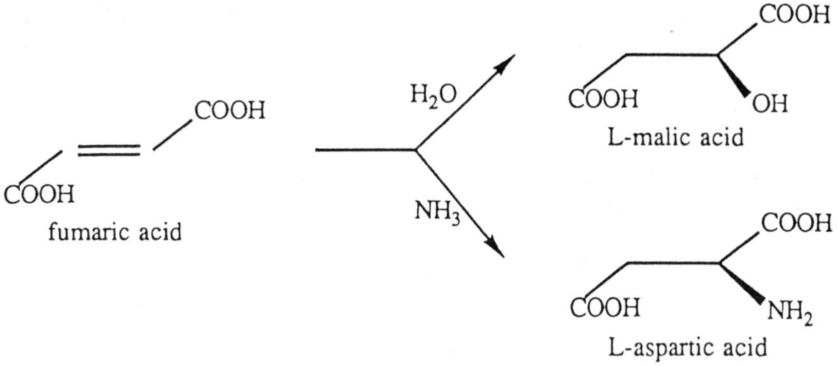

COOH
COOH
fumaric acid
H_2O
NH_3
COOH
COOH
OH
L-malic acid
COOH
COOH
NH_2
L-aspartic acid

FIGURE 28 Asymmetric biotransformations of fumaric acid.

2. *Amino Acid*

L-Serine and L-threonine dehydratases dehydrate and subsequently deaminate the amino acid to the corresponding α-keto acid. These enzymes are known to require pyridoxal-5′-phosphate as a coenzyme. They can function in a biosynthetic or catabolic manner (99). Both enzymes can cause problems for the whole-cell-based production of L-serine (100).

L. Lyase: C—N Formation

1. *Amino Acid*

A commercial process for the production of L-asparatic acid has been developed based on *Escherichia coli* strains with high aspartase activity (101–103). Aspartase catalyzes the stereospecific addition of ammonia to fumaric acid (Fig. 29). L-aspartic acid can be enzymatically decarboxylated to yield the product L-alanine (104).

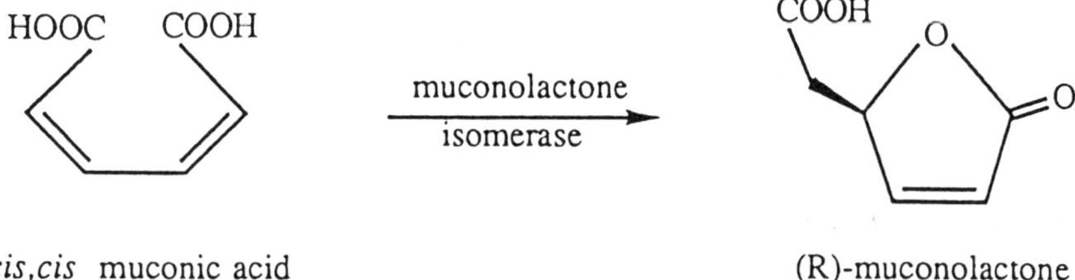

FIGURE 29 Conversion of *cis,cis*-muconic acid to muconolactone.

Various commercial routes for the production of L-phenyalanine have been developed because of the utilization of this amino acid in the dipeptide sweetener Aspartame. One route that has been actively pursued is the synthesis of L-phenylalanine from *trans*-cinnamic acid using the enzyme phenylalanine ammonia lyase (105,106). This enzyme catalyzes the reversible, nonoxidative deamination of L-phenylalanine and can be isolated from various plant and microbial sources (107,108).

M. Isomerase: Rearrangement

1. Lactone

Isomerases catalyze conversions within one molecule. For example, the *cis,cis*-muconate lactonizing enzyme (cycloisomerase) catalyzes the chiral conversion of *cis,cis*-muconic acid to (*R*)-muconolactone (Fig. 29). It is a key enzyme in the degradation of benzoate via the β-ketoadipate pathway (109). Chiral lactones could be useful as chiral synthons.

IV. CONCLUSION

It is clear that enzyme systems can and will play an ever-increasing role in the industrial production of chiral compounds. Not all classes of enzymes were represented in this review, for example, ligases; only enzymes with demonstrated industrial potential were discussed. Hydrolases are currently the most exploited group of biocatalysts, largely due to their commercial availability and relative simplicity of use. However, other groups of enzymes with greater catalytic versatility, in particular the lyases, are growing in commercial significance.

The exploitation of biotransformations to date has been hampered by a number of factors, including the nonjudicious choice of target products. Often, the large cost in time and money for the development of such processes far exceeds the final financial reward. A manufacturing issue that is often overlooked or underestimated is the eventual integration of the biotransformation step with conventional chemical process technology. Many problems can be minimized with the careful amalgamation of the appropriate disciplines required to turn an idea into a product. It is hoped that industry will learn to capitalize more on the inherent asymmetric nature of enzymes for everyone's benefit, including the environment.

ACKNOWLEDGMENTS

The author would like to thank Patricia Petras for her patience and diligence in the preparation of this manuscript.

REFERENCES

1. J. M. Thomas, Advanced catalysts: Interfacing in the physical and biological sciences, *Angew. Chem. Int. Ed. Engl. Adv. Mater.*, *28*:1079 (1989).
2. W. Keim, Nickel: An element with wide application in industrial homogeneous catalysis, *Angew. Chem. Int. Ed. Engl.*, *29*:235 (1990).
3. G. M. Whiteside and C.-H.Wong, Enzymes as catalysts in synthetic organic chemistry, *Angew. Chem. Int. Ed. Engl.*, *24*:617 (1985).
4. S. Butt and S. M. Roberts, Recent advances in the use of enzyme-catalysed reactions in organic research: The synthesis of biologically-active natural products and analogues, *Nat. Prod. Rep.*, *3*:489 (1986).
5. J. B. Jones, Enzymes in organic synthesis, *Tetrahedron.*, *42*:3351 (1986).
6. S. Butt and S. M. Roberts, Opportunities for using enzymes in organic synthesis, *Chem. Brit.*, *23*:127 (1987).
7. H. Yamada and S. Shimizu, Microbial and enzymatic processes for the production of biologically and chemically useful compounds, *Angew. Chem. Int. Ed. Engl.*, *27*:622 (1988).
8. N. J. Turner, Recent advances in the use of enzyme-catalyzed reactions in organic synthesis, *Nat. Prod. Rep.*, *6*:625 (1989).
9. A. M. Klibanov, Asymmetric transformations catalyzed by enzymes in organic solvents, *Acc. Chem. Res.*, *23*:114 (1990).
10. Y. Asano, T. Yasuda, Y. Tani, and H. Yamada, A new enzymatic method for acrylamide production, *Agric. Biol. Chem.*, *46*:1183 (1982).
11. I. Watanabe, Process for the production of acrylamide using microorganism, U.S. Patent 4,343,900, 1982.
12. H. Yamada, K. Ryuno, T. Nagasawa, K. Enomoto, and I. Watanabe, Optimum culture conditions for the production by *Pseudomonas chloraphis* B23 of nitrile hydratase, *Agric. Biol. Chem.*, *50*:2859 (1986).
13. R. L. Hinman, A role for biotech in producing chemicals?, *Bio/Tech.*, *9*:533 (1991).
14. S. M. Roberts, Synthesis of some useful optically active compounds using enzymes, *Chem. Ind.*: 384 (1988).
15. R. A. Sheldon, The industrial synthesis of optically active compounds, *Proceedings of Chiral Synthesis Symposium and Workshop*, Manchester, England, Spring Innovations, Stockport, 1989, pp. 21–29.
16. D. L. Coffen, P. B. Madan, and A. Schwartz, Glycidic acid ester and process of preparation, U.S. Patent 5,008,411, 1991.
17. D. I. Stirling, A. L. Zeitlin, and G. W. Matcham, Enantiomeric enrichment and stereoselective synthesis of chiral amines, U.S. Patent 4,950,606, 1990.
18. D. I. Stirling and G. W. Matcham, The production of chiral amine compounds via biological resolution and synthesis, *Proceedings of Chiral '90*, Manchester, England, Spring Innovations, Stockport, 1990, pp. 111–113.
19. C. R. Bayley, The development of resolution methods for the separation of racemic mixtures, *Proceedings of Chiral Synthesis Symposium and Workshop*, Manchester, England, Spring Innovations, Stockport, 1989, pp. 11–12.

20. P. J. Reider, P. Davis, P. L. Hughes, and E. J. J. Grabowski, Crystallization-induced asymmetric transformation: Stereospecific synthesis of a potent peripheral CCIT antagonist, *J. Org. Chem.*, *52*:955 (1987).
21. J. Jacques, A. Collet, and S. H. Whilen, Resolution by entrainment, *Enantiomers, Racemates and Resolutions*, Wiley, New York, 1981.
22. V. S. Martin, S. S. Woodward, T. Katsuki, Y. Yamada, M. Ikeda, and K. B. Shapless, Kinetic resolution of racemic alkylic alcohols by enantioselective eposidation. A route to substances of obsolite enantiomeric purify, *J. Am. Chem. Soc.*, *103*:6237 (1981).
23. Y. Okamato, K. Suzuki, T. Kitayama, H. Yuki, H. Kageyama, K. Miki, N. Tanaka, and N. Kasai, Kinetic resolution of racemic α methylbenzyl methacrylate: Asymmetric selective polymerization catalysed by Grignard reagent-(−)-sparteine derivative complexes, *J. Am. Chem. Soc.*, *104*:4618 (1982).
24. J. M. Brown, Catalytic kinetic resolution, *Chem. Ind.*, *12*:612 (1988).
25. T. Ohashi, S. Takahashi, T. Nagamichi, K. Yoneda, and H. Yamada, A new method for 5-(4-hydroxyphenyl) hydantoin synthesis, *Agric. Biol. Chem.*, *45*:831 (1981).
26. S. M. Roberts and N. J. Turner, The greening of chemistry, *New Sci.*, *126*:38 (1990).
27. F. S. Sariaslani, Microbial enzymes for oxidation of organic molecules, *Crit. Rev. Biotech.*, *9*:171 (1989).
28. D. T. Gibson, M. Hensley, H. Yoshioka, and T. J. Mabry, Formation of (+)-*cis*-2,3-dihydroxy-1-methyl-cyclohexa-4,6-diene from toluene by *Pseudomonas putida*, *Biochem.*, *9*:1626 (1970).
29. B. A. Finette, V. Subramanian, and D. T. Gibson, Isolation and characterization of *Pseudomonas putida* PpFl mutants defective in the toluene dioxygenase enzyme system, *J. Bacteriol.*, *160*:1003 (1984).
30. L. P. Wackett, L. D. Kwart, and D. T. Gibson, Benzylic monooxygenation catalyzed by toluene dioxygenase from *Pseudomonas putida*, *Biochem.*, *27*:1360 (1988).
31. A. G. Katopodis, K. Wimalasena, J. Lee, and S. W. May, Mechanistic studies on non-heme monooxygenase catalysis: Epoxidation, aldehyde formation, and demethylation by the omega-hydroxylation system of *Pseudomonas oleovorans*, *J. Am. Chem. Soc.*, *106*:7928 (1984).
32. G. T. Phillips, Biotransformations and their role in industrial synthesis, *Proceedings of Chiral '90*, Manchester England, Spring Innovations, Stockport, 1990, pp. 17–22.
33. D. R. Boyd, C. T. Walsh, and Y-C. J. Chen, *S*-oxygenases II—Chirality of sulfoxidation reactions, *Sulfur Drugs and Related Organic Chemicals: Chemistry, Biochemistry, and Toxicology*, Vol. 2 (L. A. Damani, ed.), Wiley, New York, 1989.
34. A. G. Katopodis, H. A. Smith, and S. W. May, New oxyfunctionalization capabilities for ω-hydroxylases: Asymmetric aliphatic sulfoxidation and branched ether demethylation, *J. Am. Chem. Soc.*, *110*:897 (1988).

35. J. V. Iersel, R. A. van der Meer, and J. A. Duine, Methylamine oxidase from *Athrobacter* P1, a bacterial copper-quinoprotein amine oxidase, *Eur. J. Biochem.*, *161*:415 (1986).
36. E. Schimizu, H. Ichise, and T. Yorifuji, Phenethylamine oxidase, a novel enzyme of *Arthrobacter globifomi*, may be a quionprotein, *Agric. Biol. Chem.*, *54*:851 (1990).
37. V. Alphand, A. Archelas, and R. Furstoss, Microbiological transformations. 13. A direct synthesis of both *S* and *R* enantiomers of 5-hexadecanolide via an enantioselective microbiological Baeyer–Villiger reaction, *J. Org. Chem.*, *72*:347 (1990).
38. M. J. Taschner and D. J. Block, The enzymatic Baeyer-Villiger oxidation: Enantioselective synthesis of lactones from mesomeric cyclohexanones, *J. Am. Chem. Soc.*, *110*:6892 (1988).
39. R. Wichmann, C. Wandrey, A. F. Buckmann, and M.-R. Kula, Continuous enzymatic transformation in an enzyme membrane reactor with simultaneous NAD (H) regeneration, *Biotech. Bioeng.*, *23*:2789 (1981).
40. R. J. Lamed and J. G. Zeikus, Novel NADP-linked alcohol-aldehyde/ketone oxidoreductase in thermophilic ethanologenic bacteria, *Biochem. J.*, *195*:183 (1981).
41. E. Keinan, E. K. Hafeli, K. K. Seth, and R. Lamed, Thermostable enzymes in organic synthesis. 2. Asymmetric reduction of ketones with alcohol dehydrogenase from *Thermoanaerobium brockii*, *J. Am. Chem. Soc.*, *108*:162 (1986).
42. H. G. W. Leuenberger, W. Boguth, E. Widmer, and R. Zell, 189. Synthese von optisch aktiven, natürchen carotinoiden und strukturell verwandten naturprodukten. I. Synthese der chiralen schlüssewerbindung (4*R*,6*R*)-4-hydroxy-2,2,6-trimethylcyclohexanon, *Helv. Chem. Act.*, *59*:1832 (1976).
43. H. Nakazawa, H. Enei, K. Kubota, and S. Okumura, Biological method of producing serine and serinol derivatives, U.S. Patent 3,871,958, 1975.
44. H. Y. Hsiao and T. T. Wei, Stabilization of tetrahydrofolic acid and serine hydroxymethyltransferase in reaction mixtures for the synthesis of L-serine, European Patent Application 84309137.2, 1984.
45. Y. Tani, Methylotrophs for biotechnology; methanol as a raw material for fermentative production, *Biotech. Gen. Eng. Rev.*, *3*:111 (1985).
46. M. Watanabe, Y. Morinaga, T. Takenouchi, and H. Enei, Efficient conversion of glycine to L-serine by a glycine-resistant mutant of almethylotroph using Co^{++} as an inhibitor of L-serine degradation, *J. Ferment. Technol.*, *65*:563 (1987).
47. V. Schirch, S. Hopkins, E. Villas, and S. Angelaccio, Serine hydroxymethyltransferase from *Escherichia coli*: Purification and properties, *J. Bacteriol.*, *163*:1 (1985).
48. D. I. Stirling, The use of aminotransferase for the production of chiral amino acids and amines, *The Industrial Production of Homochiral Compounds* (A. N. Collins and G. N. Sheldrake, eds.), Wiley, Chichester, 1992.
49. M. C. Fusee, Amino acid production from alpha-keto acid by microbiological conversion, German Patent Application DE3427495 A1, 1985.

50. B. Lawlis, W. Rastetter, and R. Snedecor, Recombinant process for preparing L-amino acids, recombinant expression vectors and transformed microorganisms for use in the process, European Patent Application No. 84305532.8, 1985.
51. J. D. Rozzell, Production of L-amino acids by transamination, European Patent Application No. 84110407.8, 1985.
52. L. L. Wood and G. J. Carlton, Process and compositions for preparing phenylalanine, European Patent Application No. 84304966.9, 1985.
53. G. J. Carlton, L. L. Wood, M. H. Updike, L. Lanty II, and J. P. Hamman, The production of L-phenylalanine by polyazetidine immobilized microbes, *Bio./Technol.*, *5*:317 (1986).
54. J. D. Rozzell, Immobilized aminotransferases for amino acid production, *M. Enzymol.*, *136*:479 (1987).
55. S. P. Crump, J. S. Meier, and J. D. Rozzell, The production of amino acids by transamination, *Biocatalysis* (D. Abramowicz, ed.), Van Nostrand-Reinhold, New York, 1990, pp. 115–133.
56. H. Ziehr, W. I. Hummel, H. Reichenbach, and M.-R. Kula, Two enzymatic routes for the production of L-phenylalanine, *3rd European Congress on Biotechnology*, Verlag Chemie, Weinheim, 1984, Vol. 1, pp. 345–350.
57. S. B. Primrose, The cloning and utilization of aminotransferase genes, European Patent Application No. 84100421.8, 1984.
58. E. Bulot and C. L. Cooney, Selective production of phenylalanine from phenylpyruvic acid using growing cells of *Corynebacterium glutamicum*, *Biotechnol. Lett.*, *7*:93 (1985).
59. P. Gramatica, Biotransformations in organic synthesis, *Chim. Oggi*, *7*:9 (1989).
60. J. B. Jones, Esterases in organic synthesis: Present and future, *Pure Appl. Chem.*, *63*:1445 (1990).
61. K. K.-C. Liu, K. Noyaki, and C.-H. Wong, Problems of acyl migration in lipase-catalysed enantioselective transformation of meso-1,3-diol systems, *Biocat.*, *3*:169 (1990).
62. S. West, Production of chiral hydroxy groups by secondary alcohol dehydrogenase, *Chim. Oggi.*, *9*:43 (1990).
63. H.-P. Meyer, J. Günther, M. Eyer, and R. Voeffray, Large scale process for enzymatic production of *R*-glycidylbutyrate, *Proceedings of Chiral Synthesis Symposium and Workshop*, Manchester, England, Spring Innovations, Stockport, 1989, pp. 31–33.
64. C. J. Sih, Q.-M. Gu, C. Fülling, S.-H. Wu, and D. R. Reddy, The use of microbial enzymes for the synthesis of optically active pharmaceuticals, *Dev. Ind. Microbiol.*, *24*:221 (1988).
65. M. Ahmar, C. Girard, and R. Bloch, Enzymatic resolution of methyl 2-alkyl-2-arylacetates, *Tetrahedron. Lett.*, *30*:7053 (1989).
66. S.-H. Wu, Z.-W. Guo, and C. J. Sih, Enhancing the enantioselectivity of *Candida* lipase catalysed ester hydrolysis via noncovalent enzyme modification, *J. Am. Chem. Soc.*, *112*:1990 (1990).

67. C.-S. Chen and C. J. Sih, General aspects and optimization of enantioselective biocatalysis in organic solvents: The use of lipases, *Angew. Chem. Int. Ed. Engl., 28*:695 (1989).
68. G. Kirchner, M. P. Scollar, and A. N. Klibanov, Resolution of racemic mixtures via lipase catalysis in organic solvents, *J. Am. Chem. Soc., 107*:7072 (1985).
69. B. Campou and A. N. Klibanov, Comparison of different strategies for the lipase-catalysed preparative resolution of racemic acids and alcohols: Asymmetric hydrolysis esterification and transesterification, *Biotech. Bioeng., 26*: 1449 (1984).
70. J.-C. Jallageas, A. Arnaud, and P. Galzy, Bioconversion of nitriles and their applications, *Adv. Biochem. Eng., 14*:1 (1980).
71. T. Nagasawa and H. Yamada, Microbial transformation of nitriles, *Trans. Biotechnol., 7*:153 (1989).
72. J. Mauger, T. Nagasawa, and H. Yamada, Synthesis of various aromatic amide derivatives using nitrile hydratase of *Rhodococcus rhodochrous* J1, *Tetrahedron, 45*:1347 (1989).
73. K. Yamamato, Y. Ueno, K. Otsubo, K. Kawakami, and K.-I. Komatsu, Productions of *S*-(+)-ibuprofen from a nitrile compound by *Acinetobacter* sp. strain AK226, *Appl. Environ. Microbiol., 56*:3125 (1990).
74. A. M. Macadam and L. J. Knowles, The stereospecific bioconversion of α-aminopropionitrile to L-alanine by an immobilized bacterium isolated from soil, *Biotechnol. Lett., 7*:865 (1985).
75. J. C. Jallageas, A. Arnaud, and P. Galzy, Preparation process of optically active α-aminated acids by biological hydrolysis of nitriles, U.S. Patent 4,366,250, 1982.
76. E. M. Meijer, W. H. J. Boesten, H. E. Shoemaker, and J. A. M. van Balken, Use of biocatalysts in the industrial production of specialty chemicals, *Biocatalysts in Organic Synthesis* (J. Tramper, H. C. van der Plas, and P. Linko, eds.), Elsevier, Amsterdam, 1985, p. 135.
77. R. Olivieri, E. Fascetti, L. Angelini, and L. Degen, Enzymatic conversion of *N* Carbonyl-D-amino acids to D-amino acids, *Enz. Microbiol. Technol., 1*:201 (1979).
78. R. Olivieri, E. Fascetti, L. Angelini, and L. Degen, Microbial transformation of racemic hydantoins to D-amino acids, *Biotech. Bioeng., 23*:2173 (1981).
79. D. J. Hardman, Biotransformation of halogenated compounds, *Crit. Rev. Biotechnol., 11*:1 (1991).
80. H. Kawasaki, K. Miyoshi, and K. Tonomura, Purification, crystallization and properties of haloacetate halohydrolase from *Pseudomonas* sp., *Agric. Biol. Chem., 45*:543 (1981).
81. A. J. Weightman, A. L. Weightman, and J. H. Slater, Stereospecificity of 2-monochloropropionate dehalogenation by the two dehalogenases of *Pseudomonas putida* PP3: Evidence for two different dehalogenation mechanisms, *J. Gen. Microbiol., 128*:1755 (1982).
82. N. Leigh, A. J. Skinner, and R. A. Cooper, Partial purification, stereo-

specificity, and stoichiometry of three dehalogenases from a *Rhizobium* species, *FEMS Microbiol. Lett.*, *49*:383 (1988).

83. S. Taylor, Chiral synthesis by biocatalysis, *Proceedings of Chiral Synthesis Symposium and Workshop*, Manchester, England, Spring Innovations, Stockport, 1989, p. 35.
84. N. Kasai, H. Shima, and K. Tsujimura, Process for producing optically active dichloropropanol using microorganisms, European Patent Application 86304184.4, 1987.
85. S. C. Bell and J. M. Turner, Bacterial catabolism of threonine. Threonine degradation initiated by L-threonine acetaldehyde-lyase (aldolase) in species of *Pseudomonas*, *Biochem. J.*, *166*:209 (1977).
86. W. Stöcklein and H.-C. Schmidt, Evidence for L-threonine cleavage and allothreonine formation by different enzymes from *Clostridium pasteurianium*: Threonine aldolase and serine hydroxymethyltransferase, *Biochem. J.*, *232*: 621 (1985).
87. H. Yamada and H. Kumagai, Synthesis of L-tyrosine and related amino acids by β-tyrosinase, *Adv. Appl. Microbiol.*, *19*:249 (1975).
88. G. Para, P. Lucciardi, and J. Barrati, Synthesis of L-tyrosine by immobilized *Escherichia intermedia* cells, *Appl. Microbiol. Biotechnol.*, *21*:273 (1985).
89. G. Para and J. Barrati, Synthesis of L-dopa by *Escherichia intermedia* cells immobilized in a polyacrylamide gel, *Appl. Microbiol. Biotechnol.*, *27*:222 (1988).
90. S. Bringer-Meyer and H. Sahm, Acetoin and Phenylacetylcarbinol formation by the pyruvate decarboxylases of *Zymomonas mobilis* and *Saccharomyces calbergensis*, *Biocat.*, *1*:321 (1988).
91. C. K. N. Tripathi, S. K. Basu, V. C. Vora, J. R. Mason, and S. J. Pirt, Continuous cultivation of a yeast strain for biotransformation of L-acetylphenylcarbinol (L-PAC) from benzaldehyde, *Biotechnol. Lett.*, *10*:635 (1988).
92. Z. Budesinky and M. Protiva, Ephedrin, *Synthetische Arzneimittel* (W. Knobloch, ed.), Springer-Verlag, Berlin, 1961.
93. J. Brussee, W. T. Loos, C. G. Kruse, A. van der Gen, Synthesis of optically active silyl protected cyanohydrins, *Tetrahedron.*, *46*:979 (1990).
94. E. Wehtje, P. Adlercreutz, and B. Mathiasson, Formation of C—C bonds by mandelonitrile lyase in organic solvents, *Biotech. Bioeng.*, *36*:39 (1990).
95. F. Effenberger, B. Hörsch, S. Förster, and T. Zeigler, Enzyme-catalyzed synthesis of (*S*)-cyanohydrins and subsequent hydrolysis to (*S*)-α-hydroxycarboxylic acids, *Tetrahedron. Lett.*, *31*:1249 (1990).
96. U. Neidermyer and M.-R. Kula, Enzyme-catalysed synthesis of (*S*)-cyanohydrins, *Angew. Chem. Int. Ed. Engl.*, *29*:386 (1990).
97. I. Takata, K. Yamamoto, T. Tosa, and I. Chibata, Immobilization of *Brevibacterium flavum* with carrageenan and its application for continuous production of L-malic acid, *Enz. Microbiol. Technol.*, *2*:30 (1980).
98. M. A. Findeis and G. M. Whitesides, Fumarase-catalyzed synthesis of L-*threo*-chloromalic acid and its conversion to 2-deoxy-D-ribase and D-*erythro*-sphingosine, *J. Org. Chem.*, *52*:2838 (1987).

99. H. C. Wong, P. Allenza, and T. C. Lessie, Hydroxy-amino acid utilization and α-ketobutyrate toxicity in *Pseudomonas cepacia*, *J. Bacteriol.*, *144*:441 (1980).
100. K. Kubota, K. Yakozeki, and H. Ozaki, Effects of L-serine dehydratase activity of L-serine production by *Corynebacterium glycinophilum* and an examination of the properties of the enzyme, *J. Ferment. Bioeng.*, *67*:341 (1989).
101. I. Chibata, T. Tosa, and T. Sato, Immobilized aspartase-containing microbial cells: Preparation and enzymatic properties, *Appl. Microbiol.*, *27*:878 (1974).
102. T. Tosa, T. Sato, T. Mori, and I. Chibata, Basic studies for continuous production of L-aspartic acid by immobilized *Escherichia coli* cells, *Appl. Microbiol.*, *27*:886 (1974).
103. T. Sato, T. Mori, T. Tosa, I. Chibata, M. Furui, K. Yamashita, and A. Sumi, Engineering analysis of continuous production of L- aspartic acid by immobilized *Escherichia coli* cells in fixed beds, *Biotechnol. Bioeng.*, *17*:1797 (1975).
104. S. Takamatsu, I. Umemura, K. Yamamoto, T. Sato, T. Tosa, and I. Chibata, Production of L-alanine from ammonium fumarate using two immobilized microorganisms, *Eur. J. Appl. Microbiol. Biotechnol.*, *15*:147 (1982).
105. B. K. Hamilton, H.-Y. Hsiao, W. Swann, D. Anderson, and J. Delente, Manufacture of L-amino acids with bioreactors, *J. Biotechnol.*, *3*:64 (1985).
106. C. T. Evans, K. Hanna, C. Payne, D. Conrad, and M. Misawa, Biotransformation of *trans*-cinnamic acid to L-phenylalanine: Optimization of reaction conditions using whole yeast cells, *Enz. Micro. Technol.*, *9*:417 (1987).
107. S. Yamada, K. Nabe, N. Izuo, K. Nakamichi, and I. Chibata, Production of L-phenylalanine from *trans*-cinnamic acid with *Rhodotorula glutinis* containing L-phenylalanine ammonia-lyase activity, *Appl. Environ. Microbiol.*, *42*:773 (1981).
108. C. T. Evans, K. Hanna, D. Conrad, W. Peterson, and M. Misawa, Production of phenylalanine ammonia-lyase (PAL): Isolation and evaluation of yeast strains suitable for commercial production of L-phenylalanine, *Appl. Microbiol. Biotechnol.*, *25*:406 (1987).
109. L. N. Ornston and R. Y. Stainer, Mechanism of β-ketoadipate formation by bacteria, *Nature*, *204*:1279 (1964).

9

STEREOSELECTIVE BIOTRANSFORMATION

Toxicological Consequences and Implications

Nico P. E. Vermeulen *Free University, Amsterdam, The Netherlands*

Johan M. te Koppele *TNO-Institute of Aging and Vascular Research, Leiden, The Netherlands*

I. INTRODUCTION

Over the last decades toxicologists have felt that the appropriate prevention of or intervention with toxic reactions requires knowledge about the molecular mechanisms by which the toxic responses of drugs and other xenobiotic chemicals occur. In the development of toxicity, different stages can be distinguished: (1) toxicokinetics (absorption, distribution, and elimination), (2) biotransformation, resulting in activation or inactivation of xenobiotics, (3) reversible or irreversible interactions with cellular or tissue components, (4) protection or repair mechanisms, and (5) nature and extent of the toxic effect for the organism.

When a toxic effect occurs, stereochemical factors may play a significant role at all stages mentioned, and therefore they will have to be considered when one is interested in understanding the role of stereochemistry in the occurrence of toxic effects (for recent reviews, see Vermeulen and Breimer, 1983; Ariens, 1984; Vermeulen, 1986; Testa, 1986; Eichelbaum, 1988; Caldwell et al., 1988; van Bladeren, 1988; Jamali et al., 1989; Williams, 1990).

Stereoisomerism manifests itself in various forms, such as those related to enantiomers, epimers, diastereoisomers, meso compounds, geometrical isomers, etc. Generally speaking, the consequences of stereoisomerism for the disposition and biological action of xenobiotics in living organisms are difficult to predict, because they are complex and of a multifactorial origin. Recent developments in analytical techniques, enabling the separation of stereoisomers with HPLC, GC, and the use of stable isotopes with GC-MS and LC-MS (Blaschke, 1986; Schill et al., 1986; Vermeulen and Testa, 1988), as well as advancements in stereocontrolled synthesis (Beld and Zwanenburg, 1986; van der Goot and Timmerman, 1988), have certainly contributed to the increased interest in stereoselective aspects of biotransformation and biological activities of drugs and other xenobiotics.

A. Stereoselectivity and Toxicity: The Role of Biotransformation

The prominent role of stereochemical factors in toxicity is clearly seen with chemotherapeutic drugs. The desired toxicity of these compounds (cytostatic effect), as well as undesired toxicity toward nontumor tissue (cytotoxic effect), depends directly on the configuration of the parent drug. For instance, platinum drugs require a *cis* and not *trans* configuration to coordinate bifunctionally to DNA, thus inhibiting DNA replication and transcription (Singh and Koropatnick, 1988). The folate antimetabolite, methotrexate, a potent inhibitor of dihydrofolate reductase, also demonstrates high stereochemical requirements related to interaction with a biological template. In a number of other cytostatic drugs, stereochemistry indirectly plays a crucial role, since substrate- or product-stereoselective biotransformation is involved. For instance, gossypol, a polyhydroxylated 2,2′-binaphthalene originating from cotton seeds (Fig. 1), was found to have in vitro and in vivo cytotoxic properties. The effect of (−)-gossypol was at least twofold greater than that of the (+)-enantiomer (Joseph et al., 1986). Multidrug-resistant MCF/Adr cells appeared more resistant to (−)-gossypol than the parent cell line, which was tentatively attributed to the inactivation of (−)-gossypol by glutathione transferases present in the MCF/Adr cells. Interestingly, (−)-gossypol was a threefold more potent inhibitor of glutathione transferase α- and π-isoenzymes than the (+) enantiomer (Benz et al., 1990).

Metabolism of the cytostatic drug cyclophosphamide (CP; Fig. 1) involves hydroxylation at the C-4 position by cytochrome P450. Subsequently, a number of detoxification reactions can occur (oxidation to the 4-keto derivative, dechloroethylation, formation of a carboxylic acid). The phosphoramide mustard resulting from spontaneous decomposition of 4-hydroxy-CP is thought to be the cytotoxic chemotherapeutic species. The chiral nature of the phosphorus atom resulted in a twofold greater therapeutic index (LD_{50}/ED_{90}) for the *S*-(−)-enantiomer (against the ADJ/PC6 plasma cell tumor in mice) without detectable differences in metabolism

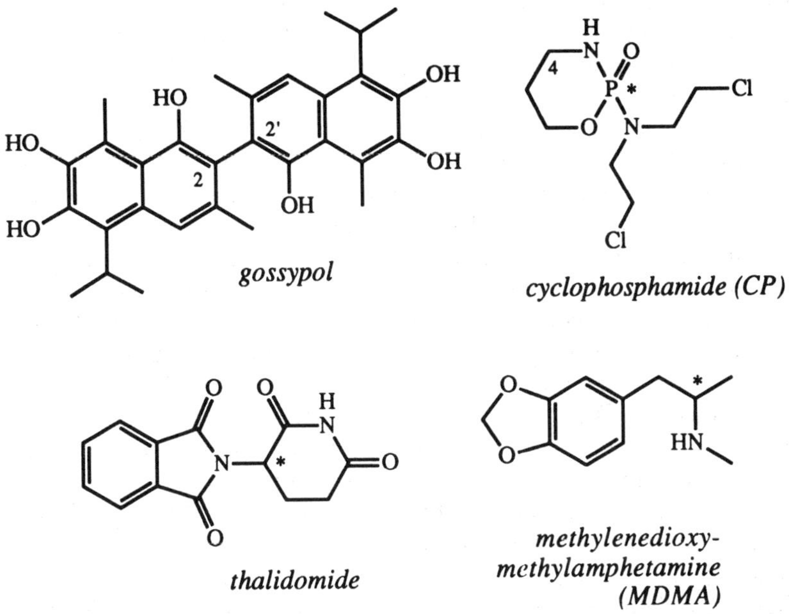

FIGURE 1 Examples of compounds that exert stereoselective toxicity. * = chiral center.

by rat liver microsomes (Cox et al., 1976). Several studies revealed considerable species differences in the metabolism of CP enantiomers, both in the initial 4-hydroxylation and in the later oxidative detoxification pathways (Farmer, 1988). Preliminary studies in man showed that the geometrical isomer iphosphamide undergoes more stereoselective metabolism than CP (Farmer, 1988). Analogues of CP were synthesized to improve the therapeutic properties. Modification of the ring structure of CP introduces a second chiral center. For instance, a methyl or phenyl substitution at the 4-position results in four stereoisomers (two *trans* isomers, *RR* and *SS*, and two *cis* isomers, *RS* and *SR*). 4-Methyl-CP isomers lacked a difference in metabolism by rat liver microsomes and chemotherapeutic efficiency against ADJ/PC6 plasma cell tumors in mice. In contrast, it was suggested that the *trans* isomers of 4-phenyl-CP are metabolized to phosphoramide mustards, whereas the *cis* diastereromers are not, in agreement with the therapeutic activity of *trans*-phenyl-CP against L1210 lymphoid leukemia in mice, which is lacking for the *cis* diastereomers (Boyd et al., 1980). Similarly, *cis*-5-fluoro-CP was hardly metabolized by mouse liver microsomes (in contrast to *trans*), corresponding to low tumor inhibitory activity (in contrast to *trans*; see Foster et al., 1981).

In addition, toxicity-related stereoselectivity can be illustrated with the classical example of thalidomide (Fig. 1), of which the underlying mechanism has as yet not been completely elucidated. At the end of the 1950s, the drug was marketed as a hypnotic agent and in the early 1960s it was withdrawn because of embryotoxic and teratogenic effects. Only in more recent years has it become clear that in various species, both thalidomide enantiomers are transformed in vivo into 1-*N*-phthaloylglutamine and 1-*N*-phthaloylglutamic acid. Reports have been published on the sensitivity of different mouse and rat strains toward thalidomide toxicity. For instance, in SWS mice only the glutamic acid metabolite derived from the *S*(−)-enantiomer was embryotoxic and teratogenic and not the one derived from the *R*(+)-isomer (Ockzenfelz et al., 1976). More recently, however, a rapid racemization of thalidomide enantiomers of α-C-substituted analogues (Blaschke and Graute, 1987) and of a phthalimide analogue of thalidomide (Schmahl et al., 1989) was suggested to occur under physiological conditions. Moreover, in rabbits both thalidomide enantiomers were found to be equally teratogenic. The suggestion that the use of *R*-thalidomide would have avoided the toxicity problems associated with the racemic drug is therefore premature at this time. Another example of stereoselective teratogenicity is presented by 2-ethylhexanoic acid, an environmental pollutant resulting from the plasticizer di-(2-ethylhexyl)-phthalate. While the *R*(−)-enantiomer was highly teratogenic (60% of the fetuses exhibited exencephaly), no teratogenic or embryotoxic effects of *S*(+)-ethylhexanoate were observed at the same dose in mice (Hauck et al., 1990).

Stereoselective neurotoxicity was reported for methylenedioxymethylamphetamine (MDMA; Fig. 1). It causes degeneration of serotonergic and, to a lesser extent, catecholaminergic neurons. MDMA is a widely abused amphetamine derivative (Ecstasy), of which the *S*(+)-enantiomer has the highest pharmacological as well as toxicological potency. Following cytochrome P450-mediated demethylation, 3,4-methylenedioxyamphetamine (MDA) is formed; it also has been shown to be a serotonergic neurotoxin. *S*(+)-MDMA undergoes this demethylation to a far greater extent than the *R*(−)-enantiomer (Fitzgerald et al., 1989). The profile of neurotoxicological reactions is identical to that of *para*-chloroamphetamine (pCA; Schmidt, 1987). Altogether, stereoselective formation of an as yet unknown metabolite of MDMA (and of MDA and pCA) seems to be involved in the stereoselective toxicity of MDMA.

Pharmacokinetic consequences of stereoselective metabolism can be illustrated by the hypnotic drug hexobarbital (HB), at present still used as a model substrate to assess cytochrome P450 enzyme activities. Large differences between the intrinsic hypnotic activities of the enantiomers of this highly lipophilic chiral drug exist; the *S*(+)-enantiomer is the most active enantiomer in most species. In addition, substrate- and product-stereoselective metabolism occurs. With a pseudoracemic mixture of HB, consisting of a mixture of *S*(+)- and N_1-trideutero-*R*(−)-HB, and GC-MS analysis (van der Graaff et al., 1985) the stereoselective disposition of HB enantiomers was studied upon simultaneous administration to the rat (Table 1). The intrinsic clearance of *R*(−)-HB (CL_{int}) was about sevenfold greater than CL_{int} of the *S*(+)-enantiomer (van der Graaff et al., 1988). A similar ninefold difference in the CL_{int} of the HB enantiomers was reported in humans. Interestingly, the observed age-related decline in stereoselectivity was attributed to a decline in various cytochrome P450 isoenzyme activities, resulting in a decrease in CL_{int}, in particular of *R*(−)-HB (Chandler et al., 1988).

Many other drugs and xenobiotics demonstrate stereoselectivity in toxicological or pharmacological effects (Tables 2 and 3) (Vermeulen, 1986; Caldwell et al., 1988; Jamali et al., 1989; Vermeulen, 1989; Eichelbaum and Gross, 1990; Williams, 1990). In addition to direct stereoselective effects of

TABLE 1 Pharmacokinetic Parameters and Recoveries of Unconjugated Metabolites in Rat Urine of *S*(+)-HB and *R*(−)-HB after p.o. and i.a. administration of 25 $mg{\cdot}kg^{-1}$. Values ($n = 6$) represent the mean ± s.e.m.

	S(+)-HB p.o.	*S*(+)-HB i.a.	*R*(−)-HB p.o.	*R*(−)-HB i.a.
$t_{1/2}$ (min)	13.4 ± 0.8	15.7 ± 1.5	16.7 ± 0.6	13.3 ± 0.4
AUC ($mg{\cdot}min^{-1}{\cdot}ml^{-1}$)	16.4 ± 2.2	366 ± 37	129 ± 22	444 ± 30
CL ($mL{\cdot}min^{-1}{\cdot}kg^{-1}$)	1621 ± 197	73.9 ± 7.5	226 ± 36	54.4 ± 2.4
V ($mL{\cdot}kg^{-1}$)		1670 ± 233		1042 ± 56
E	0.94 ± 0.01		0.68 ± 0.03	
OH-HB (% dose)	15.6 ± 0.9	20.6 ± 2.2	10.9 ± 1.3	16.9 ± 2.3
K-HB (% dose)	19.3 ± 0.8	17.3 ± 1.7	56.2 ± 1.8	54.9 ± 9.0
DMBA (% dose)	9.1 ± 1.4	7.8 ± 1.2	11.8 ± 1.9	11.6 ± 4.8
Total (% dose)	44.0 ± 1.8	45.7 ± 3.3	78.9 ± 2.9	83.4 ± 6.6

Source: Data derived from Van der Graaff et al., 1988.

I. *Halogenated methanes*
(1 chiral center)

GSH R_1 R_2 R_3 X $\xrightarrow{S_N2}$ $X^- + H^+$ GS R_1 R_2 R_3

II. *Epoxides*
(2 chiral centers, unless R=H)

cis *trans*

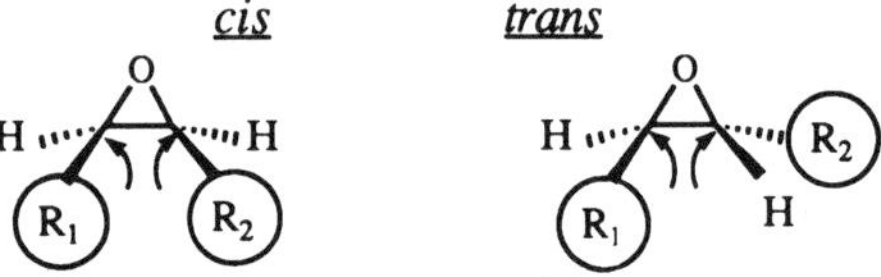

A. Two benzylic carbons ($R_1 = R_2$ = aryl)
B. One or no benzylic carbons (R_1 = aryl or alkyl; R_2 = alkyl)

III. *Olefins*
(achiral)

A. Halogenated ethenes

GSH F F F Cl — 1,2-addition → F F G-S H F Cl + F F G-S H F Cl

B. α,β-unsaturated ketones

GSH R_2 H O R_1 — Michael (1,4)-addition → R_1 H G-S R_2 O + R_1 H G-S R_2 O

FIGURE 7 Types of substrates for GST enzymes that undergo stereoselective GSH conjugation. (Taken from te Koppele and Mulder, 1991.)

mercapturic acids and no *cis* isomers were excreted (Vermeulen et al., 1980; van Bladeren et al., 1981). In alkylhalogenides rat liver cytosolic GSH transferases conjugated chiral α-phenylethyl-chloride and -bromide with a high degree of substrate stereoselectivity. *S* enantiomers were far better substrates than *R* enantiomers, and the chloride analogue showed a higher stereoselectivity than the bromide analogue, a finding that was suggested to be related to the enantioselectivity of GSH transferase isoenzymes from the α and μ family (Mangold and Abdel-Monem, 1983).

A *trans*-addition mechanism involving S_N2-substitution was also found for the enzymatic conjugation of (±)-benz[a]pyrene-4,5-oxide using rat liver cytosol as a source of transferase activity. Both positional isomers (at C_4 and C_5) were produced, however, with a preference for two diastereoisomers. Subsequent investigations with optically active (+)-4*S*,5*R*- and (−)-4*R*,5*S*-benzo[a]pyrene-4,5-oxide have shown that the rat liver cytosolic GSH transferases preferentially catalyze the formation of the 5-glutathionyl-isomer (4*S*,5*S*) from the (+)benzo[a]pyrene-4,5-oxide and the 4-glutathionyl-isomer (4*S*,5*S*) from the (−) isomer. Apparently, these GSH transferases prefer to attack the oxirane at the carbon atom with the *R* configuration (Cobb et al., 1983).

Unless different isoenzymes possess the same kind of stereoselectivity toward a substrate, it may be anticipated that a mixture of isoenzymes will exhibit a diminished stereoselective metabolism. In the case of GSH transferases, stereoselection is even more complicated since one isoenzyme may exist as a homo- or heterodimeric protein of two subunits, each of the subunits potentially having structurally different active sites apart from their GSH-binding site. In a detailed study with purified rat isoenzymes 3–3, 3–4, and 4–4, Armstrong and co-workers (Cobb et al., 1983) reported that subunit 4 was more stereoselective in the conjugation with arene and aza-arene epoxides (with more than a 99% attack at the oxirane carbon with the *R* configuration) than subunit 3. It was suggested that hydrophobic interactions between the substrate and enzyme surface distal to the oxirane ring, rather than electronic differences between the two oxirane carbons, are important in determining stereoselectivity. Recently, it was shown that cytosolic GSH transferases from rat hepatic and extra-hepatic tissues varied markedly in their stereoselectivity toward a number of polycyclic arene and alkene oxide substrates. This stereoselectivity was found to correlate roughly with known tissue differences in isoenzyme composition (π transferases preferentially attacking *S*-configured carbon atoms of K-region epoxides), and it was suggested to use this stereo- and enantioselectivity for the functional characterization of GSH-transferase isoenzymes in different tissues (Dostal et al., 1986). In Fig. 8, a schematic representation of binding interactions of some arene and alkene oxides is presented. Studies with GST 4-4 catalyzed conjugation of *trans-*

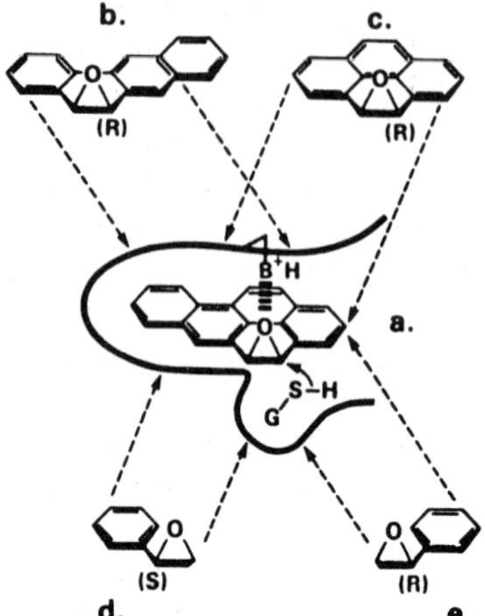

FIGURE 8 Schematic representation of binding interactions of arene oxides at the H site of GST: A, (4*R*,5*S*)-benzo[a]pyrene oxide; B, (5*S*,6*R*)-benzo[a]anthracene oxide; C, (4*R*,5*S*)-pyrene oxide; D, (7*S*)-styrene oxide, and E, (7*R*)-styrene oxide. (Taken from Dostal et al., 1986.)

and *cis*-stilbene oxide demonstrated a highly stereoselective conjugation of the carbon atom with the *R* configuration of both substrates, suggesting that the H site of this isoenzyme can accommodate phenyl rings out of the plane, perpendicular to the epoxide (Armstrong, 1987).

That stereoselective metabolism by cytosolic GSH transferases may be important from a toxicological point of view was illustrated several years ago by experiments in which the GSH transferases were bioactivating nontoxic substrates. In this context, the mutagenic behavior of *cis*- and *trans*-1,2-dichlorocyclohexane was investigated with and without metabolic activation of rat liver fractions (van Bladeren et al., 1979). By using the Ames test with *Salmonella typhimurium* cells, it was shown that, in contrast to the *trans* isomer, the *cis* isomer demonstrated mutagenic activity that was significantly increased upon the addition of 9000 g and 100,000 g supernatant of a rat liver homogenate. The explanation of this stereoselective effect was given in terms of the stereospecific action of glutathione transferases present in the liver homogenate fractions and catalyzing the substitution of the vicinal chlorine atoms according to a S_N2-substitution mechanism. Only in *cis*-1,2-dichlorocyclohexane can this conjugation reaction lead to the formation of a reactive thiiranium ion, which is suspected to be responsible for the mutagenic activity.

Apart from stereoselective bioactivation, stereoselective bioinactivation may be important too from a toxicological point of view. For instance, in contrast to the basic (α-ϵ), the human near neutral (μ) and acidic (π) classes of GSH transferases were efficient in the conjugation of benz[a]-pyrene-7,8-diol-9,10-epoxides and, in particular, the carcinogenic (+)-anti-enantiomer (and other (+)-*anti*-enantiomers of the bay-region diol-epoxides of benz[a]anthracene and chrysene) (Jernström et al., 1985; Robertson et al., 1986).

In recent studies of the rat, GSH conjugation of the chiral hypnotic drug α-bromoisovaleryl-urea was also shown to proceed with a high degree of stereoselectivity in vivo (Fig. 9, te Koppele et al., 1987). Most interestingly, the subunits 3 and 4 (μ-type isoenzymes) were particularly selective for the *R* enantiomer, whereas the subunits 1 and 2 (α-type isoenzymes) were found to be particularly active toward the opposite *S* enantiomer. Since this drug possesses the profile of a relatively ideal model substrate for in vivo applications, it has been proposed as an in vivo probe to assess the activity of different GSH transferases, including the polymorphic and toxicologically potentially important μ class of isoenzymes (te Koppele, 1988).

In recent years, the existence of a microsomal GSH transferase has also been established firmly (Morgenstern et al., 1983). Apart from the usual substrates for the cytosolic GSH transferases, compounds like hexachloro-1,3-butadiene and tetrafluoroethylene seem to be the better substrates for the microsomal enzyme. Dohn et al. (1985) were the first to demonstrate with ^{19}F-NMR the stereochemical control of an enzymatic addition reaction between the pro-chiral chlorotrifluoroethene (CTFE) and GSH. By a microsomal rat liver fraction, predominantly one diastereomeric GSH conjugate (the *S*-configured chlorotrifluoroethyl GSH conjugate; see Hargus et al., 1991) was formed, whereas a cytosolic fraction produced an equimolar fraction of two diastereomeric 2-chloro-1,1,2-trifluoroethyl GSH conjugates. Similarly, the formation of diastereomeric GSH conjugates was shown to be highly product-stereoselective in the case of a GST 4-4 catalyzed Michael addition of GSH to phenylbutenones (Zhang and Armstrong, 1990). 4-Hydroxyalkenals, toxic products of lipid peroxidation, might be anticipated to behave similarly. Further studies of the (regio- and) stereospecificities of GSH transferases might reveal information on the respective mechanisms of these reactions.

In summary, GSH transferase isoenzymes operate highly stereoselectively and this may have important consequences for the detoxification or toxification of chiral or pro-chiral substrates.

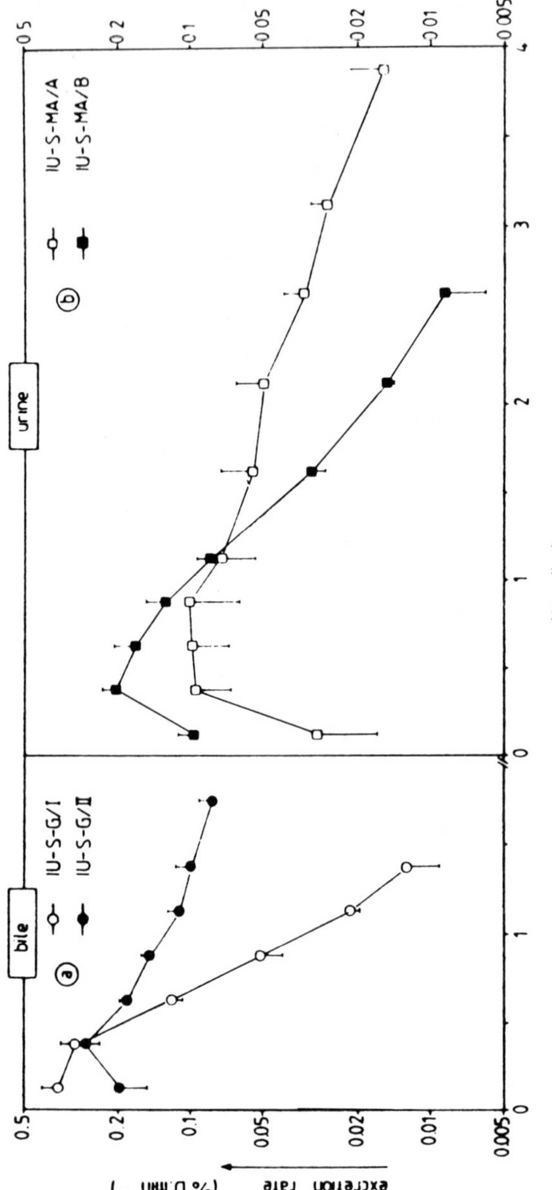

FIGURE 9 Excretion rates of BIU metabolites in bile and urine in anesthetized rats. a, diastereomeric glutathione conjugates in bile (n = 4). b, diastereomeric mercapturates in urine, (n = 3). Glutathione conjugate I and mercapturate B are derived from R(+)-BIU. The conjugates IU-S-G/II and IU-S-MA/A result from S(−)-BIU. (Taken from te Koppele et al., 1987.)

E. Epoxide Hydrolases

Epoxide-containing compounds are often rather reactive and, in many cases, they therefore exhibit strong and direct mutagenic or carcinogenic properties. Thus, the *R* and *S* enantiomers of epoxystyrene exhibit strongly different mutagenic activity in *Salmonella typhimurium* TA 100, although their chemical reactivity toward nucleophiles, such as 4-(p-nitrobenzyl)pyridine, does not differ considerably (Seiler, 1990). This difference in intrinsic mutagenic activity has been attributed to qualitative differences in binding to DNA, rather than to differences in indirect factors, such as metabolic detoxication by GSH-*S* transferases or epoxide hydrolase. Epoxide hydrolases catalyze the hydrolysis of alkene and arene oxides into dihydro-diols. For instance, the widely used anticonvulsant phenytoin, following the cytochrome P450-mediated formation of the 3′,4′-epoxide is stereoselectively hydrated by epoxide hydrolase. The (+) epoxide is metabolized to a greater extent than the (−) enantiomer (Maguire et al., 1980). Microsomal, membrane-bound (Oesch, 1979), as well as cytosolic, soluble epoxide hydrolases have been isolated and identified (Hammock et al., 1980). Substrate selectivities of both hydrolases are broad and largely overlapping. The cytosolic hydrolase with lower activity seems to possess a greater activity toward aliphatic epoxides. In contrast, the microsomal epoxide hydrolase has a higher degree of substrate stereoselectivity toward such epoxides (Belluci et al., 1989b). Mechanistically, the enzymatic hydration of epoxides most likely proceeds according to the general base-catalyzed nucleophilic addition of "activated water." A unionized histidine residue has been implicated as essential for the deprotonation of the attacking water molecule (Fig. 10, Armstrong, 1987). An electrophilic (acid-catalyzed) mechanism was ruled out by experiments with $H_2{}^{18}O$, which demonstrated regiospecific hydration at the 2-position of unlabeled styrene and nafthaleneoxides and by experiments that demonstrated the inability of the epoxides to accept nucleophiles other than water (Hanzlik et al., 1976).

Stereochemical studies have shown that the hydration of epoxides (of cyclic olefins as well as arenes) is performed by epoxide hydrolase in such a way that *trans*-diols are formed. Aliphatic epoxides have also been shown to undergo *trans*-hydration, which means that *trans*-epoxides yield *erythro*-glycols and *cis*-epoxides result in *threo*-glycols (Hammock et al., 1980).

Varying degrees of substrate enantioselectivity have been described for microsomal epoxide hydrolases. For example, for benzo[a]pyrene-4,5-oxide a 40-fold difference between the rates of hydration for the (+) and (−) enantiomers was observed in vitro (Armstrong et al., 1980). Monosubstituted epoxides (such as 1,2-epoxyhexane and its geometrical isomer,

Apart from substrate-stereoselective glucuronidation of hydroxylated hexobarbital, considerable enantioselective differences and species differences have been shown in vitro for propranolol (see the references in Vermeulen, 1986), fenoterol (Koster et al., 1986), carvedilol (Fujimaki and Hakusui, 1990), dihydrodiols of polycyclic aromatic hydrocarbons (Armstrong et al., 1988), and oxazepam (Table 4) (Sisenwine et al., 1982). (−)-Morphine has been shown to be glucuronidated exclusively at the 3-hydroxy position, whereas (+)-morphine is preferentially glucuronidated at the 6-hydroxy position. In vivo, enantioselectivity is often less pronounced than in vitro. 3-Methylcholanthrene, β-naphtoflavone, inducers of the so-called GT_1 isoenzyme, and ethanol have been shown to increase enantioselectivity in the formation of diastereomeric glucuronides from racemic oxazepam in hepatic microsomes of rabbits and to produce a large increase in enantioselectivity, whereas inducers of the GT_2 isoenzyme, such as phenobarbital, only increased oxazepam glucuronidation without affecting enantioselectivity. Inducers of forms other than GT_1 and GT_2, such as trans-stilbene-oxide and clofibric acid, generally cause increases in activities, coupled with increases in *R*/*S* ratios. Measurement of enantiomeric selectivity in the glucuronidation of racemic oxazepam was suggested to be a useful tool to characterize the nature and selectivity of inducers of the glucuronyltransferase system (Yost and Finley, 1985).

As of yet, very little is known about the active sites of these enzymes. The enzyme mechanism is probably consistent with a single displacement, S_N2-type nucleophilic attack of the aglycone on UDP-glucuronic acid in a ternary complex. Further studies seem necessary for a better understanding of the stereochemical behavior and active-site geometry of UDP-glucuronyl transferases (Armstrong, 1987; Armstrong et al., 1988; Mulder et al., 1990).

G. Sulfotransferases

Sulfotransferases catalyze the transfer of a sulphate group from 3′-phosphoadenosine 5′-phosphosulfate (PAPS) to nucleophiles such as alcohols, phenols, and amines. Several isoenzymes, grouped into two classes based on substrate selectivity toward phenols and alcohols, have been purified (Mulder et al., 1990). How the active sites of sulfotransferases participate in the catalysis is not known (Armstrong, 1987).

Generally speaking, conclusions regarding stereoselective sulfation in vivo are difficult to draw, because of competition between (stereoselective) glucuronidation and sulfation of the same substrates. Stereoselective sulfation of 4′-hydroxypropranol enantiomers was demonstrated in vitro in hepatic tissue of various species (Christ and Walle, 1985). Enantiomeric (−/+)-4′-hydroxypropranolol sulfate ratios varied from 1.07 in the rat and

TABLE 4 In vitro Glucuronidations of Oxazepam Enantiomers (nmol/hr/g liver)

Species	*S*(+)	*R*(−)	Ratio *S*/*R*
Rhesus monkey	238	596	0.4
Dog	826	43	19.1
Miniature swine	41	29	1.4
Rabbit	178	92	1.9
Rat	424	22	19.3

Source: Sisenwine et al., 1982.

0.73 in the dog to 0.62 in the hamster. Terbutaline and prenalterol, a structural analogue of terbutaline, both showed preferential sulfation for the respective (+)-enantiomers in human liver cytosol (Walle and Walle, 1990). Both the species-dependent preference as well as the degree of apparent stereoselectivity are probably due to the activity of multiple sulfotransferase isozymes present in the supernatant preparations. At least three isozymes have been isolated from rat liver. No examples are yet known, demonstrating the causal relationship between sulfation and stereoselective toxicity. Metabolic activation, however, has been shown to occur, for example, in the case of 7-hydroxymethyl-12-methylbenz[a]anthracene, the sulfation of which is followed by alkylation of DNA via the sulfate ester (Watabe et al., 1985).

H. Cysteine Conjugate β-Lyase

Several cysteine conjugates are metabolized by β-lyase to produce thiols, ammonia, and pyruvate. Cytosolic β-lyases have been isolated from the kidney, liver, and intestinal microflora. Since it has become clear that β-lyases can play a decisive role in the bioactivation of cysteine conjugates (such as those from hexachloro-1,3-butadiene or other halogenated alkenes) to nephrotoxic agents, the scientific interest in this enzyme has increased considerably (Elfarra and Anders, 1984; Commandeur et al., 1987).

As for stereoselectivity, relatively little is known as yet, except that the natural *R* configuration in the cysteine residue is a prerequisite for a cysteine conjugate to be a substrate of β-lyase. This fact has been used as a tool to ascertain the role of β-lyase in the development of nephrotoxicity by cysteine conjugates. No data are presently available with regard to the substrate-stereoselective effects of the noncysteine part of the thioether substrates of β-lyase (Commandeur and Vermeulen, 1990). Such stereoselective effects may be anticipated, however, since, for example, the regioisomeric 1,2- and 2,2-dichlorovinyl-L-cysteine conjugates have also been shown to cause strongly different β-lyase-mediated mutagenic and nephrotoxic effects (Commandeur et al., 1991).

III. STEREOSELECTIVE PROTECTION AGAINST TOXICITY

The antagonism or prevention of toxic effects may also benefit from stereochemical principles. Thus, optically active flavanones have been shown to inhibit the metabolic activation of the carcinogen benzo[a]pyrene to metabolites that bind covalently to DNA (Chae et al., 1992). Moreover, the (+) enantiomers of 3-*O*-methylcatechin and catechin have been demonstrated to protect stereoselectively against lipid peroxidation due to paracetamol (Devalia et al., 1982).

Alternatively, *R*-cysteine is utilized for the biosynthesis of cosubstrates for several conjugation reactions, such as glutathione, sulfate (PAPS), and taurine. The unphysiological isomer *S*-cysteine has been used to investigate the mechanistic aspects of physiological processes involving cysteine. In rats, the sulfoxidation rates of *R*- and *S*-cysteine, as well as the sulfation rates of the test substrate harmol sulfate, were found to be very similar, so that stereoselectivity for the amino acid does not seem to play a role in these reactions (Glazenburg et al., 1984). Since *S*-cysteine, in contrast to the *R*-isomer, did not increase the taurine concentration in serum, this type of stereoselectivity can be used to selectively enhance sulfate availability in vivo.

Glutathione, as the main intracellular nonprotein sulfhydryl, plays an important role as a cosubstrate for conjugation or a reductant in the detoxication of electrophilic compounds or radicals in organisms, as well as in the repair of various kinds of cellular injury. Consequently, increased levels of glutathione may exert beneficial effects in some cases. *N*-acetyl-*R*-cysteine, which is rapidly hydrolyzed intracellularly to *R*-cysteine, more successfully promotes glutathione synthesis in vivo than *R*-cysteine itself and is known to protect more efficiently against paracetamol-induced hepatotoxicity than *N*-acetyl-*S*-cysteine. Alternatively, 2-alkyl- or 2-aryl-substituted thiazolidine-4-*R*-carboxylic acids (Fig. 11), especially those derived from the condensation of *R*-cysteine with even alkyl carbon aldehydes, that is, aldehydes that are readily metabolizable to nontoxic products, can serve as precursors of *R*-cysteine and protect against paracetamol toxicity (Nagasawa et al., 1984). The incorporation of *R*-cysteine into GSH and not the liberation of sulfhydryl groups per se appears to be important, since analogues with a 4-*S* configuration, which are converted into unphysiological *S*-cysteine, were found to be totally ineffective as protecting agents. *R*-2-oxothiazolidine-4-carboxylate has been shown to increase brain cysteine levels and has been suggested as useful to protect, for

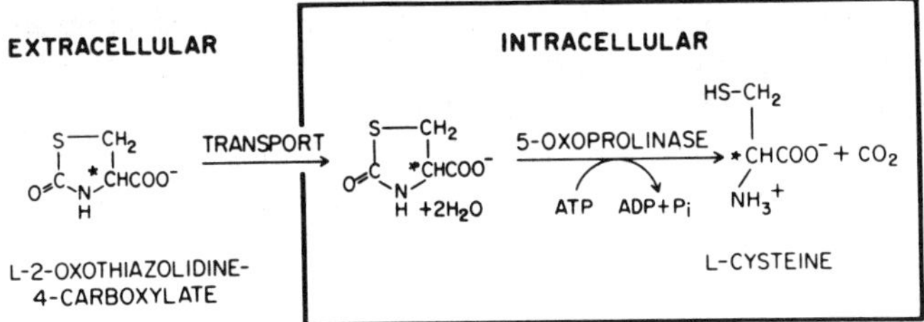

FIGURE 11 Chemical structure, transport, and metabolic degradation of 2-oxothiazolidine-4-carboxylate. * = chiral center. (Taken from Anderson et al., 1989.)

instance against the neurotoxicity of 6-hydroxydopamine (Anderson and Meister, 1989).

IV. GENERAL CONCLUSIONS

This chapter has demonstrated that it is generally recognized at present that stereochemical factors play a significant role in the disposition and pharmacological or toxicological action of drugs and other xenobiotics. As for the toxicity of xenobiotics, stereoselectivity in biotransformation (i.e., bioactivation or bioinactivation) is probably a factor of major importance. As a consequence, the biological actions of xenobiotics, in general, are difficult to predict because of the complex regulation of the processes involved.

In recent years, considerable progress has been made in the elucidation of stereochemical mechanisms of several important biotransformation enzymes, despite the fact that the multiplicity, a widely varying substrate selectivity as well as secondary metabolism (which in itself can be stereoselective), considerably complicates such studies. In principle, the same factors do complicate the correlation of data on stereoselective effects obtained in vitro with those obtained in vivo, even when it concerns the same species. The prediction of stereoselective effects from one species to another or from one chemical to another is even more difficult.

As this review also has demonstrated, new insights into stereochemical mechanisms at a molecular level [i.e., at the level of binding of substrate molecules to active (or binding) sites of isolated and purified (iso)-enzymes (or other proteins)] in some cases provided more or less simplified models or working hypotheses that might be helpful in predicting stereoselective biotransformations. Progress in research along these lines will be most promising, from not only an academic but also a more practical point of view, namely, that concerned with predictability. From a toxicological point of view, it is important to stress that knowledge on stereoselective effects at the different levels of the development of biologically active drugs or other chemicals should be obtained as early as possible. Without this knowledge, it is almost impossible to interpret the toxicodynamics and toxicokinetics of chiral compounds.

REFERENCES

Anderson, M. E. and Meister, A. (1989). Marked Increase of Cysteine Levels in Many Regions of the Brain After Administration of 2-Oxothiazolidine-4-carboxylate, *FASEB J.*, *3*:1632–1635.

Ariëns, E. J. (1984). Domestication of Chemistry by Design of Safer Chemicals: Structure-Activity Relationships, *Drug Metab. Rev.*, *15*:425–504.

Armstrong, R. N. (1987). Enzyme-Catalyzed Detoxication Reactions: Mechanisms and Stereochemistry, *CRC Crit. Rev. Biochem.*, *22*:39–89.

Armstrong, R. N., Levin, W., and Jerina, D. M. (1980). Hepatic Microsomal Epoxide Hydrolase. Mechanistic Studies of the Hydration of K-Region Arene Oxides, *J. Biol. Chem.*, *255*:4698–4705.

Armstrong, R. N., Andre, J. C., and Bessems, J. G. M. (1988). Mechanistic and Stereochemical Investigations of UDP-Glucuronosyltransferases, In *Cellular and Molecular Aspects of Glucuronidation Colloque. Inserm.* (G. Siest, J. Magdalou, and B. Burchell, eds.), John Libbey Eurotext Ltd., London, Vol. 173, pp. 51–58.

Aoyama, T., Korzekwa, K., Nagata, K., Adesnik, M., Reiss, A., Lapenson, D. P., Gillette, J., Gelboin, H. V., Waxman, D. J., and Gonzalez, F. J. (1989). Sequence Requirements for Cytochrome P-450IIB1 Catalytic Activity. Alteration of the Stereospecificity and Regioselectivity of Steroid Hydroxylation by a Simultaneous Change of Two Hydrophobic Amino Acid Residues to Phenylalanine, *J. Biol. Chem.*, *264*:21,327–21,333.

Ayesh, R., Idle, J. R., Ritchie, J. C., Crothers, J. J., and Hetzel, M. R. (1984). Metabolic Oxidation Phenotypes as Markers for Susceptibility to Lung Cancer, *Nature*, *312*:169–170.

Beld, A. J. and Zwanenburg, B. (1986). Stereocontrolled Synthesis: Modern Approaches for Preparing Bioactive Compounds. In *Innovative Approaches in Drug Research* (A. F. Harms, ed.), Elsevier, Amsterdam, pp. 331–352.

Belluci, G., Berti, G., Bianchini, R. Cetera, P., and Mastrorilli, E. (1982). Stereoselectivity in the Epoxide Hydrolase Catalyzed Hydrolysis of the Stereoisomeric 3-*tert*-Butyl-1,2-epoxycyclohexanes. Further Evidence for the Topology of the Enzyme Active Site, *J. Org. Chem.*, *47*:3105–3112.

Belluci, G., Chiappe, C., Conti, L., Marioni, F., and Pierini, G. (1989a). Substrate Enantioselection in the Microsomal Epoxide Hydrolase Catalyzed Hydrolysis of Monosubstituted Oxiranes. Effects of Branching of Alkyl Chains, *J. Org. Chem.*, *54*:5978–5983.

Belluci, G., Chiappe, C., and Marioni, F. (1989b). Enantioselectivity of the Enzymatic Hydrolysis of Cyclohexene Oxide and (±)-1-Methylcyclohexane

Oxide: A Comparison Between Microsomal and Cytosolic Epoxide Hydrolases, *J. Chem. Soc. Perkin Trans. I*:2369–2373.

Benz, C. C., Kenry, M. A., Ford, J. A. Townsend, A. J., Cox, F. W., Palayoor, S., Matlin, S. A., Hait, W. N., and Cowan, K. H. (1990). Biochemical Correlates of the Antitumor and Antimitochondrial Properties of Gossypol Enantiomers, *Mol. Pharmacol.*, *37*:840–848.

Blaschke, G. B. (1986). Chromatographic Resolution of Chiral Drugs on Polyamides and Cellulose Triacetate, *J. Liq. Chromatogr.*, *9*:341–368.

Blaschke, G. and Graute, W. F. (1987). Enantiomere des Konfigurationsstabilen C-Methylthalidomids, *Liebigs Ann. Chem.*, *1987*:647–648.

Boobis, A. R., Murray, S., Hampden, C. E., and Davies, D. S. (1985). Genetic Polymorphism in Drug Oxidation: in vitro Studies of Human Debrisoquine 4-Hydroxylase and Bufuralol 1′-Hydroxylase Activities, *Biochem. Pharmacol.*, *34*:65–71.

Boyd, V. L., Zon, G., Himes, V. L., Stalick, J. K., Mighell, A. D., and Secor, H. V. (1980). Synthesis and Antitumor Activity of Cyclophosphamide Analogues. 3. Preparation, Molecular Structure Determination, and Anticancer Screening of Racemic *cis*- and *trans*-4-Phenylcyclophosphamide, *J. Med. Chem.*, *23*: 372–378.

Caldwell, J., Winter, S. M., and Hutt, A. J. (1988). The Pharmacological and Toxicological Significance of the Stereochemistry of Drug Disposition, *Xenobiotica*, *18*:59–70.

Cashman, J. R., Olsen, L. D., and Bornheim, L. M. (1990). Enantioselective *S*-Oxygenation by Flavin-Containing and Cytochrome P-450 Monooxygenases, *Chem. Res. Toxicol.*, *3*:344–349.

Chae, Y. H., Ho, D. K., Cassady, J. J., Cook, V. M., Marcus, C. B., and Baird, W. M. (1992). Effects of Synthetic and Naturally Occurring Flavonoids on Metabolic Activation of Benzo[*a*]pyrene in Hamster Embryo Cell Cultures, *Chem. Biol. Interact.*, *82*:181–193.

Chandler, M. H. H., Acott, S. R., and Blouin, R. A. (1988). Age-Associated Stereoselective Alterations in Hexobarbital Metabolism, *Clin. Pharmacol. Ther.*, *43*:436–441.

Chang, R. L., Levin, W., Wood, A. W., Yagi, H., Tada, M., Vyas, K. P., Jerina, D. M., and Conney, A. H. (1983). Tumorigenicity of Enantiomers of Chrysene 1,2-Dihydrodiol and of the Diastereomeric Bay-Region Chrysene 1,2-Diol-3,4-epoxide on Mouse Skin and in Newborn Mice, *Cancer Res.*, *43*:192–196.

Christ, D. D. and Walle, T. (1985). Stereoselective sulfate conjugation of 4-hydroxy propranolol by different species. *Drug Metab. Disposition*, *13*:380–381.

Cobb, D., Boehlert, C., Lewis, D., and Armstrong, R. N. (1983). Stereoselective of Isozyme C of Glutathione *S*-Transferase Toward Arene and Azaarene Oxides, *Biochemistry*, *22*:805–812.

Commandeur, J. N. M., Oostendorp, R. A. J., Schoofs, P. R., Xu, B., and Vermeulen, N. P. E. (1987). Nephrotoxicity and Hepatotoxicity of 1,1-Dichloro-2,2-difluoroethylene in the Rat. Indications for Differential Mechanisms of Bioactivation, *Biochem. Pharmacol.*, *36*:4229–4237.

Commandeur, J. N. M. and Vermeulen, N. P. E. (1990). Molecular and Biochemical

Mechanisms of Chemically Induced Nephrotoxicity: A Review, *Chem. Res. Toxicol.*, *3*:171–194.

Commandeur, J. N. M., Boogaard, P. J., Mulder, G. J., and Vermeulen, N. P. E. (1991). Mutagenicity and Cytotoxicity of Two Regioisomeric Mercapturic Acids and Cysteine *S*-Conjugates of Trichloroethylene, *Arch. Toxicol.*, *65*: 373–380.

Conney, A. H. (1982). Induction of Microsomal Enzymes by Foreign Chemicals and Carcinogenesis by Polycyclic Aromatic Hydrocarbons, *Cancer Res.*, *42*:4875–4917.

Cova, D., Arnoldi, A., Colombo, R. and Rossini, I. (1986). *Toxicol. Lett.*, *30*:273–280.

Cox, P. J., Farmer, P. B., Jarman, M., Jones, J., Stec, W. J., and Kinas, R. (1976). Observations on the Differential Metabolism and Biological Activity of the Optical Isomers of Cyclophosphamide, *Biochem. Pharmacol.*, *25*:993–999.

Dayer, P., Leemann, T., Gut, J., Kronbach, T., Kupfer, A., Francis, R., and Meyer, U. (1985). Steric Configuration and Polymorphic Oxidation of Lipophilic β-Adrenoceptor Blocking Agents: in vitro–in vivo Correlations, *Biochem. Pharmacol.*, *34*:399–400.

Devalia, J. L., Ogilvie, R. L., and McLean, A. E. M. (1982). Dissociation of Cell Death from Covalent Binding of Paracetamol by Flavones in Hepatocyte System, *Biochem. Pharmacol.*, *31*:37–45.

Devanesan, P. D., Cremonesi, P., Nunnally, J. E., Rogan, E. G., and Cavalieri, E. L. (1990). Metabolism and Mutangenicity of Dibenzo[*a,e*]pyrene and the Very Potent Environmental Carcinogen Dibenzo[*a,l*]pyrene, *Chem. Res. Toxicol.*, *3*:580–586.

Dipple, A., Pigott, M. A., Agarwal, S. K., Yagi, H., Sayer, J. M., and Jerina, D. M. (1987). Optically Active Benzo[*c*]phenanthrene Diol Epoxides Bind Extensively to Adenine in DNA, *Nature*, *327*:535–536.

Distlerath, L. M., Reilly, P. E. B., Martin, M. V., Wilkinson, G. R., and Guengerich, F. P. (1985). Immunochemical Characterization of the Human Liver Cytochrome P450 Involved in Debrisoquine Hydroxylation. In *Microsomes and Drug Oxidations* (A. R. Boobis, J. Caldwell, F. De Matters, and C. R. Elcombe, eds.), Taylor & Francis, London, pp. 380–389.

Dohn, D. R., Quebbeman, A. J., Borch, R. F., and Anders, M. W. (1985). Enzymatic Reaction of Chlorotrifluoroethene with Glutathione. ^{19}-NMR Evidence for Stereochemical Control of the Reaction, *Biochemistry*, *24*:5137–5143.

Dostal, L. A., Aitio, A., Harris, C., Bhatia, A. V., Hernandez, O., and Bend, J. R. (1986). Cytosolic Glutathione *S*-Transferases in Various Rat Tissues Differ in Stereoselectivity with Polycyclic Arene and Alkene Oxide Substrates, *Drug Metab. Disposition*, *14*:303–309.

Eichelbaum, M. (1988). Pharmacokinetic and Pharmacodynamic Consequences of Stereoselective Drug Metabolism in Man, *Biochem. Pharmacol.*, *37*:93–96.

Eichelbaum, M. and Gross, A. S. (1990). The Genetic Polymorphism of Debrisoquine/sparteine Metabolism—Clinical Aspects, *Pharmacol. Ther.*, *46*:377–389.

Elfarra, A. A. and Anders, M. W. (1984). Renal Processing of Glutathione-Conjugates, *Biochem. Pharmacol.*, *33*:3729– 3732.

Farmer, P. B. (1988). Enantiomers of Cyclophosphamide and Iphosphamide, *Biochem. Pharmacol.*, *37*:145–152.

Fitzgerald, R. L., Blanke, R. V., Rosecrans, J. A., and Glennon, R. A. (1989). Stereochemistry of the Metabolism of MDMA to MDA, *Life Sci.*, *45*:295–301.

Foster, A. B., Jarman, M., Kinas, R. W., van Maanen, J. M. S., Taylor, G. N., Gaston, J. L., Parkin, A., and Richardson, A. C. (1981). 5-Fluoro- and 5-Chlorocyclophosphamide: Synthesis, Metabolism, and Antitumor Activity of the *cis* and *trans* Isomers, *J. Med. Chem.*, *24*:139–146.

Fritz, S., Lindner, W., Roots, I., Frey, B. M., and Kupfer, A. (1987). Stereochemistry of Aromatic Phenytoin Hydroxylation in Various Drug Hydroxylation Phenotypes in Humans, *J. Pharmacol. Ther.*, *241*:615–622.

Fujimaki, M. and Hakusui, H. (1990). Identification of Two Major Biliary Metabolites of Carvedilol in Rats, *Xenobiotica*, *20*:1025–1034.

Glazenburg, E., Jekel-Halsema, I. M. C., Baranczyk-Kuzma, A., Krijgsheld, K. R., and Mulder, G. J. (1984). D-Cysteine as a Selective Precursor for Inorganic Sulfate in the Rat in vivo. Effect of D-cysteine on the Sulfation of Harmol, *Biochem. Pharmacol.*, *33*:625–628.

Guengerich, F. P. (1991). Oxidation of Toxic and Carcinogenic Chemicals by Human Cytochrome P-450 Enzymes, *Chem. Res. Toxicol.*, *4*:391–407.

Hammock, B., Ratcliff, M., and Schooley, D. A. (1980). Hydration of an 18O Epoxide by a Cytosolic Epoxide Hydrolase from Mouse Liver, *Life Sci.*, *27*: 1635–1641.

Hanzlik, R. P., Edelman, M., Michaely, W. J., and Scott, G. (1976). Enzymatic Hydration of [^{18}O]-Epoxides. Role of Nucleophilic Mechanisms, *J. Amer. Chem. Soc.*, *98*:1952–1957.

Hargus, S. J., Fitzsimmons, M. E., Aniya, Y., and Anders, M. W. (1991). Stereochemistry of Microsomal Glutathione *S*-Transferase Catalyzed Addition of Glutathione to Chlorotrifluoroethene, *Biochemistry*, *30*:717–721.

Hauck, R. S., Wegner, C., Blumtritt, P., Fuhrhop, J. H., and Nau, H. (1990). Asymmetric Synthesis and Teratogenic Activity of (*R*)- and (*S*)-2-ethylhexanoic Acid, a Metabolite of the Plasticizer Di-(2-ethylhexyl)phthalate, *Life Sci.*, *46*:513–518.

Jamali, F., Mehvar, R., and Pasutto, F. M. (1989). Enantioselective Aspects of Drug Action and Disposition: Therapeutic Pitfalls, *J. Pharm. Sci.*, *78*:695–715.

Jerina, D. M., Sayer, J. M., Yagi, H., van Bladeren, P. J., Thakker, D. R., Levin, W., Chang, R. L., Wood, A. W., and Conney, A. H. (1985). In *Microsomes and Drug Oxidations* (A. R. Boobis, J. Caldwell, F. de Mattheis, and R. Elcombe, eds.), Taylor & Francis, London, pp. 310–319.

Jernström, B., Mariner, H., Meyer, D. J., and Ketterer, B. (1985). Glutathione Conjugation of the Carcinogenic and Mutagenic Electrophile (+/−)-7 beta, 8 alpha-dihydroxy-9 alpha, 10 alpha-oxy-7,8,9,10-tetra hydrobenzo[a]pyrene Catalyzed by Purified Rat Liver Glutathione Transferases, *Carcinogenesis*, *6*: 85–93.

Joseph, A. E. A., Matlin, S. A., and Knox, P. (1986). Cytotoxicity of Enantiomers of Gossypol, *Brit. J. Cancer*, *54*:511–518.

Kadlubar, F. F. and Hammons, G. J. (1987). The Role of Cytochrome P450 in the Metabolism of Chemical Carcinogens, In *Mammalian Cytochromes P-450* (F. P. Guengerich, ed.), Vol. II, CRC Press, Boca Raton, Florida, pp. 81–130.

Kaminsky, L. S., Dunbar, D. A. Wang, P. P., Beaune, P., Larrey, D., Guengerick, F. P., Schnellmann, R. G., and Sipes, I. G. (1984). Human Hepatic Cytochrome P450 Composition as Probed by in vitro Microsomal Metabolism of Warfarine, *Drug Metab. Disposition*, 12:470–476.

Koster, A. S., Frankhuijzen-Sierevogel, A. C., and Mentrup, A. (1986). Stereoselective Formation of Fenoterol-*para*-Glucuronide and Fenoterol-*meta*-Glucuronide in Rat Hepatocytes and Enterocytes, *Biochem. Pharmacol., 35*: 1981–1985.

Koymans, L. M. H., Vermeulen, N. P. E., van Acker, S. A. B. E., te Koppele, J. M., Heykants, J. J. P., Lavrijsen, K., Meuldermans, W., and Donné-Op den Kelder, G. M. (1992). A Predictive Model for Substrates of Cytochrome P450-Debrisoquine (2D6), *Chem. Res. Toxicol., 5*:211–219.

Küpfer, A., Roberts, R. K., Schenker, S., and Branch, R. A. (1981). Stereoselective Metabolism of Mephenytoin in Man, *J. Pharmacol. Exp. Ther., 218*:193–199.

Küpfer, A., Patwardhan, R., Ward, S., Schenker, S., Preisig, R., and Branch, R. A. (1984). Stereoselective Metabolism and Pharmacogenetic Control of 5-Phenyl-5-ethylhydantoin Hydroxylation in Man, *J. Pharmacol. Exp. Ther., 230*:28–33.

Maguire, J. H., Butler, T. C., and Dudley, K. C. (1980). Absolute Configuration of the Dihydrodiol Metabolites of 5,5-Diphenylhydantoin (Phenytoin) from Rat, Dog and Human Urine, *Drug Metab. Disposition, 8*:325–331.

Mangold, J. B. and Abdel-Monem, M. M. (1983). Stereochemical Aspects of Conjugation Reactions Catalyzed by Rat Liver GSH-Transferase Isozymes, *J. Med. Chem., 26*:66–73.

Mannervik, B. and Danielson, U. H. (1988). GSH-Transferases—Structure and Catalytic Function, *CRC Crit. Rev. Biochem., 23*:293–337.

Meese, C. O., Fisher, C., and Eichelbaum, M. (1988). Stereoselectivity of the 4-Hydroxylation of Debrisoquine in Man, Detected by GC-MS, *Biomed. Environ. Mass Spectrom., 15*:63–71.

Meyer, U. A., Gut, J., Kronback, T., Scoda, C., Meier, U. T., Catin, T., and Dayer, P. (1986). The Molecular Mechanism of Two Common Polymorphisms of Drug Oxidation—Evidence for Functional Changes in Cytochrome P450 Isozymes Catalyzing Bufuralol and Mephenytoin Oxidation, *Xenobiotica, 16*:449–464.

Meyer, U. A., Skoda, R. C., and Zanger, U. M. (1990). The Genetic Polymorphism of Debrisoquine/sparteine Metabolism—Molecular Mechanisms. *Pharmacol. Ther., 46*:297–310.

Morgenstern, R., Guthenberg, C., Mannervik, B., and Depierre, J. W. (1983). The Amount and Nature of GSH-Transferases in Rat Liver Microsomes Determined by Immunochemical Methods, *FEBS Lett., 160*:264–268.

Mulder, G. J., Coughtrie, M. W. H., and Burchell, B. (1990). Glucuronidation, In *Conjugation Reactions in Drug Metabolism: An Integrated Approach* (G. J. Mulder, ed.), Taylor & Francis, London, pp. 51–107.

Nagasawa, H. T., Coon, D. J. W., Muldoon, W. P., and Zera, R. T. (1984). 2-Substituted thiazolidine-4(*R*)-Carboxylic Acids as Prodrugs of L-Cysteine. Protection of Mice Against Acetaminophen Hepatotoxicity, *J. Med. Chem.*, *s27*: 591–596.

Ockzenfels, H., Köhler, F., and Meise, W. (1976). Teratogene Wirkung und Stereospezifität eines Thalidomid-Metaboliten, *Pharmazie*, *31*:492–493.

Oesch, F. (1979). Epoxide Hydratase, In *Progress in Drug Metabolism, Part 3* (J. W. Bridges and L. F. Chasseaud, eds.), Wiley, Chichester, pp. 253–302.

Ortiz de Montellano, P. R. (1989). Cytochrome P-450 Catalysis: Radical Intermediates and Dehydrogenation Reactions, *Trends Pharmacol. Sci.*, *10*:354–360.

Ortiz de Montellano, P. R., Mangold, B. L., Wheeler, C., Kunze, K. L., and Reich, N. O. (1983). Stereochemistry of Cytochrome P450 Catalyzed Epoxidation and Prosthetic Heme Alkylation, *J. Biol. Chem.*, *258*:4208–4216.

Pickett, C. B. and Lu, A. Y. H. (1989). GSH-*S*-Transferases: Gene Structure, Regulation and Biological Function, *Ann. Rev. Biochem.*, *58*:743–764.

Relling, M. V., Ayoma, T., Gonzales, F. J., and Meyer, U. A. (1990). Tolbutamide and Mephenytoin Hydroxylation by Human Cytochrome-P450s in the CYP2Cu Subfamily, *J. Pharmacol. Exp. Ther.*, *252*:442–447.

Robertson, I. G. C., Guthenberg, C., Mannervik, B., and Jernström, B. (1986). Differences in Stereoselectivity and Catalytic Efficiency of Three Human GSH-Transferases in the Conjugation of GSH with 7β_1,8α-Dihydroxy-9α,10α-oxy-7,8,9,10-tetrahydrobenzo[a]pyrene, *Cancer Res.*, *46*:2220–2224.

Rojas, M. and Alexandrov, K. (1986). In vivo Formation and Persistence of DNA Adducts in Mouse and Rat Skin Exposed to Benzo[a]pyrene-dihydrodiols and -Dihydrodiolepoxides, *Carcinogenesis*, *7*:1553–1556.

Sayer, J. M., Yagi, H., van Bladeren, P. J., Levin, W., and Jerina, D. M. (1985). Stereoselectivity of Microsomal Epoxide Hydrolase Toward Diol Epoxides and Tetrahydroepoxides Derived from Benz[a]anthracene, *J. Biol. Chem.*, *260*:1630–1640.

Schill, G., Wainer, I. W., and Barkan, S. A. (1986). Chiral Separation of Cationic Drugs on an α_1-Acid Glycoprotein Bonded Stationary Phase, *L. Liq. Chromatogr.*, *9*:641–666.

Schmahl, H. J., Heger, W., and Nau, H. (1989). The Enantiomers of the Teratogenic Thalidomide-Analogue EM 12; Chemical Stability, Stereoselective Metabolism and Renal Excretion in the Marmoset Monkey, *Toxicol. Lett.*, *45*:23–33.

Schmidt, C. J. (1987). Neurotoxicity of the Psychedelic Amphetamine, Methylene-Dioxymethamphetamine, *J. Pharmacol. Exp. Ther.*, *240*:1–7.

Seiler, J. P. (1990). Chirality-Dependent DNA-Reactivity as the Possible Cause of the Differential Mutagenicity of the Two Components in an Enantiomeric Pair of Epoxides, *Mutat. Res.*, *245*:165–169.

Singh, G. and Koropatnick, J. (1988). Differential Toxicity of *cis* and *trans* Isomers of Dichlorodiammineplatinum, *J. Biochem. Toxicol.*, *3*:223–233.

Sisenwine, S. F., Tio, C. O., Hadley, F. V., Lin, A. L., Kimmel, H. B., and Ruelius, H. W. (1982). Species-related differences in the stereoselective glucuronidation of oxazepam. *Drug Metab. Disp.*, *10*:605–608.

Speirs, C. J., Murray, S., Davies, D. S., Mabadeje, A. F. B., and Boobis, A. R. (1990).

Debrisoquine Oxidation Phenotype and Susceptibility to Lung Cancer, *Brit. J. Clin. Pharmacol.*, *29*:101–109.

Swinney, D. C., Ryan, D. E., Thomas, P. E., and Levin, W. (1987). Regioselective Progesterone Hydroxylation Catalyzed by Eleven Rat Hepatic Cytochrome P-450 Isozymes, *Biochemistry*, *26*:7073–7083.

te Koppele, J. M. (1988). Pharmacokinetics and Stereoselectivity of GSH-Conjugation, Ph.D. dissertation, State University of Leiden, The Netherlands.

te Koppele, J. M., Dogterom, P., Vermeulen, N. P. E., Meijer, D. K. F., van der Gen, A., and Mulder, G. J. (1987). α-Bromoisovalerylurea as Model Substrate for Studies on Pharmacokinetics of GSH-Conjugation in the Rat. II. Pharmacokinetics and Stereoselectivity of Metabolism and Excretion in vivo and in Perfused Liver, *J. Pharmacol. Exp. Ther.*, *239*:905–914.

te Koppele, J. M. and Mulder, G. M. (1991). Stereoselective Glutathione Conjugation by Subcellular Fractions and Purified Glutathione *S*-Transferases, *Drug Metab. Rev.*, *23*:331–354.

Testa, B. (1986). Chiral Aspects of Drug Metabolism, *TiPS*, *7*:60–64.

Testa, B. (1988). Substrate and Product Stereoselectivity in Monooxygenase-Mediated Drug Activation and Inactivation, *Biochem. Pharmacol.*, *37*:85–92.

Thakker, D. R., Levin, W., Yagi, H., Conney, A. H., and Jerina, D. H. (1982). In *Biological Reactive Intermediates—II. Part A* (R. Snyder, D. Parker, J. Kocsis, D. Jollow, G. Gibson, and C. Witmer, eds.), Plenum Press, New York, pp. 525–539.

Thijssen, H. H. W. and Baars, L. G. M. (1987). The Biliary Excretion of Acenocoumarol in the Rat: Stereochemical Aspects, *J. Pharm. Pharmacol.*, *39*: 655–657.

van Bladeren, P. J. (1988). Stereoselectivity in the Biotransformation of Crop Protectants, In *Stereoselectivity of Pesticides; Biological and Chemical Problems* (E. J. Ariëns, J. J. S. van Rensen, and W. Welling, eds.), Elsevier, Amsterdam, pp. 357–374.

van Bladeren, P. J., van der Gen, A., Breimer, D. D., and Mohn, G. R. (1979). Stereoselective Activation of Vicinal Dihalogen Compounds by GSH-Conjugation, *Biochem. Pharmacol.*, *28*:2521–2524.

van Bladeren, P. J., Breimer, D. D., van Huijgenvoort, I. A. T. C. M., Vermeulen, N. P. E., and van der Gen, A. (1981). The Metabolic Formation of *N*-acetyl-*S*-2-hydroxyethyl-L-cysteine from D_4-1,2-Dibromoethane: Relative Importance of Oxidation and GSH-Conjugation in vivo, *Biochem. Pharmacol.*, *30*:2499–2505.

van Bladeren, P. J., Sayer, J. M., Ryan, D. E., Thomas, P. E., Levin, W., and Jerina, D. M. (1985). Differential Stereoselectivity of Cytochromes P-450*b* and P-450*c* in the Formation of Naphthalene and Anthracene 1,2-oxides, *J. Biol. Chem.*, *260*:10,226–10,235.

van der Goot, H. and Timmerman, H. (1988). Sterical Isomerism: One of the Determinant Factors for Chemical and Biological Properties. In *Stereoselectivity of Pesticides: Biological and Chemical Problems. Chemicals in Agriculture,*

Vol. 1 (E. J. Ariëns, J. J. S. van Rensen, and W. Welling, eds.), Elsevier, Amsterdam, pp. 11–39.

van der Graaff, M., Vermeulen, N. P. E., Hofman, P. H., Breimer, D. D., Knabe, J., and Schamber, L. (1985). Mass Fragmentographic Assay of Pseudoracemic Hexobarbital and Its Metabolites in Blood and Urine of the Rat, *Biomed. Mass Spectrom.*, *12*:464–469.

van der Graaff, M., Vermeulen, N. P. E., and Breimer, D. D. (1988). Disposition of Hexobarbital: Fifteen Years of an Intriguing Model Substrate, *Drug Metab. Rev.*, *19*:109–164.

Vermeulen, N. P. E. (1986). Stereoselective Biotransformation: Its Role in Drug Disposition and Drug Action, In *Innovative Approaches in Drug Research* (A. F. Harms, ed.), Elsevier, Amsterdam, pp. 393–416.

Vermeulen, N. P. E. (1989). Stereoselective Biotransformation and Its Toxicological Implications, In *Xenobiotic Metabolism and Disposition* (R. Kato, R. W. Estabrook, and M. N. Cayen, eds.), Taylor & Francis, London, pp. 193–206.

Vermeulen, N. P. E., Cauvet, J., Luijten, W. C. M. M., and van Bladeren, P. J. (1980). Electron Impact and Chemical Ionization Mass Spectrometry of *cis*- and *trans*-*S*-(2-Hydroxycyclohexyl)-*N*-acetyl-L-cysteinemethyl Esters, *Biomed. Mass Spectrom.*, *7*:413–417.

Vermeulen, N. P. E. and Breimer, D. D. (1983). Stereoselectivity in Drug and Xenobiotic Metabolism, In *Stereochemistry and Biological Activity of Drugs* (E. J. Ariens, W. Soudijn, and P. B. W. M. M. Timmermans, eds.), Blackwell Scientific Publications, Oxford, pp. 33–54.

Vermeulen, N. P. E. and Testa, B. (1988). Principles of Stereoselective Analysis of Pesticides, In *Stereoselectivity of Pesticides; Biological and Chemical Problems* (E. J. Ariëns, J. J. S. van Rensen, and W. Welling, eds.), Elsevier, Amsterdam, pp. 375–407.

Wade, D. R., Airy, S. C., and Sinsheimer, J. E. (1978). Mutagenicity of Aliphatic Epoxides, *Mutat. Res.*, *58*:217–223.

Walle, T. (1985). Stereochemistry of the *in vivo* disposition and metabolism of propranolol in dog and man using deuterium-labeled pseudoracemates. *Drug Metab. Disp.*, *13*:279–282.

Walle, T. and Walle, U. K. (1990). Stereoselective Sulphate Conjugation of Racemic Terbutaline bu Human Liver Cytosol, *Brit. J. Clin. Pharmacol.*, *30*:127–133.

Watabe, T., Hiratsuka, A., Ogura, K., and Endoh, K. (1985). A Reactive Hydroxymethyl Sulfate Ester Formed Regioselectively from the Carcinogen 7,12-Dihydroxymethyl-benz[a]Anthracene, by Rat Liver Sulfotransferase, *Biochem. Biophys. Res. Commun.*, *131*:694–702.

Waxman, D. J. (1988). Interactions of Hepatic Cytochromes P-450 with Steroid Hormones. Regioselectivity and Stereospecificity of Steroid Metabolism and Hormonal Regulation of Rat P-450 Enzyme Expression, *Biochem. Pharmacol.*, *37*:71–84.

Waxman, D. J., Light, D. R., and Walsh, C. (1982). Chiral Sulfoxidations Catalyzed by Rat Liver Cytochromes P-450, *Biochemistry*, *21*:2499–2507.

White, R. E., Miller, J. P., Favreau, L. V., and Bhattacharyya, A. (1986). Stereochemical Dynamics of Aliphatic Hydroxylation by Cytochrome P-450, *J. Amer. Chem. Soc.*, *108*:6024–6031.

Williams, K. M. (1990). Enantiomers in Arthritic Disorders, *Pharmacol. Ther.*, *46*:273–295.

Wood, A. W., Ryan, D. E., Thomas, P. E., and Levin, W. (1983). Regio- and Stereoselective Metabolism of Two C_{19} Steroids by Five Highly Purified and Reconstituted Rat Hepatic Cytochrome P-450 Isozymes, *J. Biol. Chem.*, *258*: 8839–8847.

Wood, A. W., Chang, R. L., Kumar, S., Shirai, N., Jerina, D. M., Lehr, R. E., and Conney, A. H. (1986). Bacterial and Mammalian Cell Mutagenicity of Four Optically Active Bay-Region 3,4-Diol-1,2-epoxides and Other Derivatives of the Nitrogen Heterocycle Dibenz[c,h]acridine, *Cancer Res.*, *46*:2760–2766.

Yacobi, A., Lai, Chii-Ming, and Levy, G. (1984). Pharmacokinetic and pharmacodynamic studies of acute interaction between Warfarin enantiomers and Metronidazol. *J. Pharmacol. Exp. Ther.*, *231*:72–79.

Yang, S. K. (1988). Stereoselectivity of Cytochrome P-450 Isozymes and Epoxide Hydrolase in the Metabolism of Polycyclic Aromatic Hydrocarbons, *Biochem. Pharmacol.*, *37*:61–70.

Yang, S. K. and Bao, Z.-P. (1987). Stereoselective Formations of K-Region and Non-K-Region Epoxides in the Metabolism of Chrysene by Rat Liver Microsomal Cytochrome P-450 Isozymes, *Mol. Pharmacol.*, *32*:73–80.

Yost, G. S. and Finley, B. L. (1985). Stereoselective Glucuronidation as a Probe of Induced Forms of UDP-Glucuronydyltransferase in Rabbits, *Drug Metab. Disposition*, *13*:5–8.

Zhang, P. and Armstrong, R. N. (1990). Construction, expression and preliminary characterisation of chimeric class υ glutathione-S-transferases with altered catalytic properties, *Biopolym.*, *29*:159–169.

Ziegler, D. M. (1991). Unique Properties of the Enzymes of Detoxication, 1990 Bernard B. Brodie Award Lecture, *Drug Metab. Disposition*, *19*:847–852.

10

Stereoselective Transport of Drugs Across Epithelia

Ronda J. Ott and Kathleen M. Giacomini *University of California, San Francisco, San Francisco, California*

I. INTRODUCTION

Drug absorption and disposition are regulated, in part, by transport across epithelial barriers. For example, both the skin and the gastrointestinal tract consist of epithelial barriers important to drug absorption. The major organs responsible for the elimination of drugs, the liver and kidney, contain epithelial cells that play essential roles in drug elimination. The renal tubules are composed of monolayers of epithelial cells that function in the active secretion and reabsorption of endogenous substances as well as xenobiotics. In recent years, it has become increasingly clear that the liver, widely recognized as the major organ responsible for drug metabolism, is an important epithelia in drug transport. Transport systems in the canalicular and sinusoidal membrane mediate drug transport into the bile. Drug distribution may also involve transport across epithelial cells. For example, distribution into the cerebrospinal fluid may involve transport systems in the choroid plexus epithelium. Transport across the specialized capillaries of the blood brain barrier is important in the distribution of drugs to the brain. The placenta is an epithelia and transport of drugs into the fetus necessarily involves placental transport. Drug excretion into the milk occurs via the mammary gland, a specialized secretory epithelium.

In the past decade, our knowledge of transport mechanisms in epithelial cells has increased dramatically. In the 1980s considerable information about the kinetics of many transport processes was obtained. Recently, molecular biology techniques have been used to clone several important transport proteins for endogenous substances and it is likely that in the next decade other transport proteins will be cloned and sequenced. With

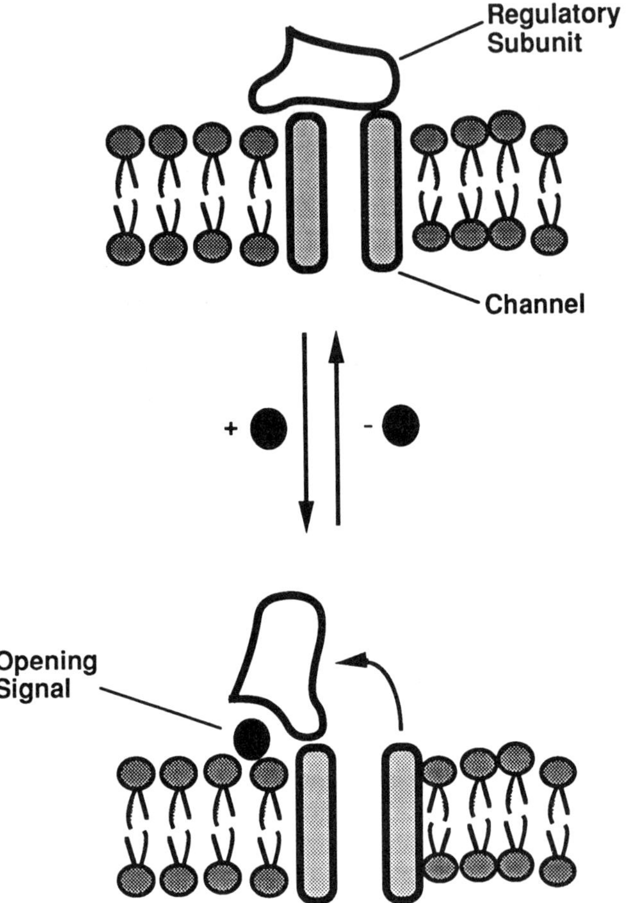

FIGURE 2 Model of an ionic channel and a regulatory subunit. Addition of an opening signal induces a conformational shift in the regulatory subunit, exposing the channel. [Adapted from Dusinsky (3).]

is an example of primary active systems (1,2). These systems are normally termed pumps, such as the Na^+K^+ ATPase pump and the H^+ ATPase pump (Fig. 6).

Secondary active transport mechanisms involve the flow of one substance against its concentration gradient coupled to the flow of another substrate down its concentration gradient. The substance moving down its concentration gradient supplies the energy, allowing the other substance to be transported against its concentration gradient. Direct evidence of such a

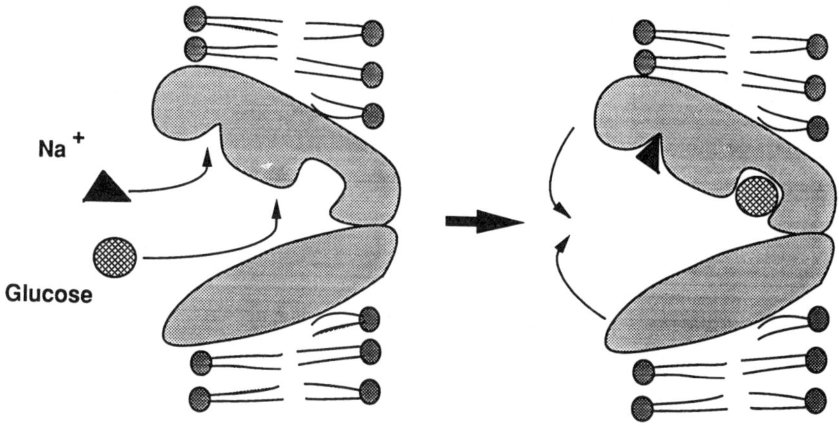

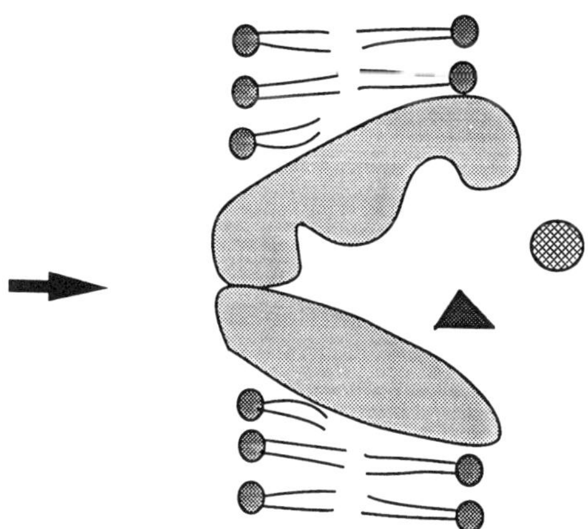

FIGURE 3 Model of the sodium-coupled glucose cotransporter. Binding site for both sodium and glucose is alternatively exposed at opposite membrane surfaces. [Adapted from Silverman (4).]

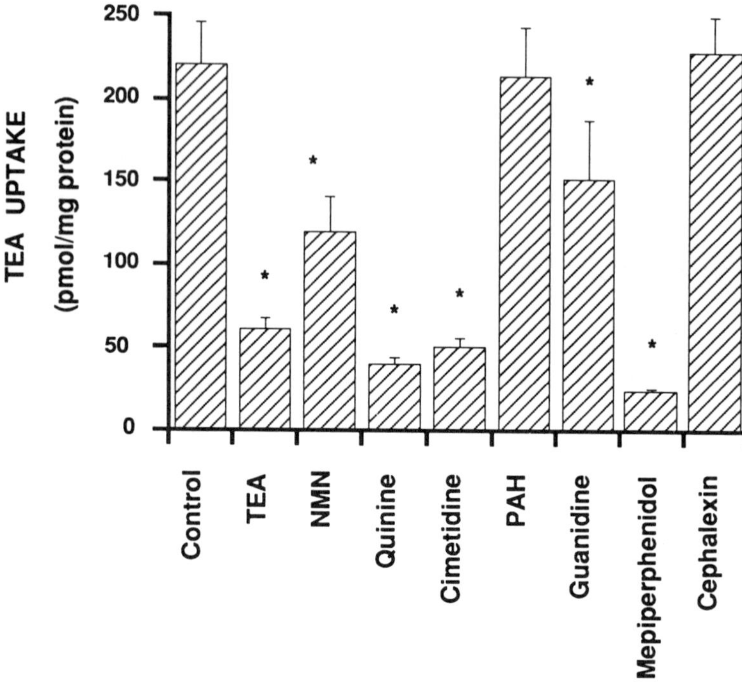

FIGURE 4 Effect of various compounds on the uptake of TEA in OK cell monolayers in the absence (control) and presence of TEA, NMN, quinine, cimetidine, *p*-aminohippuric acid (PAH), guanidine, mepiperphenidol, or cephalexin (all at 100 μM). The asterisk indicates results significantly different from the control. [Reprinted with permission from Yuan et al. (5).]

mechanism can be determined in studies in isolated plasma membrane vesicles. The organic cation tetraethylammonium (TEA) in human renal brush border membrane vesicles accumulates over time, reaching an apparent equilibrium in 60 min (Fig. 7) (7). When an outwardly directed proton gradient is present (pH in = 6.0, pH out = 7.4), the accumulation of TEA is enhanced transiently, producing an "overshoot" above its equilibrium value. This "overshoot phenomenon," reflecting the transient accumulation of TEA in response to the proton gradient, occurs because of the direct coupling of the downhill transport of protons to the uphill transport of TEA.

Many sodium-coupled systems located in the proximal tubule of the kidney exhibit an analogous mechanism (see Fig. 3). When the transport of both substances, the substrate going uphill and the coupled substrate going downhill, is flowing in the same direction, the transporter is termed

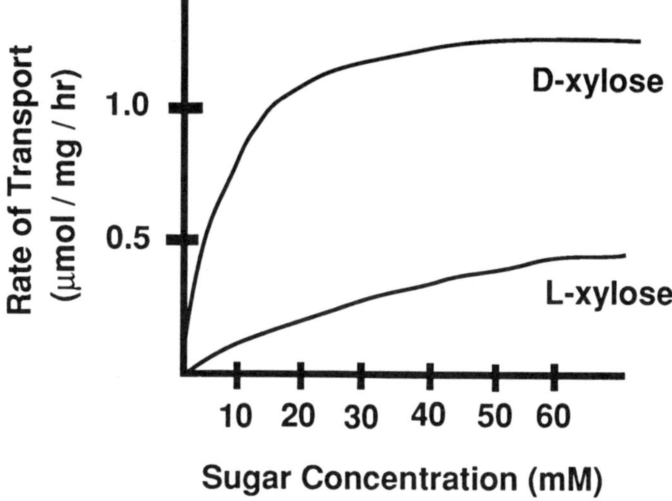

FIGURE 5 Kinetics of transport of D- and L-xylose in the yeast *Rhodotorula gracilis*. [Adapted from Dusinsky (1).]

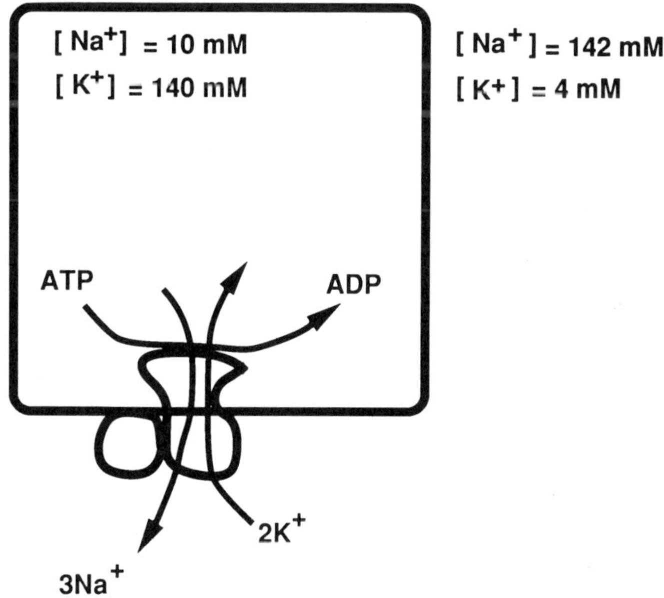

FIGURE 6 Model of the Na^+K^+ATPase pump showing the coupling ratio of Na^+ and K^+. Intracellular and extracellular concentrations of Na^+ and K^+ normally maintained by the pump are included.

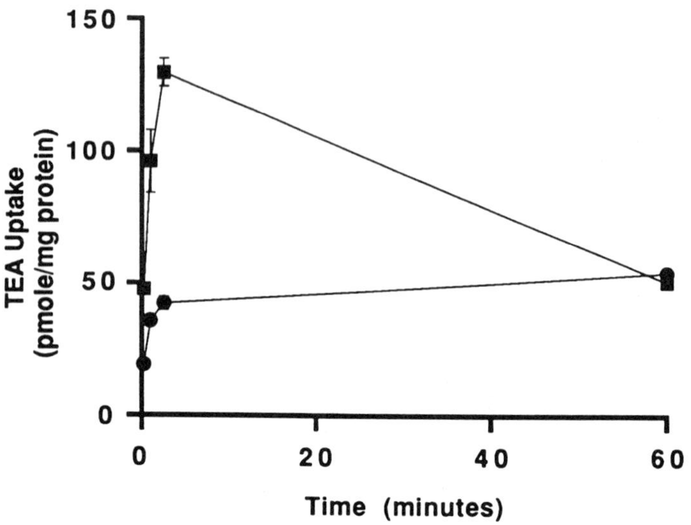

FIGURE 7 TEA uptake into human brush border membrane vesicles in the presence of an outwardly directed proton gradient (square) and in the absence of a proton gradient (circle). [Reprinted with permission from Ott et al. (7).]

a cotransporter. When the substances are transported in opposite directions across the membrane, the transporter is termed an antiporter. The passive facilitated glucose transporter in the red blood cells mentioned previously, involving a single substrate, is called a uniporter (1,2). Except for the requirement of energy, active transport systems have the same properties of facilitated diffusion systems.

For enantiomeric recognition by a protein to occur, the substrate must bind to the protein at three or more sites (8). Fewer than three interactions will not promote a stereoselective binding event (Fig. 8). When the protein of interest is a transporter, there are several stages of the transport process that could exhibit stereoselectivity (Eq. 3). Although stereoisomers would not affect the number of transporters, the J_{max} could be affected in a stereoselective manner if enantiomers produce differences in the turnover number or translocation rate of the transporter. The complex rate constant K_t, which is a function of the rate of formation and rate of breakdown of the substrate-transporter complex along with the rate of translocation of the complex, can be severely affected by stereoisomeric structural changes. First, the binding affinity may differ between enantiomers when a three-point binding interaction is involved. Second, the rate of release of the substrate from the complex may differ between enantiomers. Finally, the

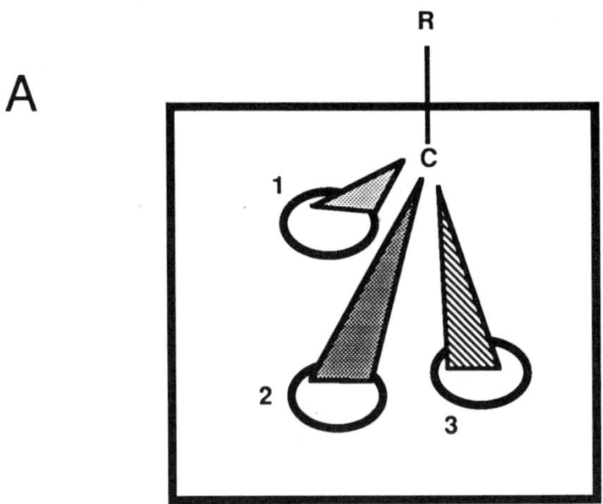

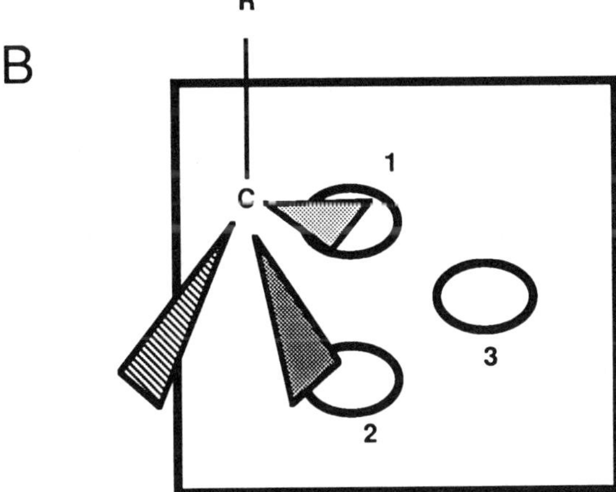

FIGURE 8 Model of substrate binding to protein showing a proper three-point binding interaction (A) and of the enantiomer showing an improper three-point binding interaction (B). [Adapted from Christensen (8).]

ability of the protein to change its conformation may be different for two enantiomers after substrate binding, thereby affecting the translocation rate of the transporter.

III. PROPERTIES OF EPITHELIAL CELLS

A major function of epithelia is to transport substances from one body cavity to another, causing the net movement of many essential ions, nutrients, and potential toxins. For this net vectorial transport to occur, the single layer of epithelial cells lining the body's cavities must have polarized membranes exhibiting different transport mechanisms. The mucosal membrane (also termed brush border, luminal, or apical) faces the outside of the body or cavity and usually has many fingerlike projections or microvilli (Fig. 9). These membrane processes increase the surface area of the cell membrane dramatically and thus facilitate the transfer of large quantities of substances. The serosal membrane (also termed basolateral, peritubular, or antiluminal) faces the blood and lies on a structural support of glycoproteins and collagen (basement membrane) that is not a significant barrier for the movement of substances. The two membrane faces are separated physically by tight junction proteins (zona occludens) that appear to prevent the movement of transmembrane proteins between the two polarized faces (9). For vectorial transport across epithelia, the substance must enter the cell, diffuse across it, and exit at the other membrane. Net movement from the basolateral through the apical membrane is called secretion and from the apical to the basolateral is referred to as absorption or reabsorption. For net flux of a nonmetabolized substance, the transport mechanisms at the two membranes must be different and active at one or both membranes. The commonly used scheme is called a pump-leak arrangement (10). For example, a substance is actively transported across one membrane into the cell, which leads to a concentration gradient across the other membrane. The exit membrane contains a transporter that allows passive facilitated diffusion out of the cell. Alternatively, simple diffusion of the substance out of the cell down its concentration gradient occurs. Cellular metabolism of the substance can also occur, complicating the above scheme. For many drugs, cellular metabolism may be important and stereoselective drug "transport" may actually represent stereoselective drug metabolism.

The major important organic electrolytes and nonelectrolytes transported by epithelial cells include sugars, amino acids, nucleosides, organic cations, and organic anions. Transport systems have significant implications for the absorption, distribution, elimination, and pharmacokinetic properties of many clinically important drugs. The major epithelial tissues

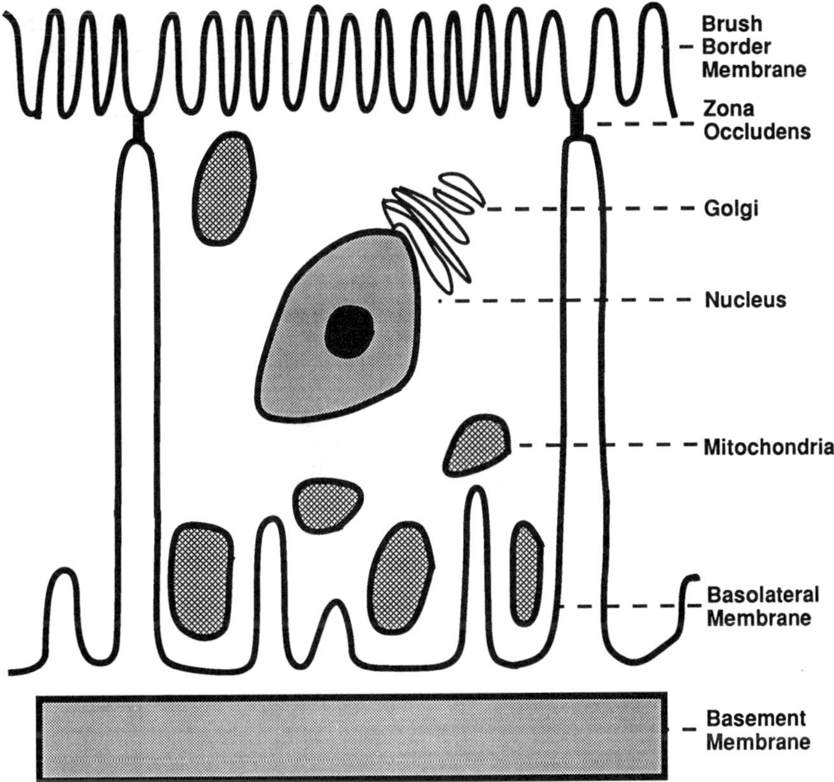

FIGURE 9 Epithelial cell showing the polarized membranes and cellular components.

important in drug absorption and disposition include the kidney, the entire gastrointestinal tract, the choroid plexus that lines the ventricles of the brain, placenta, liver hepatocytes, and the linings of the lungs (10). The remainder of this chapter will focus on the transporters located in the kidney and how the use of stereoselective mechanisms promote the specific transport of organic substances. Some discussion of transport in other epithelia is also included.

IV. KIDNEY

The major functions of the kidney include the regulation of water, electrolyte and nutrient balance of the body, and the excretion of metabolic waste products and foreign substances. These functions are carried out by a series of complex interactions involving the filtration, reabsorption, and secretion of plasma water and its constituents. The major structural and functional unit of the kidney is the nephron (Fig. 10).

Each kidney is composed of approximately 1 million nephrons (9). A complete nephron consists of a glomerular capillary network, Bowman's capsule, proximal tubule, loop of Henle, distal tubule, and collecting duct. The glomerular capillary network and Bowman's capsule are responsible for the formation of the initial ultrafiltrate of the blood. The plasma water and all non-protein-bound constituents are forced through the capillary network and into the Bowman's space formed by the capsule. Most plasma proteins and red blood cells are usually retained in the capillary. The rate of plasma filtered by the glomeruli is referred to as the glomerular filtration rate (GFR) and in a normal healthy adult is usually around 125 mL/min (9). The final urine flow averages 1 mL/min (9). Therefore, over 99% of the filtered plasma is reabsorbed in its passage through the kidney tubules. The ultrafiltrate composition changes dramatically as it flows through the

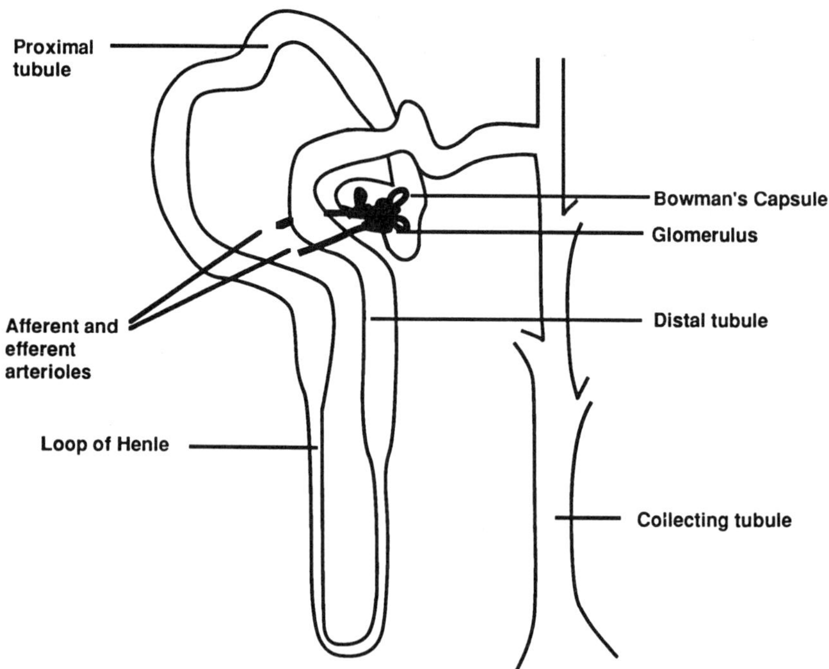

FIGURE 10 Diagram of a nephron.

extensive tubule system of the nephron. Based on the requirements of the body at any specific time, the exiting urine can be diluted to 50 mOsm/kg H_2O or concentrated to 1200 mOsm/kg H_2O, suggesting that the secretion and reabsorption of inorganic ions is an active and highly regulated process (9).

The renal excretion rate of a compound is determined by a combination of three factors: filtration at the glomeruli, tubular reabsorption, and tubular secretion (11) and is described by the following equation:

$$\text{renal excretion rate} = \text{rate of filtration} - \text{rate of reabsorption} + \text{rate of secretion} \tag{4}$$

and the renal clearance of a compound is defined by the following equation:

$$\text{renal clearance} = \frac{\text{renal excretion rate}}{\text{plasma concentration of compound}} \tag{5}$$

Therefore, the renal clearance of a compound is

$$\frac{\text{rate of filtration} - \text{rate of reabsorption} + \text{rate of secretion}}{\text{plasma concentration}} \tag{6}$$

The rate of filtration for a substance is equal to GFR*fu*C, where GFR is the glomerular filtration rate, fu the fraction of the drug unbound in the plasma, and C the plasma concentration (11). The GFR can be experimentally determined by measuring the renal clearance of inulin, an exogenous polysaccharide that is not bound to plasma proteins ($fu = 1$) and is not reabsorbed, secreted, or metabolized by the tubules of the kidney. Therefore, the renal clearance of inulin is equal to the GFR (11). By comparing the renal clearance of a drug to its clearance by filtration (GFR*fu), one can determine if the compound undergoes net reabsorption or secretion by the kidney. Therefore, if the clearance of a drug is lower than the GFR*fu, net reabsorption must have occurred. This does not exclude the possibility that simultaneous secretion of the drug also occurred. Conversely, if the clearance of a drug is higher than GFR*fu, the compound must be actively secreted by the kidney (simultaneous reabsorption is also possible, but secretion must predominate). The fact that many organic nutrients such as glucose (which is present in plasma and not bound to plasma proteins) are almost completely absent from the urine suggests active reabsorption by the kidney. Therefore, reabsorption and secretion of compounds may be active or passive, and the processes may occur simultaneously and in different sections of the nephron. For example, potassium is usually reabsorbed in the proximal tubule and the loop of Henle, whereas it may be secreted in the distal tubule and reabsorbed again in the collecting duct (9).

The renal clearance of two enantiomers will be different if any of the abovementioned processes exhibit stereoselectivity (12–14). The filtration clearance (GFR*fu) will reflect differences between enantiomers if a compound is bound stereoselectively to plasma proteins (12–14). The rate of reabsorption or secretion of enantiomers can be different if transport proteins at either the basolateral or brush border membrane exhibit stereoselective kinetics. Stereoselective cellular metabolism is a well described process and is dealt with in depth elsewhere in this book. However, the possibility of stereoselective metabolism occurring within the renal cells cannot be ignored as a potential mechanism responsible for differing renal clearance between two enantiomers.

Many methodologies have been developed to study transport processes in the renal tubules. The total renal clearances of various substances can be determined in human or whole animal studies and in isolated perfused kidneys. Transport systems along the nephron are usually located by stop-flow or microperfusion studies. Isolated perfused or nonperfused tubules can be used to determine specific transport mechanisms at individual membranes. For example, in the nonperfused tubule that has a collapsed lumen, transport reflects the process occurring at the basolateral membrane only. In the last decade, isolated plasma membrane vesicles, either of the brush border or basolateral membrane, have been widely used to study transport mechanisms at each of the polarized membranes. Vesicles are excellent for studies of drug transport since potential problems related to the metabolism of substances by the whole cells are eliminated. The use of primary or continuous renal cell lines is particularly suited to the study of the effects of long-term exposure to regulatory or toxic agents on transport proteins.

The proximal tubule is responsible for the reabsorption of over 65% of the sodium and water of the glomerular ultrafiltrate (9). Most if not all the organic nutrients such as glucose and amino acids are also reabsorbed in the proximal tubule (9). It has been experimentally determined that the renal handling of exogenous substances such as organic cations and organic anions that are generally secreted also occurs within the proximal tubule (15). Therefore, the rest of this section will be concerned with the stereoselective handling of organic substance within the proximal tubule.

A. Monosaccharides

Glucose is almost completely and actively reabsorbed within the first third of the proximal tubule of the kidney (9). Because of this and the importance of glucose and its urinary excretion in the monitoring of diabetes, much work has been conducted on the specific mechanisms responsible for glucose reabsorption. The transport of glucose across the brush border membrane is active and coupled to the downhill movement of Na^+ across the membrane by the Na^+-glucose cotransporter (Figs. 11 and 12), with the Na^+ gradient supplied by the Na^+K^+ATPase pump (16). The transport of glucose down its concentration gradient across the basolateral membrane into the blood is thought to occur by a passive uniporter similar to that seen in red blood cells (Fig. 12) (17).

This pump-leak arrangement for glucose is so efficient that only in cases such as diabetes, does any glucose escape reabsorption. The Na^+-glucose cotransporter is selectively inhibited by phloridzin (18–25), whereas the glucose transporter at the basolateral membrane is selectively inhibited by phloretin (16). Many studies have been conducted on the specificity of the two transporters, and for both transporters a D-glucopyranose ring in the chair conformation with the C1 hydroxyl in the equatorial position is required (18–25). In addition, the brush border membrane cotransporter appears to have essential requirements for free C2 and C3 hydroxyl in the equatorial positions (Fig. 13) (26).

The stereoselective requirements for the transport of sugars by these transporters are quite restrictive. The D isomers of mannose, fructose, and galactose appear to interact with both systems; however, only D-galactose, a methyl-D-glucoside, and 6-deoxy-D-glucose appear to be transported

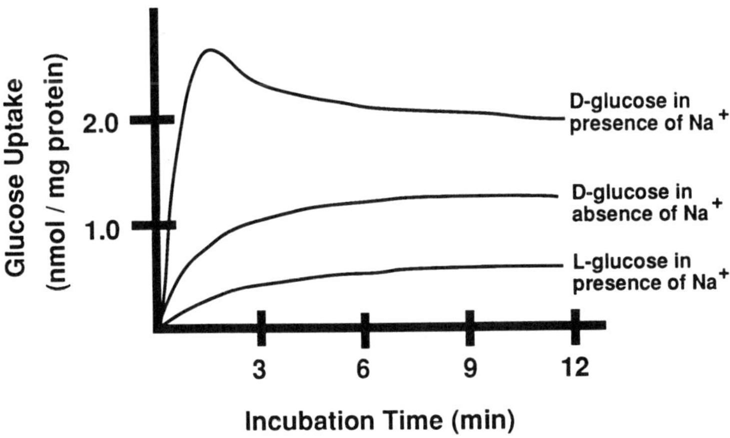

FIGURE 11 Uptake of D- and L-glucose in rat renal brush border membrane vesicles in the presence or absence of a sodium gradient [Adapted from Kinne et al. (16).]

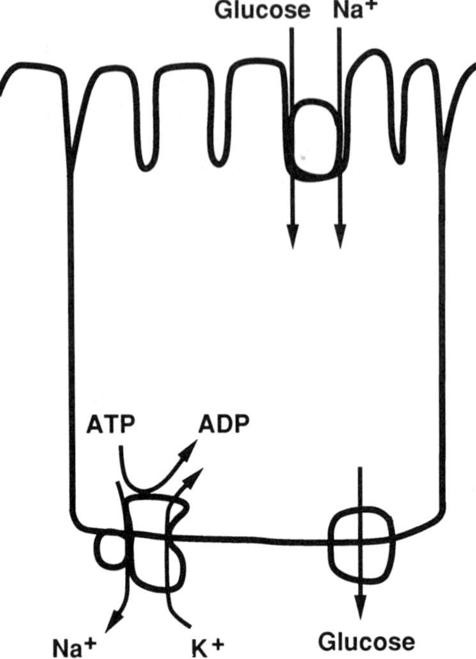

FIGURE 12 Model of transepithelial flux of glucose in the proximal tubule.

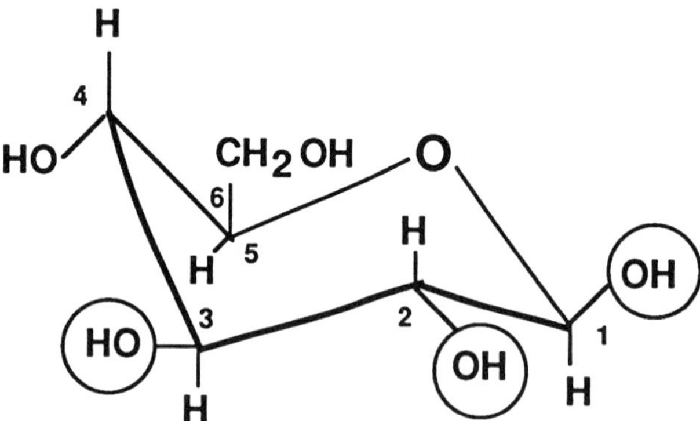

FIGURE 13 Structural requirements of monosaccharide transport across the brush border membrane of the proximal tubule. Hydroxyl groups in circles are essential for transport. [Adapted from Deetjen and von Baeyer (26).]

by the Na^+ system (18–25). The L isomer of glucose is not transported across either membrane by mediated mechanisms and appears to cross the membrane by simple diffusion (Fig. 11). There has been one report of L-arabinose interacting with the basolateral uniporter (27). Therefore, the stereoselective requirements for sugar transport in the kidney are very restrictive and only a few structural changes to the ribose ring are allowed.

B. Amino Acids

Amino acids, like glucose, are very efficiently reabsorbed by the early proximal tubule via a pump-leak arrangement, with a Na^+-cotransporter located in the brush border membrane (Fig. 14) (28). However, there are some additional complicating factors affecting amino acid transport. Brush border membrane aminopeptidases can split peptides within the lumen, increasing the concentration of amino acids after filtration (29). Under certain conditions, such as hereditary hyperamino aciduria, amino acids can undergo net secretion in the tubules by mechanisms not completely understood, involving both active and leak pathways (29). In addition, there are three different types of amino acids: acidic, basic, and neutral, and there are amino acids with aliphatic, aromatic, amide, and sulfur-containing side chains. There is one transporter for acidic amino acids, one for basic amino acids, and at least two transporters for neutral amino acids

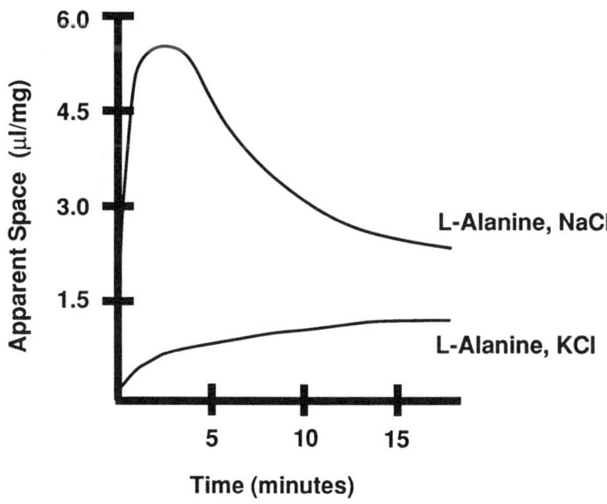

FIGURE 14 Uptake of L-alanine in rabbit renal brush border membrane vesicles in the presence and absence of an inwardly directed sodium gradient. [Adapted from Mircheff et al. (28).]

TABLE 1 Stereoselectivity of Alanine Uptake[a]

Incubation	Uptake (pmol/30 sec/mg protein)
L-alanine	2.48 ± 0.41
D-alanine	2.45 ± 0.43
L-alanine + NaCl	8.39 ± 1.15
D-alanine + NaCl	3.41 ± 0.36

[a]Rabbit renal brush border membrane vesicles were incubated for 30 sec at room temperature.
Source: Fass et al. (38).

in the brush border membrane (29–37). The naturally occurring L isomers are transported through the brush border membrane at a much greater rate than the D forms (Table 1) (38). However, at high concentrations, the D isomers can inhibit the transport of L isomers and vice versa. Therefore, the D isomers do appear to be actively transported by these systems, albeit with a much lower affinity (38). The acidic amino acid carrier appears to exhibit pronounced stereoselectivity. D-glutamate has almost no affinity for this transporter; however, D-aspartate inhibits the transport of both L-glutamate and L-aspartate (29).

C. Organic Cations

The renal handling of organic cations and bases has received considerable attention during the last decade due to the identification of many clinically relevant drug–drug interactions occurring in the kidney. The active tubular secretion of organic cations was discovered in the 1940s (39); however, because of their detrimental effects on the cardiovascular system (dopamine, norepinephrine, neostigmine, histamine, and epinephrine), mechanisms of organic cation transport were not elucidated until the advent of in vitro methodologies, including the isolation of plasma membrane vesicles and of perfused and nonperfused tubules.

In 1981, a model for the transepithelial secretion of organic cations was proposed by Holohan and Ross (40). Organic cations such as TEA and N^1-methylnicotinamide (NMN) are transported into the proximal tubule cell across the peritubular membrane by a facilitated, passive electrogenic carrier-mediated system and accumulate within the cell because of the favorable electrochemical gradient across that membrane (40). The transport of organic cations into the lumen of the tubule involves a secondarily active system driven by a counterflow of protons. This system is referred to as the organic cation/proton antiporter (7,40–46). The Na^+/H^+ antiporter, also located in the brush border membrane, is thought to be largely responsible for the proton gradient that exists from lumen to cell. The Na^+ gradient across the apical membrane is produced by the Na^+K^+ATPase system in the basolateral membrane (40). The complete model for organic cation secretion is detailed below (Fig. 15). Microperfusion studies, isolated tubules, and membrane vesicle studies have localized this transport system to the proximal tubule, with more transport occurring in the first segment (47).

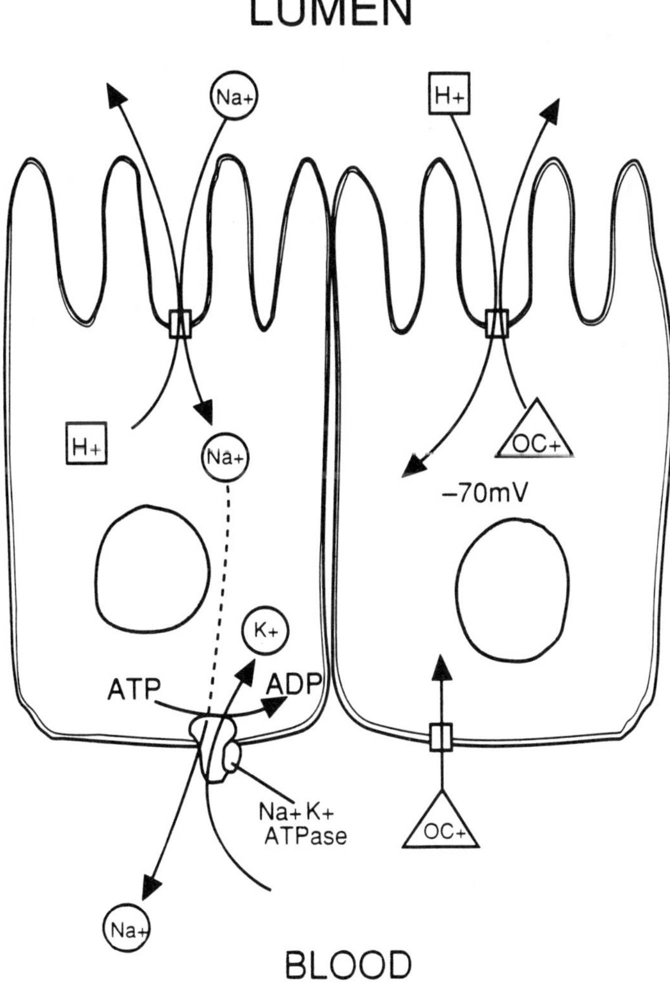

FIGURE 15 Model of the transepithelial flux of organic cations in the proximal tubule.

Many exogenous, pharmacologically active, clinically important organic cations including cimetidine, procainamide, quinine, quinidine, triamterene, and pindolol are actively secreted (renal clearance > GRF*fu) (48–52). Drug interactions between many of these substances have been documented in vivo in humans. For example, cimetidine has been shown to decrease the renal clearance of procainamide, pindolol, ranitidine, cephalexin, and amiloride (53–55). Competitive inhibition of the organic cation/proton antiporter at the brush border membrane has been suggested as the mechanism for these drug–drug interactions. Consistent with this mechanism, in vitro studies of the organic cation/proton antiporter demonstrate the competitive inhibition between many of these organic cations (Fig. 16) (5). This, however, does not preclude interactions at the basolateral membrane.

Stereoselective biotransformation of drugs has long been observed. However, the possibility of stereoselective renal excretion of exogenous compounds has only been recently proposed (56,57). Pindolol, a β-adrenoreceptor blocking agent, is actively secreted (renal clearance > GRF*fu) by the organic cation transport system, and recent clinical studies have suggested that it is stereoselectively eliminated. The $S(-)$ enantiomer of pindolol had a 30% higher renal clearance than the $R(+)$ form (Fig. 17) (56).

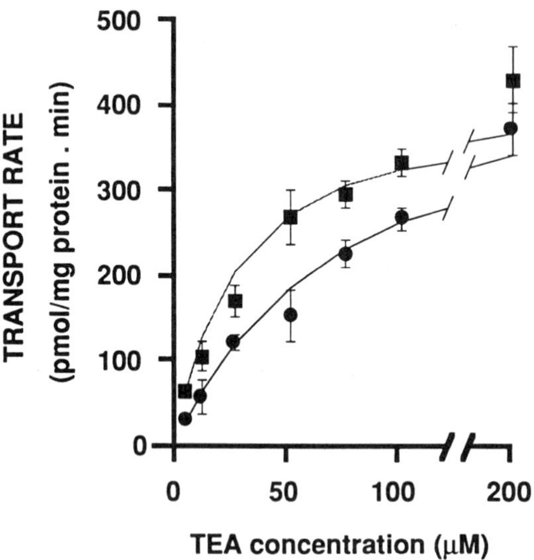

FIGURE 16 Effect of increasing TEA concentration in the absence (square) or presence (circle) of 100 μM NMN on the rate of TEA uptake in OK cell monolayers. [From Yuan et al. (5).]

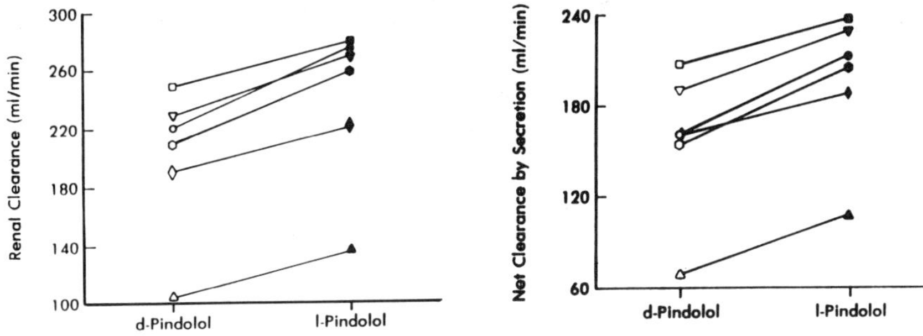

FIGURE 17 Renal clearance (left) and net clearance by secretion (right) of *l*- and *d*-pindolol in six subjects [From Hsyu and Giacomini (56).]

Possible mechanisms responsible for this stereoselective renal clearance of pindolol appear to be stereoselective renal metabolism or stereoselective renal secretion (stereoselective binding to plasma proteins was not observed). Recently, Somogyi et al. investigated the effect of coadministration of cimetidine on the renal clearance of the two enantiomers of pindolol (58). Cimetidine significantly reduced the renal clearance of both enantiomers, but reduced the renal clearance of the *R* enantiomer by a greater extent than the *S* enantiomer. These data are consistent with the stereoselective renal elimination mechanism for pindolol, with the *S* enantiomer being preferentially cleared.

Although diastereoisomers, both quinine and quinidine, have similar physical properties (Fig. 18). In clinical studies, the renal clearance of quinidine was fourfold greater than that of quinine (57). No stereoselective differences in plasma protein binding were observed. The renal filtration and passive reabsorption of these two diastereoisomers should be similar since the compounds have similar octanol-water partition coefficients and pKa values (57). Therefore, stereoselective active renal secretion may be the mechanism responsible for the observed differences in the renal clearances of quinine and quinidine.

Further clinical examples of potential stereoselective renal secretion of organic cations have been recently reported. A major metabolite of verapamil (D-617) is actively secreted by the kidney (59). Upon coadministration of cimetidine, the renal clearance of the *S*-D-617 isomer was significantly decreased, whereas the clearance of the *R*-D-617 metabolite was unaffected by cimetidine administration (59). Stereoselective renal secretion was suggested as the mechanism of this effect. However, it is not known if this metabolite is actually secreted by the organic cation transport system. The renal clearance of unbound *S*(+) disopyramide was

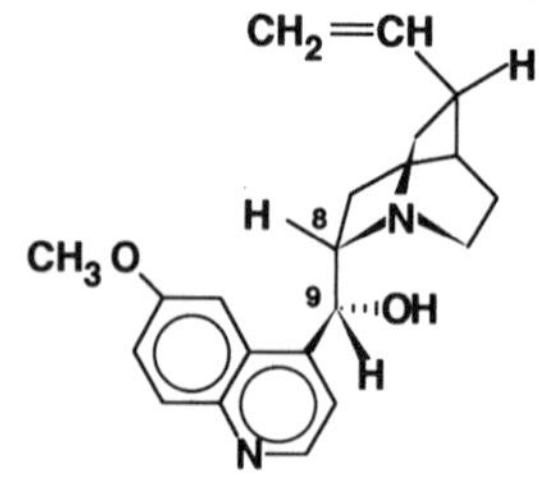

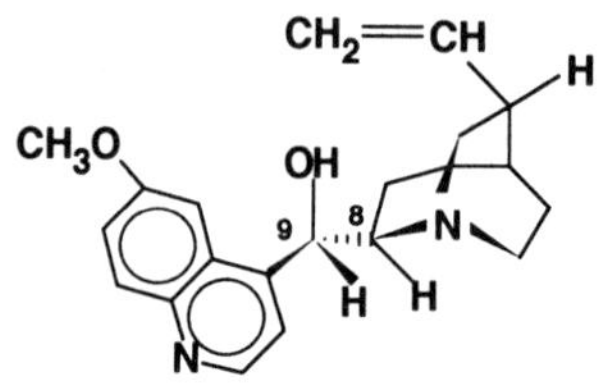

QUININE (8S,9R)

QUINIDINE (8R,9S)

FIGURE 18 Structures of quinine and quinidine.

much greater than that of the $R(-)$ enantiomer (60). In addition, stereoselective plasma protein binding may produce stereoselective renal clearance of the enantiomers of disopyramide (61).

The results of these studies suggest that there may be stereoselective secretion by the renal organic cation transport system. However, the potential concomitant stereoselective renal metabolism of these drugs was not addressed in these studies. The use of isolated plasma membranes eliminates the metabolic machinery of the renal cell (62), thereby allowing the direct examination of the interactions of enantiomers with organic cation transport proteins located at individual membranes. We examined the effect of $R(+)$ and $S(-)$ pindolol and the diastereoisomers quinine and quinidine on the uptake of NMN in renal brush border membrane vesicles from the rabbit (63). Our rationale was that if these substances interacted stereoselectively with the organic cation/proton antiporter, the K_I or IC_{50} values should reflect this interaction. All four compounds significantly inhibited the transport of NMN in brush border membranes vesicles (Fig. 19). The IC_{50} values of the inhibitors were $R(+)$ pindolol = 0.14 μM, $S(-)$ pindolol = 0.12 μM, quinine = 2.5 μM, and quinidine = 2.4 μM. However, there was no significant difference in the IC_{50} values for the enantiomers of

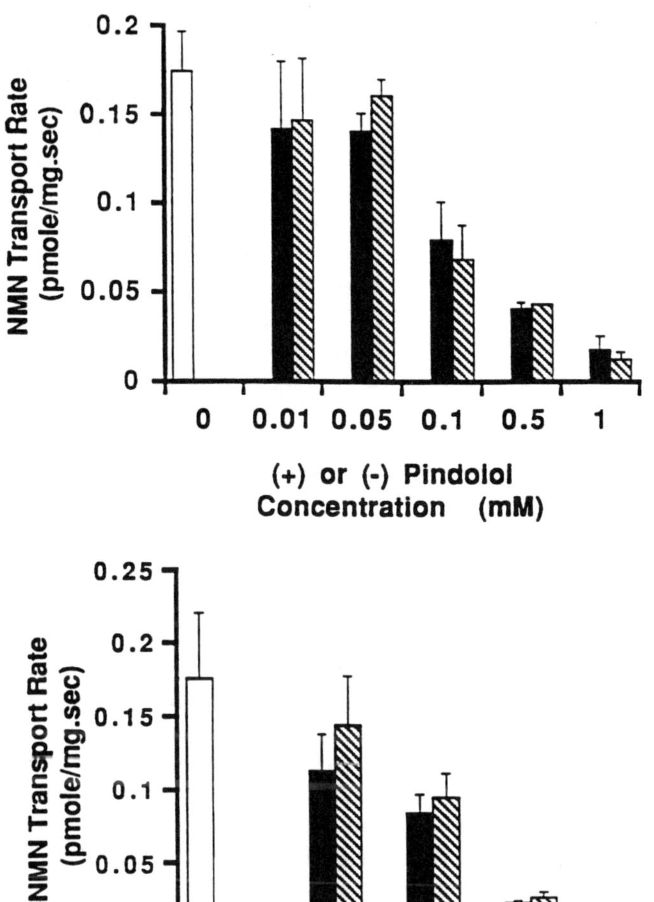

FIGURE 19 Uptake of NMN in rabbit brush border membrane vesicles in the presence of various concentrations of quinine and quinidine (lower panel) and *d*- and *l*-pindolol (upper panel). The clear bar in each panel is the control, the black bars represent either quinine (lower panel) or *d*-pindolol (upper panel) and the shaded bars represent either quinidine (lower panel) or *l*-pindolol (upper panel). [From Ott et al. (63).]

pindolol or the IC_{50} values of quinine and quinidine, suggesting no stereoselective interaction with the organic cation/proton antiporter. This same approach was undertaken by Bendayan et al. (64) using dog renal brush border membrane vesicles. They determined that the uptake of NMN was inhibited by quinine with a K_I of 7 μM and quinidine with a K_I of 0.7 μM. Although not stated if these values were significantly different, the authors suggest that the organic cation/proton antiporter demonstrates a degree of stereospecificity.

Wong et al. examined the effect of quinine and quinidine on the uptake of the organic cation amantidine (65), using rat renal cortical slices and proximal tubule suspensions. They observed a stereoselective inhibition of amantidine, with quinine being the significantly more potent inhibitory species. This is in direct conflict with the previous studies, which demonstrated either no stereoselective inhibition between the diastereoisomers or that quinidine is the more potent species in terms of renal transport.

Clearly, the results to date using in vitro methods are conflicting. Of the three studies detailed, each produced significantly different findings and conclusions regarding the potential stereoselectivity of the organic cation transport system. These conflicting data may be a result of species differences in transport mechanisms or in differences in experimental methods. Although the method employed by Wong et al. includes the basolateral membrane (proximal tubule suspensions and cortex slices), no study has specifically addressed the potential interactions occurring at the basolateral membrane. Further studies are required to determine potential stereoselective interactions for this system. The studies until now have relied on the inhibition potencies (IC_{50}) of enantiomers. Studies directly examining the transport of enantiomers are needed; they may reveal differences in binding and translocation rates.

D. Organic Anions

The active secretion of organic anions by the proximal tubule of the kidney has been studied extensively. *p*-Aminohippuric acid has been the model compound of organic anions for many decades. Transporters in the two membranes of proximal cells appear to be arranged opposite to that observed for organic cations (66,67). The active step is located in the basolateral membrane, but the exact mechanism is still very controversial. Some studies have suggested that the active transport of organic anions involves a Na^+ cotransport mechanism, whereas others have suggested organic anion exchange with endogenous organic anions (66,67). The brush border membrane is thought to contain a downhill facilitated carrier-mediated step in exchange for urate (66,67). Probenecid has become the model inhibitor of the organic anion transporter (68). Organic anions transported by this system include salicylates, penicillins, and cephalosporins (68,69). Although much work has been conducted on the structural specificity of this system for organic anions, there have been few reports detailing the stereoselectivity of this system, possibly due to the fact that both model compounds for this system are achiral. Ullrich et al. demonstrated that the *N*-a-acetyl-D isomers of many amino acids have a much higher affinity for the basolateral membrane than the L forms (70). Cox et al. observed no stereoselective renal handling of the nonsteroidal anti-inflammatory agent, ibuprofen, in rat isolated perfused kidney (71). However, Ahn et al. (72) demonstrated that $R(-)$ ibuprofen is cleared more rapidly in the isolated perfused rat kidney than its $S(+)$ enantiomer, suggesting differences in tubular transport or renal drug metabolism. These conflicting results suggest that further studies are required to determine potential stereoselective interactions of the organic anion transport system.

E. Reabsorption of Drugs in the Kidney

As mentioned previously, both the secretion and reabsorption of compounds can occur simultaneously and in different areas of the kidney, complicating the net handling of the drugs by the kidney. The kidney appears to actively reabsorb most endogenous nutrients, including inorganic ions, sugars, amino acids, and vitamins. However, in general for many exogenous compounds, reabsorption appears to occur by simple diffusion and is dependent on the degree of ionization of the drug, lipophilicity of the drug, and urine pH and flow. The driving force for the passive reabsorption of drugs is reabsorption of water. As water is reabsorbed along the nephron, the concentration of drug within the lumen increases, causing the required concentration gradient to be imposed across the membrane. Drug reabsorption will then occur passively by the laws of simple diffusion.

V. OTHER EPITHELIA

A. Small Intestine

The absorption of sugars and amino acids appears to follow the same basic mechanisms as previously described for the proximal tubule of the kidney (73–77). An additional D-fructose-facilitated carrier mechanism has been located in the brush border membrane that is stereoselective and specific (78). It is Na^+-independent and does not accept D- or L-glucose. After transport into the cell, the intracellular fructose concentration is decreased by metabolism, which promotes further passive reabsorption through the brush border membrane (78).

The absorption of most exogenous compounds occurs by a first-order simple diffusion process. Therefore, the physicochemical properties of the drug, molecular weight, degree of lipophilicity, and concentration gradient will determine the extent of absorption (11). Also important is the pH of the lumen solution, which determines the percent of the nonionizable drug. Additional factors determining the extent of absorption include metabolism of the drug in the gut lumen due to enzymatic or chemical degradation, gastric emptying, and local blood flow (11). A few exogenous drugs are transported in the small intestine by carrier-mediated systems. 5-Fluorouracil and methotrexate are structural analogs of endogenous compounds and are thought to be transported by those systems (11). A carrier-mediated mechanism has also been observed for aminocephalosporins such as cefadroxil and cephradine (79). These drugs show a saturable and inhibitable transport mechanism. These compounds are also inhibited by dipeptides, suggesting that certain drugs may have some affinity for the dipeptide transport system in the small intestine. This system transports dipeptides coupled to an inwardly directed proton gradient, independent of a sodium gradient. It has recently been documented that the *cis* isomer of ceftibuten is markedly concentrated by this system, whereas the *trans* isomer is not readily absorbed in the small intestine and is not stimulated by a proton gradient (80). Therefore, it appears that some exogenous compounds may be absorbed through the small intestine by carrier mechanism and a report suggests some limited stereoselectivity.

B. Liver

Studies have demonstrated multiple transport carriers in hepatocytes for organic cations, organic anions, and neutral charged compounds (81–83). For biliary excretion to occur, compounds must transverse two different membranes of the hepatocyte, the sinusoidal and canaliculi membranes. Many large positively charged (monovalent and bivalent) lipophilic compounds with a minimum molecular-weight threshold in the range of 200–600 daltons are actively transported into the bile from the blood through hepatocytes (81–83). These include procainamide ethobromide, vecuronium, *N*-methylimipramine, and tubocurarine. Therefore, the pharmacokinetics of anticholinergics, neuromuscular blocking agents, psychotropic amines, and local anesthetics can potentially be affected by biliary excretion. Also, bile acids and salts, steroid hormones, bilirubin, fatty acids, and cardiac glycosides have been demonstrated to be transported by system in the liver cells (81–83). Proton cotransport along with sodium dependent and independent mechanisms have been suggested for transport. Because the major structural feature affecting the selectivity of transport appears to be lipophilicity, only one study showing stereoselectivity has been documented (84). Clearly, more work is needed in this area in not only the determination of ionic driving forces, but also potential stereoselective structural requirements.

VI. SUMMARY

There are many examples of stereospecific epithelial transport. Especially significant are the transport of endogenous compounds and nutrients, such as glucose and amino acids. Less well understood and identified are those transport processes involving exogenous compounds, including racemic drugs. To clearly understand the pharmacokinetics and pharmacodynamics of drugs, transport processes involving absorption, distribution, and excretion must be clearly delineated. Further work on the various aspects of stereoselective transport needs to be done for those compounds in which transport processes greatly affect the handling of these drugs.

REFERENCES

1. M. Hofer, *Transport Across Biological Membranes*, Pitman, Boston, 1981.
2. W. D. Stein, *Channels, Carriers, and Pumps*, Academic Press, San Diego, Calif., 1990.
3. W. P. Dubinsky, Jr., The physiology of epithelial chloride channels, *Hosp. Practice*, 24:69–82 (1989).
4. M. Silverman, Molecular biology of the Na+-D-glucose cotransporter, *Hosp. Practice*, 24:180–204 (1989).
5. G. Yuan, R. J. Ott, C. Salgado, and K. M. Giacomini, Transport of organic cations by a renal epithelial cell line (OK), *J. Biol. Chem.*, *266*:8978–8986 (1991).
6. M. A. Kasanicki and P. F. Pilch, Regulation of glucose transporter function, *Diab. Care*, *13*:219–227 (1990).
7. R. J. Ott, A. C. Hui, G. Yuan, and K. M. Giacomini, Organic cation transport in human renal brush border membrane vesicles, *Am. J. Physiol.*, *261*:F443–F451 (1991).
8. H. N. Christensen, *Biological Transport*, W. A. Benjamin, London, 1975.
9. A. J. Vander, *Biological Transport*, McGraw-Hill, New York, 1985.
10. C. E. Stirling, Epithelial transport, *Textbook of Physiology* (H. D. Patton et al., eds.), W. B. Saunders, Philadelphia, Pa., 1047–1060, 1989.
11. M. Rowland and T. N. Tozer, *Clinical Pharmacokinetics*, Lea and Febiger, Philadelphia, Pa., 1980.
12. T. F. Blaschke and K. M. Giacomini, Pharmacodynamic consequences of

enantioselective drug disposition, *Pharmacology* (M. J. Rand and C. Raper, eds.), Elsevier, Amsterdam, 1987.

13. D. E. Drayer, Pharmacodynamic and pharmacokinetic differences between drug enantiomers in humans: An overview, *Clin. Pharmacol. Ther.*, *40*:125–133 (1986).
14. R. H. Levy and A. V. Boddy, Stereoselectivity in pharmacokinetics: A general theory, *Pharm. Res.*, *8*:551–556 (1991).
15. I. M. Weiner, Organic acids and bases and uric acid, *The Kidney, Physiology and Pathophysiology* (D. W. Seldin and G. Giebisch, eds.), Raven Press, New York, 1703–1724, 1985.
16. R. Kinne, H. Murer, E. Kinne-Saffran, M. Thees, and G. Sachs, Sugar transport by renal plasma membrane vesicles, *J. Mem. Biol.*, *21*:375–395 (1975).
17. M. Silverman, Glucose reabsorption in the kidney, *Can. J. Physiol. Pharmacol.*, *59*:209–224 (1981).
18. R. Kinne, Properties of the glucose transport system in the renal brush border membrane, *Curr. Top. Mem. Trans.*, *8*:209–267 (1976).
19. P. S. Aronson and B. Sacktor, The Na^+ gradient-dependent transport of D-glucose in renal brush border membranes, *J. Biol. Chem.*, *250*:6032–6039 (1975).
20. J. E. G. Barnett, W. T. S. Jarvis, and K. A. Munday, Structural requirements for active intestinal sugar transport, *Biochem. J.*, *109*:61–67 (1968).
21. A. Kleinzeller, The specificity of the active sugar transport in kidney cortex cells, *Biochim. Biophys. Acta*, *211*:264–276 (1970).
22. M. Silverman, The chemical and steric determinants governing sugar interactions with renal tubular membranes, *Biochim. Biophys. Acta*, *332*:248–262 (1974).
23. K. J. Ullrich, G. Rumrich, and S. Kloss, Specificity and sodium dependence of the active sugar transport in the proximal convolution of the rat kidney, *Pflugers Arch.*, *351*:35–48 (1974).
24. H. Glossman and D. M. Neville, Phlorizin receptors in isolated kidney, *J. Biol. Chem.*, *247*:7779–7789 (1972).
25. P. S. Aronson and B. Sacktor, Transport of D-glucose by brush border membranes isolated from the renal cortex, *Biochim. Biophys. Acta*, *356*:231–243 (1974).
26. P. Deetjen and H. von Baeyer, Renal handling of D-glucose and other sugars, *Textbook of Nephrology* (S. G. Massry and J. Glassock, eds.), Williams and Wilkins, Baltimore, Md., 88–91, 1989.
27. M. Silverman, M. A. Aganon, and F. P. Chinard, Specificity of monosaccharide transport in dog kidney, *Am. J. Physiol.*, *218*:743–750 (1970).
28. A. K. Mircheff, I. Kippen, B. Hirayama, and E. M. Wright, Delineation of sodium-stimulated amino acid transport pathways in rabbit kidney brush border vesicles, *J. Mem. Biol.*, *64*:113–122 (1982).
29. S. Silbernagl, Amino acids and oligopeptides, *The Kidney: Physiology and Pathophysiology* (D. W. Seldin and G. Giebisch, eds.), Raven Press, New York, 1677–1701, 1985.

30. J. Lerner, Acidic amino acid transport in animal cells and tissues, *Comp. Biochem. Physiol.*, *87B*:443–457 (1987).
31. K. Sigrist-Nelson, H. Murer, and U. Hopfer, Active alanine transport in isolated brush border membranes, *J. Biol. Chem.*, *250*:5674–5680 (1975).
32. A. M. Lynch and J. D. McGivan, Evidence for a single common Na+-dependent transport system for alanine, glutamine, leucine and phenylalanine in brush-border membrane vesicles from bovine kidney, *Biochim. Biophys. Acta*, *899*:176–184 (1987).
33. J. Evers, H. Murer, and R. Kinne, Phenylalanine uptake in isolated renal brush border vesicles, *Biochim. Biophys. Acta*, *426*:598–615 (1976).
34. P. D. McNamara, C. T. Rea, and S. Segal, Lysine uptake by rat renal brush-border membrane vesicles, *Am. J. Physiol.*, *251*:F734–F742 (1986).
35. M. R. Hammerman and B. Sacktor, Transport of amino acids in renal brush border membrane vesicles, *J. Biol. Chem.*, *252*:591–595 (1977).
36. M. R. Hammerman and B. Sacktor, Na+-dependent transport of glycine in renal brush border membrane vesicles: Evidence for a single specific transport system, *Biochim. Biophys. Acta*, *686*:189–196 (1982).
37. B. Stieger, G. Stange, J. Biber, and H. Murer, Transport of L-cysteine by rat renal brush border membrane vesicles, *J. Mem. Biol.*, *73*:25–37 (1983).
38. S. J. Fass, M. R. Hammerman, and B. Sacktor, Transport of amino acids in renal brush border membrane vesicles, *J. Biol. Chem.*, *252*:583–590 (1977).
39. B. R. Rennick, Renal tubule transport of organic cations, *Am. J. Physiol.*, *240*: F83–F89 (1981).
40. P. D. Holohan and C. R. Ross, Mechanisms of organic cation transport in kidney plasma membrane vesicles: 2. ΔpH studies, *J. Pharmacol. Exp. Ther.*, *216*:294 298 (1981).
41. W. H. Dantzler, O. H. Brokl, and S. H. Wright, Brush-border TEA transport in intact proximal tubules and isolated membrane vesicles, *Am. J. Physiol.*, *256*:F290–F297 (1989).
42. P.-H. Hsyu and K. M. Giacomini, The pH gradient-dependent transport of organic cations in the renal brush border membrane: Studies with acridine orange, *J. Biol. Chem.*, *262*:3964–3968 (1987).
43. H. Maegawa, M. Kato, K.-I. Inui, and R. Hori, pH sensitivity of H^+/organic cation antiport system in rat renal brush-border membranes, *J. Biol. Chem.*, *263*:11,150–11,154 (1988).
44. C. Rafizadeh, F. Roch-Ramel, and C. Schali, Tetraethylammonium transport in renal brush border membrane vesicles of the rabbit, *J. Pharmacol. Exp. Ther.*, *240*:308–313 (1987).
45. S. H. Wright, Transport of N^1-methylnicotinamide across brush border membrane vesicles from rabbit kidney, *Am. J. Physiol.*, *249*:F903–F911 (1985).
46. S. H. Wright and T. M. Wunz, Transport of tetraethylammonium by rabbit renal brush-border and basolateral membrane vesicles, *Am. J. Physiol.*, *253*: F1040–F1050 (1987).
47. T. D. McKinney, Heterogeneity of organic base secretion by proximal tubules, *Am. J. Physiol.*, *243*:F404–F407 (1982).

48. M. Takano, K.-I. Inui, T. Okano, and R. Hori, Cimetidine transport in rat renal brush border and basolateral membrane vesicles, *Life Sci.*, *37*:1579–1585 (1985).
49. S. H. Wright and T. M. Wunz, Amiloride transport in rabbit renal brush-border membrane vesicles, *Am. J. Physiol.*, *256*:F462–F468 (1989).
50. L. G. Gisclon, F. M. Wong, and K. M. Giacomini, Cimetidine transport in isolated luminal membrane vesicles from rabbit kidney, *Am. J. Physiol.*, *253*: F141–F150 (1987).
51. T. D. McKinney and M. E. Kunnemann, Procainamide transport in rabbit renal cortical brush border membrane vesicles, *Am. J. Physiol.*, *249*:F532–F541 (1985).
52. T. D. McKinney and M. E. Kunnemann, Cimetidine transport in rabbit renal cortical brush-border membrane vesicles, *Am. J. Physiol.*, *252*:F525–F535 (1987).
53. A. Somogyi, New insights into the renal secretion of drugs, *TIPS*, *8*:354–357 (1987).
54. J. van Crugten, F. Bochner, J. Keal, and A. Somogyi, Selectivity of the cimetidine-induced alterations in the renal handling of organic substrates in humans. Studies with anionic, cationic and zwitterionic drugs, *J. Pharmacol. Exp. Ther.*, *236*:481–487 (1986).
55. M. Muirhead, F. Bochner, and A. Somogyi, Pharmacokinetic drug interactions between triamterene and ranitidine in humans: Alterations in renal and hepatic clearances and gastrointestinal absorption, *J. Pharmacol. Exp. Ther.*, *244*:734–739 (1988).
56. P.-H. Hsyu and K. M. Giacomini, Stereoselective renal clearance of pindolol in humans, *J. Clin. Invest.*, *76*:1720–1726 (1985).
57. D. A. Notterman, D. E. Drayer, L. Metakis, and M. M. Reidenberg, Stereoselective renal tubular secretion of quinidine and quinine, *Clin. Pharmacol. Ther.*, *40*:511–517 (1986).
58. A. A. Somogyi, F. Bochner, and B. C. Sallustio, Stereoselective inhibition of pindolol renal clearance by cimetidine in humans, *Clin. Pharmacol. Ther.*, *5*: 379–387 (1992).
59. G. Mikus, M. Eichelbaum, C. Fischer, S. Gumulka, U. Klotz, and H. K. Kroemer, Interactions of verapamil and cimetidine: Stereochemical aspects of drug metabolism, drug disposition and drug action, *J. Pharmacol. Exp. Ther.*, *253*:1042–1048 (1990).
60. J. J. Lima, H. Boudoulas, and B. J. Shields, Stereoselective pharmacokinetics of disopyramide enantiomers in man, *Drug Met. Disp.*, *13*:572–577 (1985).
61. K. M. Giacomini, W. L. Nelson, R. A. Pershe, L. Valdivieso, K. Turner-Tamiyasu, and T. F. Blaschke, *In vivo* interactions of the enantiomers of disopyramide in human subjects, *J. Pharm. Biopharm.*, *14*:335–356 (1986).
62. C. R. Ross and P. D. Holohan, Transport of organic anions and cations in isolated renal plasma membranes, *Ann. Rev. Pharmacol. Toxicol.*, *23*:65–85 (1983).
63. R. J. Ott, A. C. Hui, F.-M. Wong, P.-H. Hsyu, and K. M. Giacomini, Interactions of quinidine and quinine and (+) and (−) pindolol with the

organic cation/proton antiporter in renal brush border membrane vesicles, *Biochem. Pharmacol.*, *41*:142–145 (1991).
64. R. Bendayan, E. M. Sellers, and M. Silverman, Inhibition kinetics of cationic drugs on *N*′-methylnicotinamide uptake by brush border membrane vesicles from the dog kidney cortex, *Can. J. Physiol. Pharmacol.*, *68*:467–475 (1990).
65. L. T. Y. Wong, D. D. Smyth, and D. S. Sitar, Stereoselective inhibition of amantidine accumulation by quinine and quinidine in rat renal proximal tubules and cortical slices, *J. Pharmacol. Exp. Ther.*, *255*:271–275 (1990).
66. J. B. Pritchard, Luminal and peritubular steps in renal transport of *p*-aminohippurate, *Biochim. Biophys. Acta*, *906*:295–309 (1987).
67. G. Burckhardt and K. J. Ullrich, Organic anion transport across the contraluminal membrane-dependence on sodium, *Kid. Int.*, *36*:370–377 (1989).
68. J. V. Moller and M. I. Sheikh, Renal organic anion transport system: Pharmacological, physiological and biochemical aspects, *Pharmacol. Rev.*, *34*:315–358 (1983).
69. K. J. Ullrich, G. Rumrich, and S. Kloss, Contraluminal organic anion and cation transport in the proximal renal tubule: V. Interactions with sulfamoyl- and phenoxy diuretics, and with β-lactam antibiotics, *Kid. Int.*, *36*:78–88 (1989).
70. K. J. Ullrich, G. Rumrich, T. Wieland, and W. Dekant, Contraluminal para-aminohippurate (PAH) transport in the proximal tubule of the rat kidney: VI Specificity: Amino acids, their *N*-methyl-, *N*-acetyl-, and *N*-benzoylderivatives: glutathione- and cysteine conjugates, di- and oligopeptides, *Pflugers Arch.*, *415*:342–350 (1989).
71. P. G. F. Cox, W. M. Moons, F. G. M. Russel, and C. A. M. van Ginneken, Renal handling and effects of *S*(+)-ibuprofen and *R*(−)-ibuprofen in the rat isolated perfused kidney, *Brit. J. Pharmacol.*, *103*:1542–1546 (1991).
72. H.-Y. Ahn, F. Jamali, S. R. Cox, D. Kittayanond, and D. E. Smith, Stereoselective disposition of ibuprofen enantiomers in the isolated perfused rat kidney, *Pharm. Res.*, *8*:1520–1524 (1991).
73. R. K. Crane, Intestinal absorption of sugars, *Physiol. Rev.*, *40*:789–825 (1960).
74. I. Bihler and R. Cybulsky, Sugar transport at the basal and lateral aspect of the small intestinal cell, *Biochim. Biophys. Acta*, *298*:429–437 (1973).
75. S. G. Schultz and P. F. Curran, Coupled transport of sodium and organic solutes, *Physiol. Rev.*, *50*:637–718 (1970).
76. R. L. Preston, J. F. Schaeffer, and P. F. Curran, Structure-affinity relationships of substrates for the neutral amino acid transport system in rabbit ileum, *J. Gen. Physiol.*, *64*:443–467 (1974).
77. H. N. Christensen, Organic ion transport during seven decades: The amino acids, *Biochim. Biophys. Acta*, *779*:255–269 (1984).
78. K. Sigrist-Nelson and U. Hopfer, A distinct D-fructose transport system in isolated brush border membrane, *Biochim. Biophys. Acta*, *367*:247–254 (1974).
79. H. Sezaki and T. Kimura, Carrier-mediated transport in drug absorption, *Topics in Pharmaceutical Sciences* (D. D. Breimer and P. Speiser, eds.), Elsevier, Amsterdam, 133–142, 1983.

2:1 expected at high absorption rates. Consistent with this analysis is our observation that a sustained-release dosage form of racemic verapamil yields an *R*:*S* ratio of about 4:1, whereas an immediate release dosage form yields an *R*:*S* ratio of about 3:1 (plots B and C in Fig. 4). The sustained-release dosage form with its intentionally lower drug release rate (average release rate approximately 30 mg/hr) has a lower absorption rate and, as predicted, a greater *R*:*S* ratio than the immediate release dosage form (average release rate approximately 40 mg/hr).

Modulation of the *R*:*S* ratio about some characteristic average value that is dependent on the particular design of a verapamil formulation is to

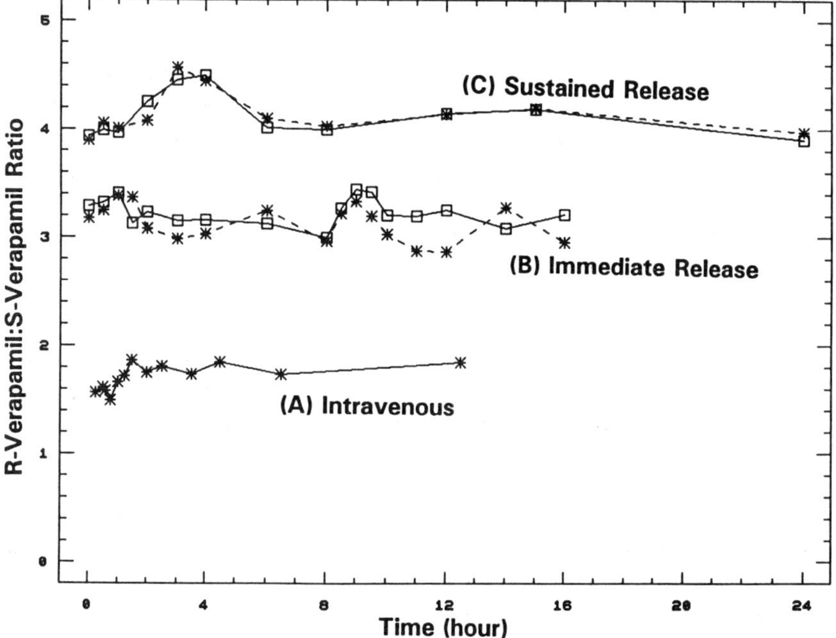

FIGURE 4 Mean *R*-verapamil:*S*-verapamil ratio observed following intravenous and oral dosing of racemic verapamil. A, After intravenous administration of a single 15-mg dose in 8 volunteers (personal communication from A. Rasymas, Univ. of Toronto, Canada). B, After oral administration of two different 120-mg immediate release formulations dosed every 8 hr to 22 normal volunteers in a cross-over design study and measured at steady state over two dosing intervals (-*-, test formulation; □, reference formulation). C, After oral administration of two different lots of a 180-mg once daily sustained-release formulation to 48 normal volunteers in a cross-over design study and measured at steady state (-*-, new manufacturing site; □, reference manufacturing site).

be expected. At this time, little information has been published that allows careful examination of how sensitive the relative plasma concentrations of *R*- and *S*-verapamil actually might be to small changes in the release characteristics of an oral formulation. One would expect that sustained-release formulations, which have slower *average* release rates but where the instantaneous release rate often fluctuates substantially over the release period, would very likely show formulation-dependent variation in the *R*:*S* ratio with time. However, the possibility that immediate release products show analogous variability in their release rate cannot be ruled out, and such formulation-specific variability may be the source of the slightly different time profiles for the *R*:*S* ratio observed with the two different immediate release formulations shown in Fig. 4 (plots B). As more studies of this type are conducted on a wider range of formulations, we will be able to confirm whether particular formulations can be associated with a signature *R*:*S* ratio.

The possibility that the modulation in the *R*:*S* ratio is simply an artifact of the formulation and not mediated through the absorption rate and presentation rate of drug to the liver has been addressed in our laboratory. Figure 5 shows results from a study in which 24 subjects were given the same sustained-release formulation of racemic verapamil at two different doses. In one treatment period, the dose was 120 mg/day to steady state, and in another it was 480 mg/day, also to steady state. The fourfold difference in absorption rates associated with the two different doses produced dramatically different profiles for the *R*:*S* ratios during the dosing interval. As predicted in the preceding discussions, the greatest *R*:*S* ratio was observed with the slower average absorption rate (low dose). However, the lowest *R*:*S* ratios were also observed with that low dose (at the beginning and end of the dosing interval). The mechanism(s) responsible for this wider fluctuation in the *R*:*S* ratio over the dosing interval at a lower dose have not yet been identified.

We are just beginning to understand and address, in a systematic fashion, the interactions in the pharmacokinetic processes that control the relative concentrations of two enantiomers in vivo. The above discussion, which is based on verapamil as a model substrate, is too simplistic for predicting enantiomer ratios in general. Until now the discussion has concentrated on the relative bioavailabilities of the two enantiomers, that is, the fraction of the dose that escaped first-pass metabolism. We have ignored other pharmacokinetic processes. If absorption is not the dominant process, drug distribution and systemic clearance will have a greater role in determining the relative systemic concentrations of the two enantiomers and the resulting *R*:*S* ratio. For verapamil, the distribution phase is rapid and the primary systemic clearance mechanisms are the same ones

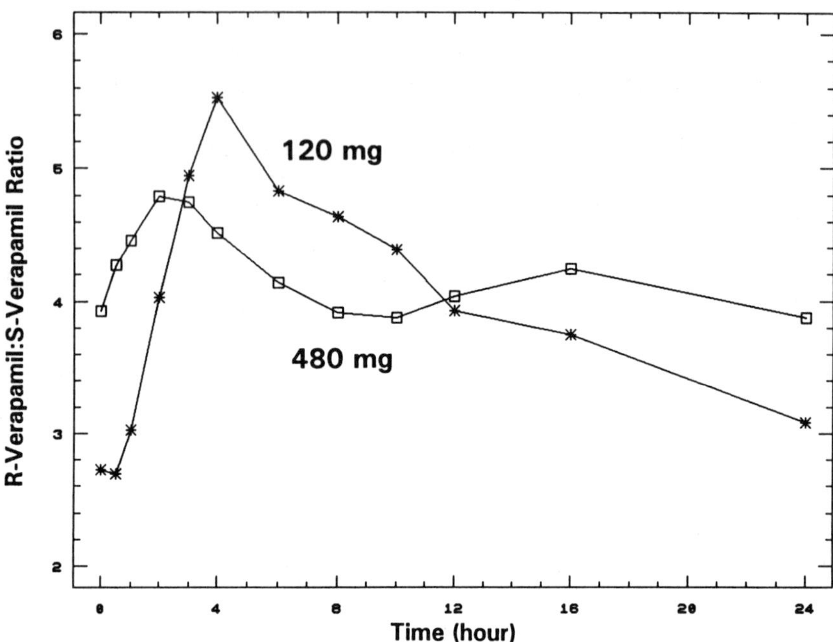

FIGURE 5 Mean *R*-verapamil:*S*-verapamil ratios observed at steady state during once daily dosing of either 120-mg (*) or 480-mg (□) racemic verapamil in 24 normal volunteers.

responsible for the first-pass effect. These limitations probably permitted our predictions above to be relatively close qualitatively. For drugs that have substantial multicompartmental kinetics, or where the primary systemic elimination pathway is different from that responsible for the first-pass effect (e.g., when renal clearance dominates elimination), the likely patterns have not yet been explored, or the important factors identified.

V. CLINICAL STUDIES: BIOEQUIVALENCE OF VERAPAMIL FORMULATIONS

Due to (1) the differential pharmacokinetics of the enantiomers of verapamil within each formulation, (2) differences in the rate of metabolism between various oral formulations, (3) large interindividual variability in drug effect (as measured by the PR interval), and (4) large interindividual variability in plasma verapamil concentration, the assessment of the bioequivalency of different verapamil formulations is not a trivial task (1,4,11).

Traditionally, the presence of enantiomers in the administered dosage form and the possibility that differential metabolic kinetics may influence systemic concentrations of the enantiomers have been ignored in trying to establish the bioequivalence of two racemic drug formulations, even for verapamil (17). Recently, however, verapamil has attracted attention as a model compound for a group of chiral drugs in which it may be necessary to use stereoselective assays to unambiguously evaluate bioequivalence. Verapamil is a prototype of a chiral drug with a high first-pass clearance (i.e., there is rapid hepatic metabolism), where the high clearance metabolic pathways display saturable kinetics and the saturable pathways also show stereoselectivity. In addition, because verapamil is an example in which the two enantiomers have different pharmacologic potencies, knowledge of the concentrations of the individual enantiomers is required to predict the total effect.

Other drugs warranting serious consideration for stereospecific bioequivalence evaluations are (1) chiral drugs shown to have a potency that depends on the route of administration; (2) chiral drugs undergoing hepatic metabolism and available in formulations with obviously different release rates; and (3) chiral drugs used in formulations that contain other chiral molecules as excipients, binders, or a retarding matrix that may promote stereoselective release rates from the dosage form.

Our first stereospecific bioequivalence evaluation using verapamil formulations was a pilot study conducted in 24 healthy male volunteers. The study compared the bioavailability of two sustained-release dosage forms and an immediate release formulation of racemic verapamil in a three-way cross-over study. Table 2 summarizes the results observed for verapamil, Table 3 the results for norverapamil. First, pharmacokinetic

TABLE 2 Achiral and Chiral Verapamil Bioequivalence Comparisons of Pellatized Capsule and Gel Tablet Formulations of Verapamil

	R, S Ratio (90% CI) power	*R* Ratio (90% CI) power	*R* Ratio (90% CI) power
AUC	94.3 (100.2, 88.5) 0.99	80.2 (92.0, 68.4) 0.72	72.3 (82.4, 52.3) 0.75
Cmax	91.4 (99.0, 83.7) 0.99	84.7 (99.7, 69.7) 0.50	79.0 (91.5, 56.5) 0.61
Tmax	73.2 (84.0, 62.3) 0.84	75.0 (81.7, 62.9) 0.88	76.9 (87.6, 56.2) 0.75
Cmin	108.7 (121.3, 96.2) 0.71	71.4 (86.7, 56.3) <0.50	76.4 (98.1, 54.8) <0.50

TABLE 3 Achiral and Chiral Norverapamil Comparisons Following Pellatized Capsule and Gel Tablet Formulations of Verapamil

	R, *S* Ratio (90% CI) power	*R* Ratio (90% CI) power	*R* Ratio (90% CI) power
AUC	96.8 (102.8, 90.8) 0.99	87.8 (97.3, 78.2) 0.88	62.6 (112.6, 87.4) 0.74
Cmax	99.2 (106.7, 91.7) 0.99	87.3 (98.8, 75.8) 0.74	92.3 (113.9, 86.1) 0.64
Tmax	81.5 (92.3, 70.8) 0.84	78.7 (88.1, 69.2) 0.89	100.0 (113.0, 87.0) 0.71
Cmin	100.0 (111.8, 88.3) 0.77	74.8 (88.3, 61.4) 0.62	58.9 (119.7, 80.4) <0.50

parameters generated from concentration data obtained with a nonstereospecific assay were used to compare the verapamil and norverapamil bioavailabilities. The two sustained-release formulations were concluded to have verapamil and norverapamil pharmacokinetic parameters that were generally not statistically different. However, when the pharmacokinetic parameters generated from concentration data obtained with a chiral assay were compared for a 12-subject subset of that study population, statistically significant differences were noted in some of the pharmacokinetic parameters, especially the parameters describing concentrations of *S*-verapamil (Cmax, Cmin, and AUC).

Also of interest to us was the observation that a simple regression analysis of the relationship between the nonstereospecific assay results and prolongation of the PR interval indicated that the concentration-effect relationship identified for the immediate release formulation differed from the relationships identified for each of the sustained-release formulations (Fig. 6). The difference approached statistical significance for all comparisons ($p = 0.1$) and exceeded it for some ($p < 0.05$). A similar finding has been reported recently by others (13). This formulation dependence disappeared once the effect was regressed against either of the individual enantiomer concentrations (data not shown). This pattern is directly analogous to the paradoxical difference in potency observed with verapamil when administered intravenously vs. orally that was discussed earlier in this chapter. The current example differs from the intravenous-oral case only in the magnitude of the effect. That is because the two types of oral formulations (sustained-release and immediate release) do not differ in their enantiomer ratio by as much as they do in a comparison of the intravenous and oral routes (Fig. 4).

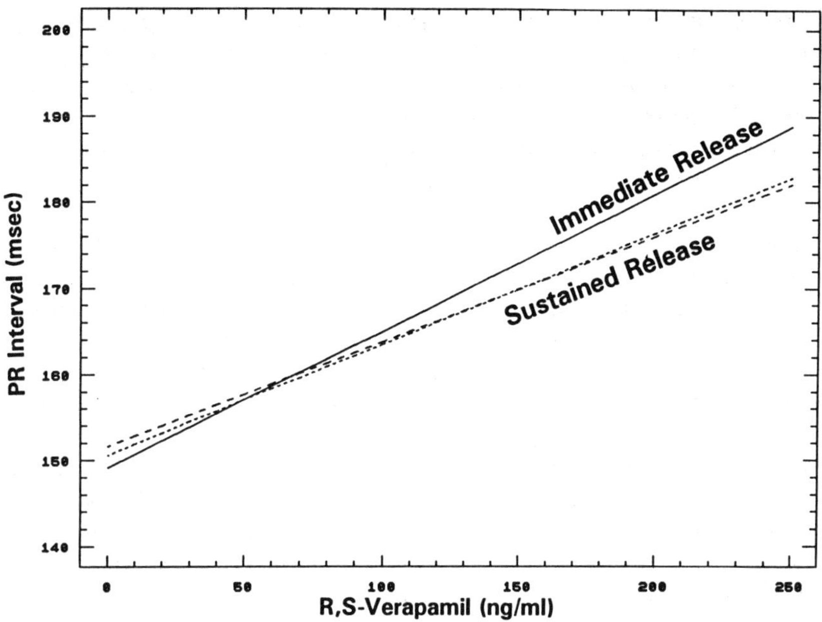

FIGURE 6 Mean relationship observed between prolongation of the PR interval and verapamil concentration measured in 24 normal volunteers using a nonstereospecific verapamil assay. The immediate release formulation was given 80 mg every 8 hr; both sustained-release formulations were given 240 mg once daily. Concentrations and PR intervals were measured at steady state.

Evaluation of the apparently conflicting results from the nonstereospecific and stereospecific assays in this pilot study forced us to recognize two pitfalls commonly overlooked in interpreting nonstereospecific data for chiral compounds. First, it is possible to conclude, using nonstereospecific data, that two formulations have no significant bioavailability differences, while concluding the opposite on the basis of data from a chiral assay, namely, that the formulations do show differences in bioavailability. Second, use of a nonstereospecific assay tends to reduce the discriminating power of a study to identify differences if they do exist. The observation identified in the first point is just one direct result of the process identified as the second point.

The nature of the pitfalls can be illustrated with an example. If we examine the simplest possible case, that is, consider a drug in which only one of the enantiomers has pharmacological activity, then inclusion of an inactive enantiomer in the measurement of the "active" entity just serves to add noise to the measurement. As with any measurement, this noise will make it a bit more difficult to identify slight changes in the concentrations of the active enantiomer if, or when, they occur. The noise factor has the least impact when the active enantiomer is a large fraction of the assay signal and the largest impact when the active enantiomer is a small fraction of the total signal.

For instance, if the active enantiomer constitutes 90% of the total drug on the average, then a 10% change in the concentration of the active enantiomer will be manifested as approximately a 9% change in total drug concentration (if we assume that the enantiomer concentrations do not change in opposite directions). In this case, there is little effect from the noise because most of the measurement is due to the active enantiomer. However, if 90% of the drug is inactive enantiomer, then the 10% fraction that is active could entirely disappear without being detected. Just normal assay variability can be great enough to mask such a systematic loss if the 90% of the assay measurement that is noise has 10% assay variability associated with it. In this example, a 100% decrease, or increase, in bioavailability (and pharmacological activity) could occur without the measured achiral concentration showing a detectable difference. Only in pharmacokinetic environments where there are coincidental, approximately equivalent fractional changes in both the active and inactive enantiomers can there be reliable detection of changes in the concentration of an active enantiomer that constitutes only a small fraction of the total drug.

This masking or "dilution" of sensitivity to change and loss in discriminating power is a well-known and widely lamented consequence of combining random noise with a signal of interest in many fields. The detrimental impact of this interpretational pitfall increases as the fraction of the total drug that is the enantiomer of interest decreases. However, the consequences of this pitfall for properly interpreting pharmacokinetic comparisons derived from nonstereospecific assays of chiral drugs are not yet widely appreciated.

This pitfall is especially relevant to studies attempting to compare the rates and extent of drug absorption of two dosage formulations of a chiral drug where both data sets are generated from nonstereospecific assay results. Such exercises raise the prospect of comparing one noise-laden signal to a second noise-laden signal. As long as the noise contribution is relatively uniform, the probability of finding a favorable conclusion (e.g., bioequivalence) is increased. However, the double dose of noise also increases the likelihood that the favorable result is a false finding.

As we have discussed, the interpretational complexities that can be encountered when using a nonstereospecific assay for a chiral compound can range from trivial to overwhelming when one of the enantiomers is inactive. Interpretation becomes almost impossible when both enantiomers have similar types of activities but different potencies (e.g., verapamil), or when the changes in the concentrations of the two enantiomers do not closely track each other in terms of direction or magnitude of change (e.g., verapamil).

The pilot study described above indicates that a conclusion of bioequivalence based on nonstereospecific data can be contradicted once chiral data are available. Although the shortcomings of using nonstereospecific data are clear, it is not as clear what other standard can replace it in actual practice. The obvious option, for all active enantiomers to simultaneously meet traditional bioequivalence guidelines, may be unrealistic. A study addressing this issue was conducted in our laboratory. Forty-four healthy male volunteers were studied in a two-way cross-over investigation comparing two sustained-release formulations from different manufacturing sites. The formulations were administered once daily to steady state. Plasma concentrations of the enantiomers of verapamil and norverapamil were determined on the final three days of each treatment period (days 6–8). On days 1–7, verapamil was administered with food, and on day 8, it was administered under fasting conditions.

The mean concentration-time profiles for each of the four analytes, and for each of the two formulations, generated by this study are shown in Fig. 7. The pharmacokinetic comparisons derived from this study for all four analytes are summarized in Table 4. As can be seen from Fig. 7 and Table 4, a four-way bioequivalence assessment proved both feasible and practical. This study also demonstrated that similar formulations will produce similar concentration-time profiles for all the enantiomers in the plasma (even given lot-to-lot variability, manufacturing site-to-site variability, and shelf-time variability).

Subsequent studies in our laboratory have reproduced the above results and demonstrated that similar formulations will produce similar *R*:*S* ratio vs. time profiles (plots C in Fig. 4), whereas dissimilar formulations produce *R*:*S* ratio vs. time profiles that appear to differ in subtle but discernible ways (plots B in Fig. 4). Since differences in the *R*:*S* ratio vs. time profiles were seen even as the formulations met the rigorous bioequivalency criteria for each of the four analytes, the *R*:*S* ratio may be a powerful tool in helping to identify and understand the interaction between absorption rate and differential saturation of metabolic enzymes.

Figure 8 illustrates another pharmacokinetic issue that needs to be considered when selecting a study design or interpreting nonstereospecific data for a chiral drug. In Fig. 8, the mean *R*:*S* ratios from 12

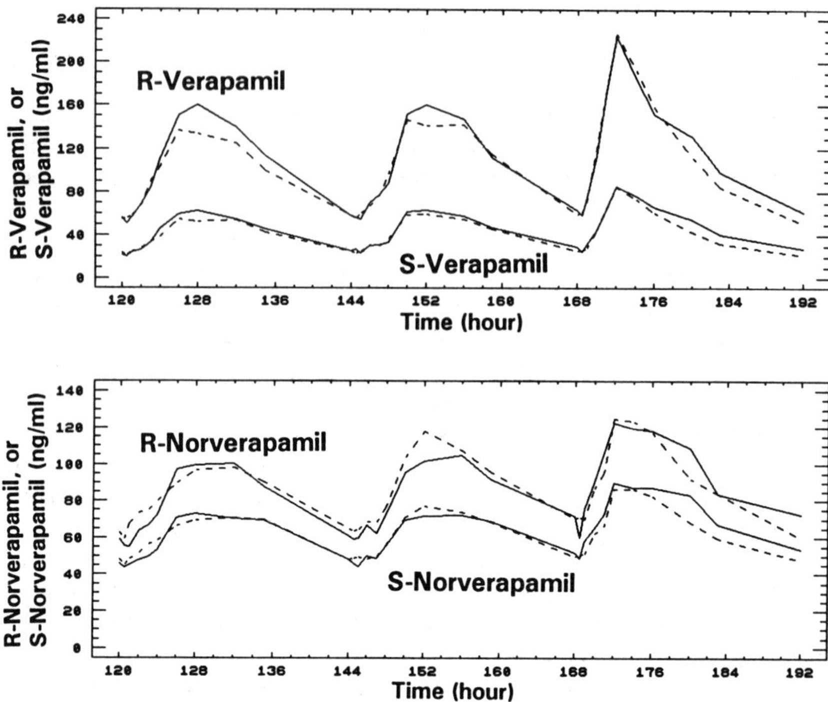

FIGURE 7 Stereospecific comparison of concentration-time profiles of verapamil and norverapamil obtained from two lots of sustained-release verapamil manufactured at different manufacturing sites (---, new manufacturing site; —, reference manufacturing site). Sustained-release racemic verapamil was administered 240 mg once daily to the 44 normal volunteers in a cross-over design study; plasma samples were collected at steady state on days 7–8. (On days 6 and 7 verapamil was administered with food, and on day 8 it was administered while fasting.)

subjects are shown following a single dose of sustained-release verapamil. The pattern shown in Fig. 8 differs substantially from all *R*:*S* profiles acquired following verapamil doses at steady state (e.g., plots B and C in Fig. 4 and both profiles in Fig. 5). Part of the differences may be related to the tendency for verapamil to decrease its clearance with chronic dosing (18), and part of them may be related to there being some "initial" enantiomer ratio present during steady-state dosing that reduces the impact of the freshly absorbed drug. The relative roles of these and other processes may become apparent as more sophisticated models of verapamil kinetics are developed. However, for now, the data indicate that care must be taken when using either single-dose or steady-state results to

TABLE 4 Verapamil and Norverapamil Chiral Bioequivalence Comparisons for Two Lots of Sustained-Release Racemic Verapamil Tablets Manufactured at Different Sites

	R-verapamil ratio (90% CI) power	*S*-verapamil ratio (90% CI) power	*R*-norverapamil ratio (90% CI) power	*S*-norverapamil ratio (90% CI) power
AUC	94.6 (100.6, 87.3) 0.95	94.4 (102.9, 85.7) 0.97	105.6 (113.2, 96.0) 0.99	97.7 (108.0, 94.2) 0.99
Cmax	92.7 (101.5, 83.1) 0.94	103.6 (106.7, 85.8) 0.87	106.7 (115.3, 97.6) 0.96	103.9 (111.1, 96.4) 0.99
Tmax	101.2 (112.0, 90.7) 0.97	96.4 (107.5, 86.5) 0.87	98.9 (111.4, 84.4) 0.66	91.0 (104.7, 77.5) 0.67
Cmin	98.2 (106.5, 88.4) 0.94	92.2 (104.3, 79.8) 0.75	103.9 (116.6, 92.2) 0.76	96.9 (106.6, 86.9) 0.90

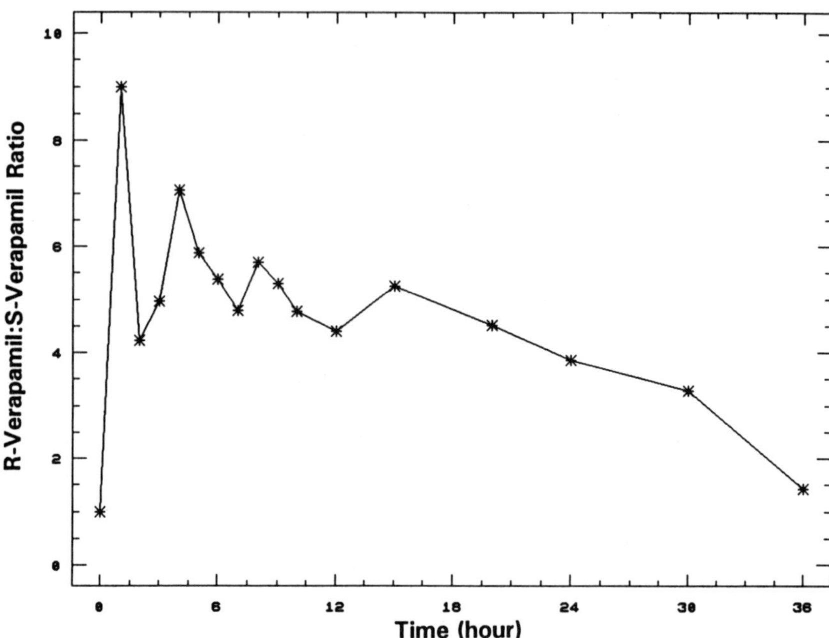

FIGURE 8 *R*-verapamil:*S*-verapamil ratio observed following a single 180-mg dose of sustained-release racemic verapamil. (The comparable relationship for the same formulation, at the same dose, but at steady state is shown as curves (C) in Fig. 3.)

extrapolate beyond their particular study conditions. This cautionary message certainly applies to both the pharmacokinetics and pharmacodynamics of verapamil, where not only the overall kinetics but also the relative contributions of the enantiomers are altered. It is not known whether other chiral drugs display similar phenomena.

Use of a pharmacodynamic endpoint to show therapeutic equivalence of two formulations as an alternative method of establishing bioequivalence has been suggested in many settings. Verapamil is a candidate for this approach because it has the readily detectable effect of prolonging the PR interval. Although each case for a pharmacodynamic endpoint must be argued on its own merits, our laboratory has identified several difficulties in using the PR interval as a surrogate for chiral assays of verapamil and norverapamil. First, study subjects are generally selected from a healthy young population whose PR intervals should not be particularly sensitive to verapamil concentrations. Second, in most such subjects the dynamic range in PR interval due to verapamil is generally limited to between 150 and 220 msec. The maximum permitted effect for this parameter is only about 50% greater than the baseline. Third, the baseline PR interval varies from subject to subject and fluctuates within a subject from day to day; the additional circadian variation in the PR interval requires that a standardized daily regimen be established and maintained. Fourth, prolongation of the PR interval is a surrogate endpoint; it is neither a measure of the therapeutic antihypertensive effect of the drug, nor a measure of the principal toxicities found in the patient population. Fifth, we have found that prolongation of the PR interval shows a saturable response (Fig. 9), indicating that it is least sensitive to differences in the enantiomer concentrations of verapamil in the region of peak concentration, the region of primary interest in terms of both safety and efficacy. Finally, neither the

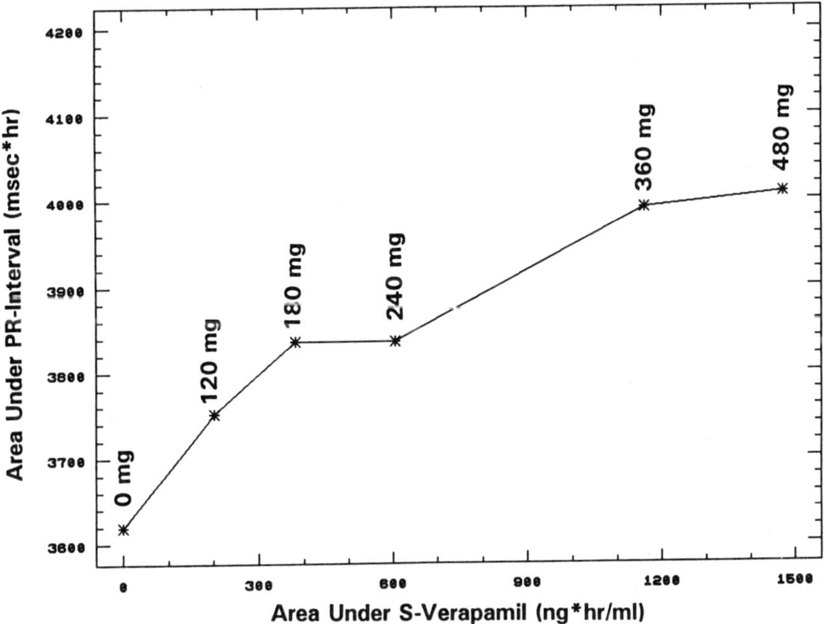

FIGURE 9 Pharmacodynamic relationship observed at steady state in 24 subjects for various doses of a sustained-release (once daily) verapamil formulation. The concentration measure used is the area under the *S*-verapamil concentration over the 24-hr dosing interval; the effect measure used is the area under the PR interval curve (measured with Holter monitor) over the 24-hr dosing interval. The doses of sustained-release verapamil that resulted in each concentration-effect pair are also shown.

appropriate size for confidence intervals nor what constitutes a clinically relevant difference has been generally agreed on for prolongation of the PR interval.

VI. CONCLUSIONS

Verapamil is a chiral drug that undergoes stereoselective metabolism in the liver. The relative bioavailabilities of the enantiomers of verapamil following oral administration of the racemic drug are dependent on the rate of presentation of verapamil to the liver by the portal vein, and hence the rate of absorption of verapamil from the gastrointestinal tract. The difference in the systemic concentrations of the enantiomers of verapamil appears to be explained by their different bioavailabilities; the different bioavailabilities are apparently a consequence of the relative affinities of the metabolizing enzymes for the two enantiomers (as reflected in their K_m values). Once absorption is completed and during chronic dosing, other stereoselective pharmacokinetic processes (i.e., protein binding, distribution volume, and plasma clearance) appear to control the relative concentrations of *R*- and *S*-verapamil. Additional research is required to understand these interactions.

The important paradoxical finding that verapamil potency depends on route of administration (intravenous vs. oral) also holds when dissimilar oral formulations are compared. The paradox is avoided by using a stereospecific assay to follow enantiomer concentrations and to associate the effect with concentrations of the appropriate enantiomer(s). Both the route of administration and type of oral formulation are important influences on the relative systemic concentrations of the two enantiomers. In addition, the amount of time elapsed post-dose, the amount of verapamil administered, and whether a single dose or chronic dosing is involved also have a major impact on the enantiomer ratio.

It is feasible and practical to establish the bioequivalence of oral verapamil products using traditional comparisons of the four active entities (the enantiomers of verapamil and those of norverapamil). The results from such a comparison may not be in agreement with comparisons based on a nonstereospecific assay, especially because of the reduced power of the nonstereospecific assay to detect differences in the lower-concentration analytes (*S*-verapamil and *S*-norverapamil). Using PR interval prolongation as a surrogate endpoint is not a suitable alternative to assaying enantiomer concentrations because of its lack of specificity as a safety or efficacy parameter, its limited dynamic range, and its decreased sensitivity as concentration increases.

Verapamil is a useful model for evaluating the relevance of stereospecific assays and enantiospecific data for racemic drugs in the clinical and regulatory environment. However, verapamil is just one of many racemic drugs, a specific example with high hepatic clearance pathways that are saturable and stereoselective. Generalizing these findings to other chiral drugs will require additional animal and clinical investigations and sophisticated simulation studies.

REFERENCES

1. R. G. McAllister and E. B. Kirsten, The pharmacology of verapamil. IV. Kinetic and dynamic effects after single intravenous and oral doses, *Clin. Pharmacol. Ther.*, *31*:418–426 (1982).
2. K. Chatterjee, J.-L. Rouleau, and W. W. Parmley, Medical management of patients with angina. Has first-line management changed?, *J. Amer. Med. Assoc.*, *252*:1170–1176 (1984).
3. H. Echizen, B. Vogelgesang, and M. Eichelbaum, Effects of *d,l*-verapamil on atrioventricular conduction in relation to its stereoselective first-pass metabolism, *Clin. Pharmacol. Ther.*, *38*:71–76 (1985).
4. D. McTavish and E. M. Sorkin, Verapamil: An updated review of its pharmacodynamic and pharmacokinetic properties and therapeutic use in hypertension, *Drugs*, *38*:19–76 (1989).
5. H. P. Dustan, Calcium channel blockers potential medical benefits and side effects, *Hypertension*, *13* (Suppl. I):I137–I140 (1989).
6. K. Satoh, T. Yanagisawa, and N. Taira, Coronary vasodilator and cardiac effects of optical isomers of verapamil in the dog, *J. Cardiovasc. Pharmacol.*, *2*: 309–318 (1980).
7. M. Eichelbaum, G. Mikus, and B. Vogelgesang, Pharmacokinetics of (+)-, (−)- and (±)-verapamil after intravenous administration, *Br. J. Clin. Pharmacol.*, *17*:453–458 (1984).
8. G. Mikus, M. Eichelbaum, C. Fisher, S. Gumulka, U. Klotz, and H. K. Kroemer, Interaction of verapamil and cimetidine: Stereochemical aspects of drug metabolism, drug disposition, and drug action, *J. Pharmacol. Exp. Ther.*, *253*:1042–1048 (1990).
9. R. H. Levy and A. V. Boddy, Stereoselectivity in pharmacokinetics: A general theory, *Pharmaceut. Res.*, *8*:551–556 (1991).
10. A. S. Gross, B. Heuer, and M. Eichelbaum, Stereoselective protein binding of verapamil enantiomers, *Biochem. Pharmacol.*, *37*:4623–4627 (1988).
11. B. Vogelgesang, H. Echizen, E. Schmidt, and M. Eichelbaum, Stereoselective first-pass metabolism of highly cleared drugs: Studies of the bioavailability of L- and D-verapamil examined with a stable isotope technique, *Br. J. Clin. Pharmacol.*, *18*:733–740 (1984).
12. M. Eichelbaum, P. Birkel, E. Grube, U. Gütgemann, and A. Somogyi, Effects of verapamil on PR-intervals in relation to verapamil plasma levels following single i.v. and oral administration and during chronic treatment, *Klin. Wochenschr.*, *58*:919–925 (1980).

13. S. Harder, P. Thürmann, M. Siewert, H. Blume, T. Huber, and N. Rietbrock, Pharmacodynamic profile of verapamil in relation to absolute bioavailability: Investigations with a conventional and a controlled-release formulation, *J. Cardiovasc. Pharmacol.*, *17*:207–212 (1991).
14. J. A. Plumb, R. Milroy, and S. B. Kaye, The activity of verapamil as a resistance modifier in vitro in drug resistant human tumour cell lines is not stereo-specific, *Biochem. Pharmacol.*, *39*:787–792 (1990).
15. A. K. Rasymas, H. Boudoulas, and J. J. MacKichan, Determination of verapamil enantiomers in serum following racemate administration using HPLC, *J. Liq. Chromatogr.* In press.
16. H. K. Kroemer, H. Echizen, H. Heidemann, and M. Eichelbaum, Predictability of the in vivo metabolism of verapamil from in vitro data: Contribution of individual metabolic pathways and stereoselective aspects, *J. Pharmacol. Exp. Ther.*, *260*:1052–1057 (1992).
17. J. G. Devane, M. Kavanagh, and J. G. Kelly, Dose proportionality and steady-state bioequivalence of verapamil sustained-release pellet-filled capsules, *Cur. Ther. Res.*, *50*:720–730 (1991).
18. J. G. Wagner, A. P. Rocchini, and J. Vasiliades, Prediction of steady-state verapamil plasma concentrations in children and adults, *Clin. Pharmacol. Ther.*, *32*:172–181 (1982).

12

Enantioselective Binding of Drugs to Plasma Proteins

Terence A. G. Noctor* *McGill University, Montreal, Canada*

I. INTRODUCTION

After systemic introduction, most drugs undergo some degree of reversible binding to plasma proteins. Two plasma proteins are responsible for the majority of drug binding: serum albumin (that will be abbreviated to HSA in the case of human serum albumin and BSA for bovine serum albumin), which accounts for the bulk of drug binding and is especially important in the binding of acidic solutes, and α_1-acid glycoprotein (AGP), to which weakly basic drugs preferentially bind. As, in general, only the free or unbound fraction of a drug is able to traverse cell membranes to exert a therapeutic effect, or to undergo enzymatic metabolism, factors influencing the binding of a drug to plasma proteins may significantly affect its pharmacokinetics and pharmacodynamics (1). This is especially true when the drug is highly bound or has low intrinsic clearance (in which case, total hepatic clearance is proportional to the free fraction). A clear understanding of the plasma protein binding behavior of a particular therapeutic agent is therefore fundamental to its safe and rational use.

In the case of chiral drugs, the possibility arises that the individual stereoisomers may bind to plasma proteins with differing affinities, resulting in differing free fractions. As over 60% of the drugs in current clinical use are chiral (2) and the majority of synthetically derived chiral drugs are administered as mixtures of the constituent stereoisomers (most commonly the racemate), this potentiality is significant.

**Present affiliation*: University of Sunderland, Sunderland, England

TABLE 2 Drugs Exhibiting Enantioselective Binding to Isolated HSA

Drug	n_1k_1 ($\times 10^{-3}$ M^{-1})[a]	Ref.
(*R*)-warfarin	429	(50)
	330	(19)
	228	(31)
	150	(37)
(*S*)-warfarin	730	(50)
	440	(19)
	496	(31)
	240	(37)
(*R*)-oxazepam hemisuccinate	7	
(*S*)-oxazepam hemisuccinate	257	(29)
(*R*)-oxazepam acetate	11	
(*S*)-oxazepam acetate	55	(58)
(−)-Ketoprofen	620	
(+)-Ketoprofen	1140	(51)
(*R*)-oxazepam glucuronide	ND	
(*S*)-oxazepam glucuronide	55	(52)
S-(+)-MK 571	1910	
R-(−)-MK 571	1222	(53)
(*S*)-α-(4-Chlorophenoxy)-propionic acid	31	
(*R*)-α-(4-Chlorophenoxy)-propionic acid	27	(54)
(*R*)-flurbiprofen	193	
(*S*)-flurbiprofen	344	(55)
(+)-Pirprofen	38	(56)
	78	(57)
(−)-Pirprofen	37	(56)
	136	(57)

[a] = binding affinity at primary binding site.
ND = not determined.

The source of the enantioselectivity in binding of warfarin to HSA is assumed to be the primary site to which it binds, that is, site I. However, Veronich et al. (38) proposed that the enantioselectivity in binding of warfarin to HSA arose from differential binding at secondary sites. In contrast to this, a recent study performed in this laboratory (39) has determined that the enantioselective binding site for warfarin on HSA accounts for over 70% of the total binding of the drug to albumin. This observation would suggest that it is indeed binding site I at which the enantioselectivity in the binding of warfarin enantiomers is expressed.

Other chiral coumarin anticoagulant drugs have been determined to exhibit higher enantioselectivity in binding to site I of HSA than warfarin.

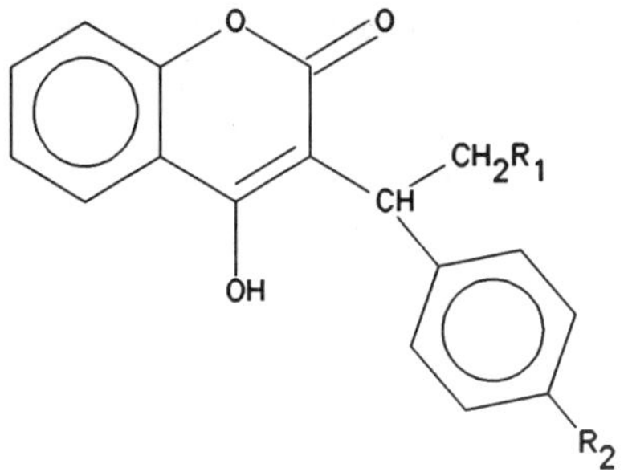

Drug	R_1	R_2
Warfarin	$COCH_3$	H
Acenocoumarol	$COCH_3$	NO_2
Phenprocoumon	CH_2CH_3	H

FIGURE 1 Coumarin anticoagulants binding to site I of human serum albumin.

The *R*-enantiomer of acenocoumarol (Fig. 1) has an association constant ($1.6 \times 10^5\ M^{-1}$) some 2.5 times greater than its antipode, determined by affinity chromatography on an HSA-sepharose filled column (37). Phenprocoumon (Fig. 1), another coumarin anticoagulant thought to bind at site I, also exhibits enantioselective binding to HSA (34), with the *S*(−)-enantiomer exhibiting an approximately twofold higher affinity than its antipode. This leads to free fractions of the *S*(−)- and *R*(+)-enantiomers of 0.0072 and 0.0107, respectively (40).

Brown et al. (34) have attempted to explain the higher enantioselectivity in the binding of phenprocoumon, compared to warfarin, using circular dichroism (CD) studies. The binding of the enantiomers of phenprocoumon differed in the orientation of the bulky substituents (side chains) attached to the chiral center, which were proposed to contribute significantly to the overall binding of coumarins to HSA. On the other hand, the side chains of the enantiomers of warfarin were found to bind in the same fashion, with the coumarin nuclei binding in differing orientations.

More recently, Fitos and Simonyi (41) used the work of Brown et al. to attempt to explain the observation that the *S*-enantiomer of warfarin is able to induce an allosteric enhancement in the binding of certain (*S*)-benzodiazepines at site II, whereas its antipode apparently is not. Their hypothesis was that a particular orientation of binding of the coumarin nucleus was responsible for the allosteric coupling between the two binding sites. However, the enantiomers of phenprocoumon behaved chromatographically in the same way as those of warfarin; the *S*-enantiomer alone exerted an allosteric increase in the binding of certain benzodiazepines. These results indicate that the coupling between binding sites is dependent on rather more than simply the orientation of the coumarin nucleus, or side chains.

Recent work by this group (39), using a stationary phase for HPLC based on HSA, has suggested that the enantioselectivity in binding to HSA of certain nonsteroidal antiinflammatory drugs (NSAIDs), namely, ketoprofen and suprofen, is expressed at site I. This hypothesis is based on the observation that although octanoic acid is able to displace approximately 70% of NSAID binding to HSA, this occurs without any concomitant loss of enantioselectivity. It is therefore apparent that the site which is responsible for the quantitative majority of binding of the NSAIDs [site II according to Sjöholm et al. (7)] is not the site responsible for the enantioselectivity in binding of these compounds. The minor binding site of ketoprofen and other NSAIDs is site I (7,9), which would therefore suggest that this is the site at which the enantioselective binding of the NSAIDs is expressed.

B. Enantioselective Binding of Drugs to Binding Site II of Serum Albumin

The occasional high enantioselectivity of the second major drug binding site of serum albumin (the indole-benzodiazepine site, site II) has led Müller (4) to describe HSA as a "silent receptor." Compounds thought to bind to site II include tryptophan, diazepam, and octanoate (59). The enantioselective binding of tryptophan to albumin is thought to arise from differential attachment of the isomers to this site (60).

The major binding site of the NSAID etodolac on HSA is site I, which accounts for only a small fraction of the total enantioselectivity in binding exhibited by this drug. The bulk of the enantioselectivity in binding of etodolac to HSA is apparently due to an almost enantiospecific attachment of the *S*-enantiomer to site II (49).

However, the most striking examples of enantioselective drug binding to albumin have arisen from the interaction between chiral 1,4-benzodiazepines and site II. Accordingly, a large body of research has addressed this topic.

1. Enantioselective Binding of Chiral 1,4-Benzodiazepines to Site II of Serum Albumin

Initial studies on the stereochemical aspects of the interaction between 1,4-benzodiazepin-2-ones (BDZs; see Fig. 2) and HSA were reported by Müller and Wollert (29,60), who first speculated on the site of the high

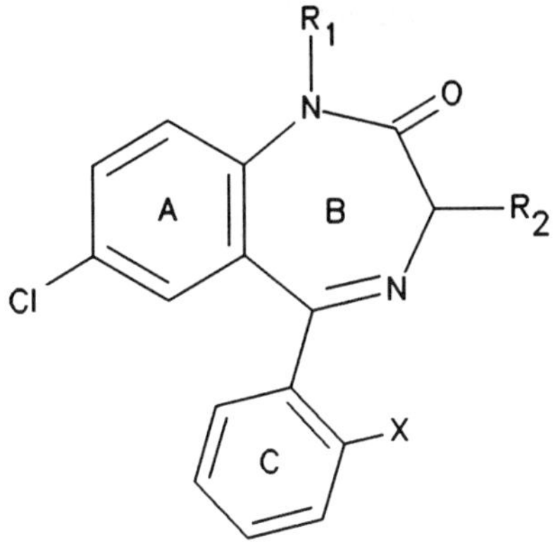

	R_1	R_2	X
Oxazepam	H	OH	H
Oxazepam Hemisuccinate	H	$OCO(CH_2)_2COO^-$	H
Lorazepam	H	OH	Cl
Lorazepam Acetate	H	$OCOCH_3$	Cl
Diazepam	CH_3	H	H

FIGURE 2 1,4-benzodiazepin-2-one drugs binding to site II of human serum albumin.

enantioselectivities often observed for the binding of these compounds. Interestingly, Müller and Wollert discovered that even prochiral benzodiazepines (e.g., diazepam; see Fig. 2) could exhibit characteristic extrinsic Cotton effects on binding to site II, emphasizing the high degree of stereoselectivity of the site: A particularly favored conformation of the ligand was selectively bound, or induced, by the protein. Subsequently, many reports of enantioselectivity in the binding of 1,4-benzodiazepines have appeared (e.g., 61–70). The various aspects of enantioselective binding of BDZs to serum albumin were expertly reviewed in 1979 by Alebic-Kolbah et al. (70).

The binding of the chiral BDZ oxazepam hemisuccinate (OXH; see Fig. 2) to HSA is a particularly striking example of enantioselective protein binding (29). As indicated in Table 2, the ratios of the affinity constants of the two enantiomers when binding to HSA may be as high as 50, depending on the conditions of observation. BSA, on the other hand, exhibits only a very small enantioselectivity in comparison to HSA, the ratio of the binding constants of *d*- and *l*-OXH being only in the region of 0.8 (29).

Studies employing CD spectroscopy of HSA-BDZ complexes have been used in an attempt to characterize the nature of the enantioselectivities observed. All BDZs studied, chiral or achiral, exhibited similar, biphasic CD spectra. The position of the bands in the CD spectra of the HSA-bound chiral compounds was only slightly shifted from those observed with the free ligands (29). These data were interpreted as supporting an extrinsically derived Cotton effect (69,70). Based on their chiroptical studies, Müller and Wollert formed several conclusions: First, both achiral and chiral BDZs bind to a common site; second, BDZ binding does not influence the protein conformation; third, both enantiomers of chiral BDZs bind in similar molecular conformations; and, finally, the primary point of attachment to albumin within the BDZ structure is the phenyl ring fused to the heptaatomic heterocycle (ring A, Fig. 2). Subsequent studies have shown that, in fact, HSA preferentially binds one of the two possible conformations of achiral BDZs (the *M* form), which is that adopted by the *S*-enantiomers of 3-OR substituted compounds, which invariably display higher binding affinities (68–70). Both enantiomers bind with the 3-substituent oriented in a quasiequatorial position, however; for the *R*-enantiomers, this necessitates adoption of the lesser favored *P* conformation. Other conclusions of Müller and Wollert have since been challenged: Binding of BDZs does apparently affect protein conformation (q.v.) (72); binding to other sites on HSA (for both *M* and *P* conformers) is revealed by closer examination of the CD data (66); and, finally, the importance of ring A in protein binding has been contested, nmr studies having indicated that the nitrogen atoms of the heptaatomic ring B play a more vital role (70).

It is clear that the binding of BDZs to HSA is not a simple, unimodular process. Recent studies using immobilized HSA indicate that although many BDZs share a common binding site, this accounts for only approximately 50% of the total binding observed, and that the remaining binding occurs at a number of disparate sites, determined by the molecular structure of the particular BDZ (73). However, the marked enantioselectivity in the binding of many BDZs to albumin remains an impressive example of this phenomenon.

C. Enantioselective Binding to Other Drug Binding Sites on Serum Albumin

The binding of several compounds to albumin may not be satisfactorily described by binding to either sites I or II. In addition to these sites, Fehske (6) proposes three further specific binding areas: for digitoxin, bilirubin, and fatty acids. Sjöholm (7,9) would add to this list a specific site for the binding of tamoxifen.

Enantioselectivity in binding to minor binding sites has not yet been investigated. This is primarily due to a lack of specific chiral markers for the new sites. Should they actually exist, there is no reason to assume that the potential for enantioselectivity in ligand binding does not occur.

D. Enantioselective Interactions Between Drug Binding Sites on Serum Albumin

One of the striking characteristics of serum albumin is its large conformational mobility. Indeed, this factor has prevented the accurate determination of the tertiary structure of the protein until very recently (74), due to the consequent difficulty in obtaining good x-ray crystallographic data. Many solutes that bind to albumin have been demonstrated to induce reciprocal conformational changes in the protein [e.g., bilirubin; see (75)]. It is therefore not surprising that drugs that bind at one distinct area of the protein are able to influence the binding of second compounds at other, apparently remote sites. In this respect, the work of Simonyi and coworkers (41,71,76–79), on the enantioselective aspects of such interactions, is prominent. The most studied enantioselective allosteric interaction, between binding sites I and II, involves the *S*-enantiomers of warfarin and certain 3-substituted 1,4-benzodiazepin-2-ones (Fig. 2). Mutual enhancements or decreases in binding are observed, the actual effect being critically controlled by the molecular structure of the BDZ. The presence of an o-chlorine atom on the ring C seems to intensify the enhancements in binding caused by the addition of (*S*)-warfarin. Conversely, alkyl substitution at N(1) eliminates the phenomenon. The nature of the C(3) substituent may augment or diminish the effect. (*R*)-warfarin has a much reduced effect on the binding of BDZs, compared to its antipode (78).

To date, the most pronounced allosteric effect observed is the enhancement in the binding of (*S*)-lorazepam acetate (Fig. 2) induced by (*S*)-warfarin (76). The *S*-enantiomer of phenprocoumon has a similar influence on the binding of BDZs to (*S*)-warfarin, with which it shares a binding site (site I) (41).

We have also been able to detect allosteric interactions between the two major drug binding sites of HSA, using an immobilized protein-based stationary phase for HPLC (26,39,73).

An unresolved question, which is difficult to explain using the site-oriented model of drug binding to albumin, is why these effects *are* enantioselective, when the enantiomers of warfarin (for instance) are thought to occupy the same site. Also, it is not apparent why other drugs that bind to sites I and II do not display mutual allosteric enhancements in binding. Developments in the methodologies available for the study of such phenomena (affinity chromatography, improved spectroscopic techniques) or a revision of current drug-HSA binding models (11) may provide the answers to these puzzles.

E. Immobilized HSA as a Tool in Protein Binding Studies

We have recently reported on the synthesis and characterization of a chiral stationary phase (CSP) for HPLC, based on immobilized HSA (the HSA-CSP) (24–26,39,73,80). This phase, because it broadly retains the binding characteristics and conformational mobility of free HSA, provides a powerful tool for the examination enantioselective aspects of ligand binding to albumin (80). Due to the inherent enantioselectivity of HSA, many chiral ligands are separated into their isomeric forms when injected onto the HSA-CSP. The basis therefore exists, by using specific modifiers added to the column mobile phase, to probe the source of enantioselectivity in the binding of drugs to HSA. For instance, the effect of one enantiomer added to the mobile phase on the chromatographic retention of its antipode may be examined. In the case of oxazepam hemisuccinate, the enantiomers were found to bind independently (25), whereas the enantiomers of ibuprofen competed for the same site(s) (24). Application of mathematical treatments to the chromatographic data can quickly provide quantitative data on the binding of individual enantiomers (39). The HSA-CSP promises to be a useful agent in the examination of enantioselective binding to HSA in the future.

III. ENANTIOSELECTIVE BINDING OF DRUGS TO α_1-ACID GLYCOPROTEIN

α_1-acid glycoprotein (AGP), which has a molecular weight of 40,000, is normally present in plasma at concentrations approximately 100 times lower than those of HSA. The actual levels of AGP found in plasma are subject to a very large variation in health, as well as disease. The primary physiological role of AGP is uncertain, but levels are increased in many acute conditions, suggesting an immunological or tissue repair function (81). AGP is an acidic protein due to its high content of sialic acid residues, although binding of basic drugs appears to be driven by hydrophobic rather than electrostatic forces. Removal of the sialic acid residues of AGP does not significantly reduce drug binding (82). At therapeutic concentrations, it is rare for drugs to bind solely to AGP (81).

Drug binding to AGP is thought to occur at a single hydrophobic pocket, or cleft, within the protein section of the molecule (8,83). The enantioselectivity of drug binding at this site was examined in 1987 by

Brunner and Müller (84). Various drugs have been found to bind enantioselectively to AGP. The laevorotatory enantiomer of propranolol (Fig. 3B) is preferentially bound to AGP (42,85), the ratio of the free fractions (−:+) being 0.79. This enantioselectivity is apparent over a range of AGP concentrations (85,86). Interestingly, the enantioselectivity in binding of propranolol to HSA is inverted, compared to that of AGP (42). The significant difference in the tissue distribution of the enantiomers of

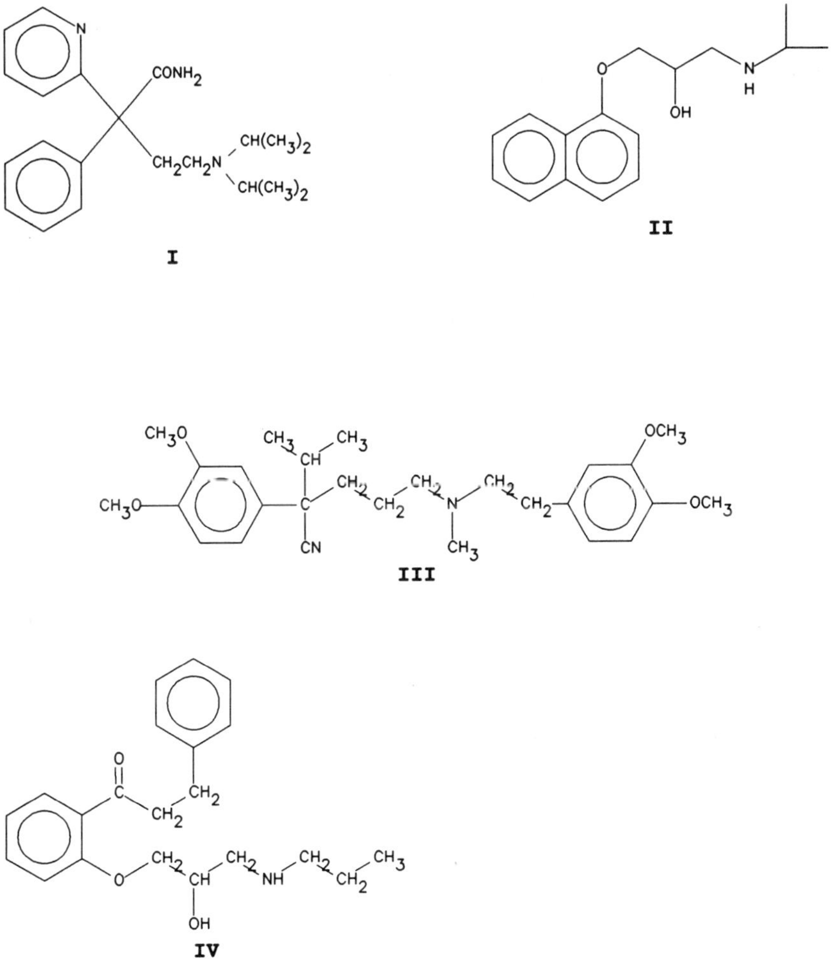

FIGURE 3 Drugs that display enantioselective binding to α_1-acid glycoprotein. I, disopyramide; II, propranolol; III, verapamil; and IV, propafenone.

propranolol in the rat is due to enantioselective binding to AGP (87). The enantioselective binding of β-adrenoreceptor antagonists, in general, significantly affects their actions (88).

The *S*-enantiomer of the antiarrhythmic drug propafenone (Fig. 3D) was found to bind to a single class of binding sites on AGP with greater affinity than its antipode (89). Both enantiomers were also significantly bound in a nonspecific fashion.

Another antiarrhythmic drug, disopyramide (Fig. 3A), is also enantioselectively bound to AGP (90,91). Disopyramide is a rare example, in that AGP almost exclusively accounts for the binding of the drug in plasma (92). The *S*-enantiomer of disopyramide was found to be the form more highly bound to human AGP and plasma. Again, both enantiomers appear to compete for a common binding site (93). The enantiomers of the major metabolite of disopyramide, mono-*N*-dealkyldisopyramide, are able to displace the enantiomers of the parent drug from this binding site in a competitive fashion (94). The plasma concentration of AGP in humans was found to have an enantioselective influence on the unbound fraction of the isomers of disopyramide, and consequently on other pharmacokinetic parameters, such as nonrenal clearance (95).

The more pharmacologically potent enantiomer of verapamil (Fig. 3C), the (−)-form, is bound more loosely to AGP than (+)-verapamil (43,96). The free fractions of the (+)- and (−)-enantiomers, in a 0.55-g/L solution of AGP, were 0.079 and 0.142, respectively (43). The free fractions of the enantiomers of verapamil in plasma samples from human patients were higher than those anticipated from in vitro studies, after oral dosage only. Samples from patients who had been administered the drug intravenously showed similar enantiomeric free fractions to those observed in solution studies on AGP. This apparent inconsistency was explained as being due to competition for the binding site(s) of the enantiomers of verapamil by drug metabolites. These attain higher concentrations following oral dosage, due to an enantioselective first-pass metabolism effect (43).

The concentration of AGP in the plasma of healthy volunteers correlated significantly with the binding ratio (bound/free) of the enantiomers, as well as the racemic form, of methadone. The average free fractions for (+)- and (−)-methadone in plasma were 0.100 and 0.142, respectively (97). Two variants of AGP, orosomucoids 2A and F1, were also found to play important determining roles in the binding of methadone enantiomers (97). The authors of the study point out that the levels of AGP variants should be considered in protein binding studies.

Other drugs reported to bind enantioselectively to AGP include the local anesthetics prilocaine, mepivacaine, and bupivicaine (3); chloroquine and desethylchloroquine (45); quinine/quinidine (3); isoprotenerol (98); vinca alkaloids (99); and acenocoumarol (37).

Although AGP is the most quantitatively important protein in the plasma binding of many basic drugs, it usually exhibits only low enantioselectivity. This may have measurable pharmacokinetic and pharmacodynamic consequences, as is the case with, for instance, disopyramide, verapamil, and the β-blockers. More generally, however, clinically important effects due to enantioselective binding of drugs to AGP are much less frequently observed than for HSA.

IV. ENANTIOSELECTIVE BINDING TO OTHER PLASMA PROTEINS

Although serum albumin and α_1-acid glycoprotein account for the majority of the drug binding capacity of the plasma proteins, other minor proteins may be important in the binding of certain drugs. The possibility of enantioselective binding to these minor proteins has not, in general, been explored. Usually, the clinical importance of binding to such proteins is negligible, but occasionally it may be pharmacologically significant. Lipoproteins, for instance, have the capability to bind basic, or neutral, lipophilic compounds, such as amitryptyline and nortryptyline (100). Although the plasma levels of lipoproteins are small, relative to those of HSA and AGP, they may be the major binding component of plasma for certain drugs, such as probucol, cyclosporin A, and ticlodipine (101). The importance of lipoproteins in the binding of a wider range of drugs may be enhanced in disease states such as hyperlipoproteinemia. To date, no work has been published that considers the possibility of enantioselectivity in binding to lipoproteins. The binding of drugs to lipoproteins closely resembles the partition between an aqueous buffer and chloroform, which may suggest that the process is essentially dissolution of the drug in a lipoprotein core, rather than specific attachment to a classical binding site. Perhaps then enantioselective binding would be unexpected in this instance.

Complement C3 appears to be important in the plasma binding of imipramine (102). Again, no systematic studies on the potential differential binding of enantiomers to this protein have appeared.

V. ENANTIOSELECTIVE DRUG BINDING TO PLASMA

Although it is important to understand the contributions of the individual components of plasma to drug binding, in the clinical situation, the interplay between plasma proteins, endogenous ligands, and drugs often results in behavior far different from in vitro observations using isolated proteins. For this reason, many studies consider the binding of enantiomers to whole plasma, rather than to isolated proteins. In such cases,

TABLE 3 Free Fractions of Drug Enantiomers in Human Plasma at Therapeutic Drug Levels

Drug	(% unbound)		Ref.
Acenocoumarol	(*S*):2.0	(*R*):1.8	(103)
Chloroquine	(−):51.5	(+):33.4	(104)
Disopyramide	(*R*):34.0	(*S*):22.2	(105)
	(*R*):36.6	(*S*):10.9	(106)
Fenfluramine	(−):2.9	(+):2.8	(107)
Glifumide	(*R*)	(*S*)	(108)
Ibuprofen	(*S*):0.643	(*R*):0.419	(109)
Indacrinone	(*R*):0.9	(*S*):0.3	(110)
Methadone	(−):12.4	(+):9.2	(111)
Mexiletine	(*S*):28.3	(*R*):19.8	(112)
Moxalatam	(*R*):53.0	(*S*):33.0	(113)
Nilvadipine	(−):0.9	(+):1.0	(114)
Oxazepam glucuronide	(*R*):45.0	(*S*):13.0	(52)
Pentobarbital	(*R*):36.6	(*S*):26.5	(115)
Phenprocoumon	(*R*):1.07	(*S*):0.72	(40)
Pindolol	(*R*)45.0	(*S*)45.0	(116)
Propoxyphene	(+):1.8	(−):1.89	(117)
Propranolol	(+)12:2	(−):10.9	(85)
	(+):20.3	(−):17.6	(88)
Quinidine	Quinine:14.0	Quinidine:13.0	(118)
Tocainide	(−):86–91	(+):83–89	(119)
Verapamil	(−):11.0	(+):6.4	(120)
	(−):22.4	(+):16.3	(121)
	(−):13.6	(+):9.6	(122)
Warfarin	(*R*):1.2	(*S*):0.9	(32)

the observed behavior may be due to the (possibly opposing) contributions of several proteins and is often modified by the presence of endogenous materials, particularly fatty acids and bilirubin. Tables 3 and 4 contain examples of enantioselectivity in the binding of drugs to plasma.

VI. EFFECT OF DISEASE STATE ON ENANTIOSELECTIVE PLASMA PROTEIN BINDING

The vast majority of studies on drug-protein binding have employed solutions of isolated plasma proteins or serum samples from healthy volunteers. Few data are currently available on the potential of enantioselective disease-induced changes in plasma protein binding. This is the

TABLE 4 Drugs Exhibiting Enantioselective Binding to Human Plasma at Therapeutic Concentrations

Drug	Dissociation constant ($\times 10^6 M^{-1}$)	Ref.
(−)-Alprenolol	0.7	
(+)-Alprenolol	2.1	(98)
(+)-Disopyramide	0.610	
(−)-Disopyramide	0.990	(94)
(*S*)-glifumide	0.4	
(*R*)-glifumide	4.4	(108)
(−)-Isoproterenol	0.9	
(+)-Isoproterenol	4.0	(98)
(+)-Propranolol	0.026	(87)
	1.4	(98)
(−)-Propranolol	0.140	(87)
	0.4	(98)
	4.0	(98)

case despite the fact that the patient treated in the clinic may present a very different environment to the drug than that existing in the rather idealized situation of in vitro experiments.

Although the level of serum albumin tends to remain relatively constant in healthy individuals, the concentration of the other major plasma protein, AGP, may vary significantly. In disease states, the levels of either or both major proteins may be vastly diminished or elevated. Particularly significant are those conditions in which the relative levels of these proteins are greatly modified. Clearly, drugs such as acenocoumarol (37), chloroquine (45), and propranolol (42), which exhibit inverse enantioselectivities in their binding to HSA and AGP, as well as significantly different enantiomeric pharmacologies, will demonstrate the most dramatic effects in this context.

Hyperalbuminemia is a relatively rare condition, but hypoalbuminemia occurs frequently in severe hepatic and renal disorders. In disorders such as these, there may also be relatively elevated serum concentrations of free fatty acids, which may compete with drugs for binding sites on HSA. The *S*-enantiomer of the glucuronide conjugate of the 1,4-benzodiazepin-2-one drug oxazepam binds to site II on HSA, whereas its antipode binds in an apparently nonspecific manner. The elevated levels of fatty acids found in renally impaired patients can therefore cause a displacement of oxazepam glucuronide (52), which may be enantioselective.

Rabbits that had been rendered uremic, by administration of uranyl nitrate, displayed an enantioselective decrease in the plasma protein binding of 2-phenylpropionic acid, the effect being more pronounced for the highly bound *R*-enantiomer (123). This behavior was repeated in a small group of uremic patients administered racemic flurbiprofen. The enantiomeric ratio of free (*R*):(*S*) flurbiprofen increased from 1.1 in healthy individuals to 1.4 in the uremic group (55).

Pathophysiology may have an indirect influence on plasma protein binding. For instance, high levels of the major metabolite of disopyramide, mono-*N*-dealkyldisopyramide, relative to the parent compound, occur in patients with renal insufficiency and also in those with elevated hepatic function (94). Enantioselective competition between the parent drug and metabolite for the same binding site on AGP were postulated to give rise to relatively higher levels of free (+)-disopyramide than in normal subjects (94).

Although the physiology of the patient may remain largely unchanged by his or her pathological condition, other coadministered agents may interfere with the enantioselective protein binding of a particular drug. This interference may be mediated directly, through competition for binding sites, or via allosteric interactions. The result may be supranormal free concentrations of the displaced compound. When the displaced drug has a small therapeutic index, the consequences of this may be clinically important. A typical example of this relatively unexamined phenomenon is the greater displacement of (*S*)-warfarin, compared to its enantiomer, by phenylbutazone, which leads to a significantly greater volume of distribution for (*S*)-warfarin (124). However, of greater clinical importance in this case is the enantioselective inhibition of the metabolism of (*S*)-warfarin by phenylbutazone (124).

Several studies have been reported (e.g., 125,126) concerning enantioselective drug interactions with highly protein-bound species, such as warfarin. As the free fractions of the individual enantiomers of the displaced compound were not determined, it is impossible to evaluate the full clinical significance of enantioselective displacement from plasma proteins. However, it may be anticipated that the enantioselective displacement of highly protein-bound drugs, with high eudismic ratios, will be clinically significant.

VII. INTERETHNIC DIFFERENCES IN ENANTIOSELECTIVE PLASMA PROTEIN BINDING OF DRUGS

The importance of genetic factors in pharmacokinetic and pharmacodynamic processes is becoming more widely recognized, particularly in the field of drug biotransformation (127). Recently, it has been determined that interethnic differences in response to drugs may be related to racial differences in plasma protein concentrations (128). For instance, Chinese subjects are more sensitive to the β-antagonist effects of propranolol than Caucasians, in part due to lower plasma protein binding in the former group (129). Zhou et al. (130) recently compared the free fractions of the enantiomers of propranolol in Chinese and Caucasian subjects. Among the Chinese, the binding of both enantiomers was significantly lower than in Caucasians, there being a greater proportion of free (−)-propranolol (the eutomer) in the Chinese subjects. This, the authors surmise, could account, in part, for the greater sensitivity of Chinese to propranolol. The lower plasma protein binding in Chinese subjects can be ascribed directly to lower plasma concentrations of AGP (128). Other chiral drugs that bind primarily to AGP (other β-blockers, lignocaine, verapamil) have higher free concentrations in Chinese subjects (128), although enantioselective aspects have yet to be explored.

VIII. SPECIES DIFFERENCES BETWEEN ENANTIOSELECTIVITY IN PLASMA PROTEIN BINDING

This chapter has addressed primarily the enantioselective plasma protein binding of drugs in humans. However, proteins from other species are sometimes used in binding studies, and the conclusions made extended to the human situation. Although this extrapolation may occasionally be valid, quantitative, and indeed qualitative, differences in enantioselective plasma protein binding may occur between species. For instance, rat plasma binds benzodiazepines (131) and warfarin (32) with similar enantioselectivities to those observed for human plasma, but displays no enantioselectivity toward disopyramide (105). The *S*-enantiomer of disopyramide was more highly bound than the *R*-isomer to serum protein in humans, the gorilla, and the pig. The reverse was true in bovine and ovine serum. These enantioselectivities were due to differences in association constants. No enantioselective binding was observed for disopyramide with serum from horses and goats (106).

The binding of a new, potent leukotriene D4 antagonist, MK-571, in the plasmas of 12 different species was recently examined (53). The species studied were subdivided into those in which the *S*-enantiomer bound more tightly (e.g., human, baboon, cow, dog), those that displayed greater binding of the *R*-enantiomer (rat, guinea pig, sheep), and those that displayed no enantioselective binding (rabbit, hamster, mouse).

Differences have also been observed between plasma proteins isolated from different species. Bovine albumin is non-enantioselective for phenprocoumon. HSA displays a marked preference for the *S*-enantiomer, whereas rodent albumins seem to bind (*R*)-phenprocoumon more strongly (40). A similar qualitative difference exists between HSA and BSA in the binding of warfarin enantiomers. HSA preferentially binds (*S*)-warfarin, whereas BSA binds the *R*-form more tightly (19).

It is clear that despite apparently small differences in the primary structure of serum albumins from different species and gross similarities in the binding of racemates, the examination of the binding behavior of enantiomers reveals many dissimilarities. This observation would imply that the molecular processes involved in the binding of small ligands to serum proteins are highly dependent on the structure and conformation of both participants in the interaction.

IX. CONCLUSION

The present chapter has sought to illustrate the growing awareness of the potential for enantiomers of drugs to be bound to plasma proteins to different extents. In general, the clinical consequences of enantioselectivity in the plasma protein binding of drugs will be minor, but in certain cases, the effects are significant. The examination of enantioselective binding to proteins has been facilitated by recent advances in chiral analysis. In addition, the use of immobilized plasma proteins as stationary phases for HPLC provides a subtle, rapid, and precise means to probe drug enantiomer-biopolymer interactions. In turn, the systematized study of enantioselective binding offers a particularly delicate insight into the molecular processes involved in the binding of drug ligands to proteins.

REFERENCES

1. U. Kragh-Hansen, Molecular aspects of ligand binding to serum albumin, *Pharmacol. Rev.*, *33*:17 (1981).
2. G. T. Tucker and M. S. Lennard, Enantiomer specific pharmacokinetics, *Pharmacol. Ther.*, *45*:309 (1989).
3. W. E. Müller, Stereoselective plasma protein binding of drugs, *Drug Stereochemistry: Analytical Methods and Pharmacology* (I. W. Wainer and D. E. Drayer, eds.), Marcel Dekker, New York, 1988, p. 227.
4. W. E. Müller and U. Wollert, Human serum albumin as a 'silent receptor' for drugs and endogenous substances, *Pharmacol.*, *19*:56 (1979).
5. G. Sudlow, D. J. Birkett, and D. N. Wade, The characterisation of two specific drug binding sites on human serum albumin, *Mol. Pharmacol.*, *11*:824 (1975).
6. K. J. Fehske, W. E. Müller, and U. Wollert, The location of drug binding sites in human serum albumin, *Biochem. Pharmacol.*, *30*:687 (1981).
7. I. Sjöholm, B. Ekman, A. Kober, I. Ljungstedt-Påhlman, B. Seiving, and T.

Sjödin, Binding of drugs to human serum albumin: XI. The specificity of three binding sites as studied with albumin immobilized in microparticles, *Mol. Pharmacol.*, *16*:767 (1979).

8. F. Brunner and W. E. Müller, Prazosin binding to human α_1-acid glycoprotein (orosmucoid), human serum albumin, and human serum. Further characterisation of the 'single drug binding site' of orosmucoid, *J. Pharm. Pharmacol.*, *37*:305 (1985).
9. I. Sjöholm, The specificity of drug binding sites on human serum albumin, *Drug-Protein Binding* (M. M. Reidenberg and S. Erill, eds.), Praeger, New York, 1986, p. 36.
10. K. J. Fehske, U. Schläfer, U. Wollert, and W. E. Müller, Characterization of an important drug binding area on human serum albumin, including the high-affinity binding sites of warfarin and azapropazone, *Mol. Pharmacol.*, *21*:387 (1982).
11. U. Kragh-Hansen, Evidence for a large and flexible region of human serum albumin possessing high-affinity binding sites for salicylate, warfarin and other ligands. Relation between drug-binding sites of human serum albumin, *Mol. Pharmacol.*, *34*:160 (1988).
12. R. Brodersen, B. Honoré, and G. Larsen, Serum albumin: A non-saturable carrier, *Acta Pharmacol. Toxicol.*, *54*:129 (1984).
13. B. Honoré, Conformational changes in human serum albumin induced binding, *Pharmacol. Toxicol.*, *66*(Suppl. 2):1 (1990).
14. F. Karush, The interaction of optically isomeric dyes with bovine serum albumin, *J. Phys. Chem.*, *56*:70 (1952).
15. F. Karush, The interaction of optically isomeric dyes with human serum albumin, *J. Am. Chem. Soc.*, *76*:5536 (1954).
16. R. H. McMenamy and J. L. Oncley, The specific binding of L-tryptophan to serum albumin, *J. Biol. Chem.*, *233*:1436 (1958).
17. R. H. McMenamy, Albumin binding sites, *Albumin Structure, Function and Uses* (V. M. Rosener, M. Oratz, and M. A. Rothschild, eds.), Pergamon Press, Oxford, 1977, p. 143.
18. B. Sébille and N. Thuaud, Determination of tryptophan-human serum albumin binding from retention data and separation of tryptophan enantiomers by high-performance liquid chromatography, *J. Liq. Chromatogr.*, *3*:299 (1980).
19. C. Lagercrantz, T. Larsson, and I. Denfors, Stereoselective binding of the enantiomers of warfarin and tryptophan to serum albumin from some different species studies by affinity chromatography on columns of immobilized serum albumin, *Comp. Biochem. Physiol.*, *69C*:375 (1981).
20. I. Fitos and M. Simonyi, Investigation of binding of tryptophan enantiomers to human serum albumin, *Acta Biochim. Biophys. Hung.*, *21*:237 (1986).
21. J. P. Monti, M. Sarrazin, C. Briand, and A. Crevat, Etude par résonance magnétique nucléaire de l'interaction sérum albumine humaine-tryptophane, *J. Chim. Phys.*, *74*:942 (1977).
22. I. W. Wainer, Proposal for the classification of high-performance liquid

chromatographic chiral stationary phases: How to choose the right column, *Trends Anal. Chem.*, *6*:125 (1987).
23. S. R. Perrin, Fast liquid chromatography for the resolution of chiral compounds, *Chirality*, *3*:188 (1991).
24. T. A. G. Noctor, G. Félix, and I. W. Wainer, Stereochemical resolution of enantiomeric 2-aryl propionic acid non-steroidal anti-inflammatory drugs on a human serum albumin based high-performance liquid chromatography chiral stationary phase, *Chromatographia*, *31*:55 (1991).
25. E. Domenici, C. Bertucci, P. Salvadori, S. Motellier, and I. W. Wainer, Immobilized serum albumin: Rapid HPLC probe of stereoselective protein binding interactions, *Chirality*, *2*:263 (1990).
26. E. Domenici, C. Bertucci, P. Salvadori, and I. W. Wainer, Use of a human serum albumin-based chiral stationary phase for high performance liquid chromatography for the investigation of protein binding: Detection of the allosteric interaction between warfarin and benzodiazepine binding sites, *J. Pharm. Sci.*, *80*:164 (1991).
27. F. M. Belpaire, M. G. Bogaert, and M. Rosseneu, Binding of beta-adrenoreceptor blocking drugs to human serum in a α1-acid glycoprotein and to human serum, *Eur. J. Clin. Pharmacol.*, *22*:253 (1982).
28. E. Janchen and W. E. Müller, Stereoselectivity in protein binding and drug distribution, *Topics in Pharmaceutical Sciences* (D. Breimer and P. Speiser, eds.), Elsevier, New York, 1983, p. 109.
29. W. E. Müller and U. Wollert, High stereospecificity of the benzodiazepine binding site on human serum albumin, *Mol. Pharmacol.*, *11*:52 (1975).
30. E. M. Sellers and J. Koch-Weser, Interaction of warfarin stereoisomers with human albumin, *Pharmacol. Res. Commun.*, *7*:331 (1975).
31. J. H. M. Miller and G. A. Smail, Interaction of the enantiomers of warfarin with human serum albumin, peptides and amino acids, *J. Pharm. Pharmacol.*, *29*:33P (1977).
32. A. Yacobi and G. Levy, Protein binding of warfarin enantiomers in serum of humans and rats, *J. Pharmacokinet. Biopharm.*, *5*:123 (1977).
33. K. Veronich, G. White, and A. Kapoor, Effects of phenylbutazone, tolbutamide and clofibric acid on binding of racemic warfarin and its enantiomers to human serum albumin, *J. Pharm. Sci.*, *68*:1515 (1979).
34. N. A. Brown, E. Janchen, W. E. Müller, and U. Wollert, Optical studies on the mechanism of the interaction of the enantiomers of the anticoagulant drugs phenprocoumon and warfarin with human serum albumin, *Mol. Pharmacol.*, *13*:70 (1977).
35. M. Otagiri, Y. Otagiri, and J. H. Perrin, Some fluorescent investigations of the interaction between the enantiomers of warfarin and human albumin, *Int. J. Pharmaceut.*, *2*:283 (1979).
36. B. Sebille, N. Thuaud, J. P. Tillement, and J. Brienne, Effect of long chain free fatty acids on the conformation of the human serum albumin warfarin binding site as studied by circular dichroism, *Int. J. Biol. Macromol.*, *6*:175 (1984).

37. I. Fitos, J. Visy, A. Magyar, J. Kajtar, and M. Simonyi, Inverse stereoselectivity in the binding of acenocoumarol to human serum albumin and α_1-acid glycoprotein, *Biochem. Pharmacol., 38*:2259 (1989).
38. K. Veronich, G. White, and A. Kapoor, Effects of phenylbutazone, tolbutamide and clofibric acid on binding of racemic warfarin and its enantiomers to human serum albumin, *J. Pharm. Sci., 68*:1515 (1979).
39. T. A. G. Noctor, D. S. Hage, and I. W. Wainer, Allosteric and competitive displacement of drugs from human serum albumin, as revealed by high-performance liquid affinity chromatography, on a human serum albumin-based stationary phase, *J. Chromatogr.*
40. W. Schmidt and E. Jahnchen, Species-dependent stereospecific serum protein binding of the oral anti-coagulant drug phenprocoumon, *Experientia, 34*:1323 (1978).
41. I. Fitos and M. Simonyi, Stereoselective effect of phenprocoumon enantiomers on the binding of benzodiazepines to human serum albumin, *2nd International Symposium on Chiral Discrimination*, Rome, 1991.
42. U. K. Walle, T. Walle, A. Bal, and L. S. Olanoff, Stereoselective binding of propranolol to human plasma, α_1-acid glycoprotein and albumin, *Clin. Pharmacol. Ther.*, 718 (1983).
43. A. S. Gross, B. Heuer, and M. Eichelbaum, Stereoselective protein binding of verapamil enantiomers, *Biochem. Pharmacol., 37*:4623 (1988).
44. D. Siebler and A. Kinawa, Binding of racemic indoprofen and its enantiomers to human serum albumin, *Arzneimittelforschung, 39*:659 (1989).
45. D. Ofori-Adjei, B. Ericsson, B. Linström, and F. Sjöquist, Protein binding of chloroquine enantiomers and desethylchloroquine, *Br. J. Clin. Pharmacol., 22*:356 (1986).
46. H. Büch, J. Knabe, W. Buzello, and W. Rummel, Stereospecificity of anaesthetic activity, distribution, inactivation and protein binding of the optical antipodes of two *N*-methylated barbiturates, *J. Pharmacol. Exp. Ther., 175*:709 (1970).
47. I. Marle, C. Pettersson, and T. Arvidsson, Determination of binding affinity of enantiomers to human serum albumin, *J. Chromatogr., 456*:323 (1988).
48. S. Iwakawa, H. Spahn, L. Z. Benet, and E. T. Lin, Stereoselective binding of the glucuronide conjugates of carprofen enantiomers to human serum albumin, *Biochem. Pharmacol., 39*:949 (1990).
49. N. Muller, C. Wioland, F. Lapicque, C. Monot, E. Payan, R. Dropsy, and P. Netter, Stereoselective binding of etodolac to human serum albumin, *2nd International Symposium on Chiral Discrimination*, Rome, 1991.
50. F. J. Diana, K. Veronich, and A. Kapoor, Binding of nonsteroidal anti-inflammatory agents and their effect on binding of racemic warfarin and its enantiomers to human serum albumin, *J. Pharm. Sci., 78*:195 (1989).
51. S. Rendic, T. Alebic-Kolbah, R. Kajfez, and V. Sunjic, Stereoselective binding of (+)- and (−)-ketoprofen to human serum albumin, *Il Farmaco, Ed. Sci., 35*: 51 (1980).
52. F. D. Boudinot, C. A. Homon, W. J. Jusko, and H. W. Ruelius, Protein binding

of oxazepam and its glucuronide conjugates to human albumin, *Biochem. Pharmacol.*, *34*:2115 (1985).
53. J. H. Lin, F. A. de Luna, E. H. Ulm, and D. J. Tocco, Species dependent enantioselective plasma protein binding of MK-571, a potent leukotriene D4 antagonist, *Drug Metab. Dispos.*, *18*:484 (1990).
54. R. I. Nazareth, T. D. Sokoloski, D. T. Witiak, and A. T. Hopper, Biological significance of serum albumin binding parameters determined *in vitro* for clofibrate-related hypolipemic drugs, *J. Pharm. Sci.*, *63*:203 (1974).
55. M. P. Knadler, D. C. Brater, and S. D. Hall, Plasma protein binding of flurbiprofen: Enantioselectivity and influence pathophysiological status, *J. Pharmacol. Exp. Ther.*, *249*:378 (1989).
56. M. Otagiri, K. Masuda, T. Imai, Y. Imamura, and M. Yamasaki, Binding of pirprofen to human serum albumin studied by dialysis and spectroscopy techniques, *Biochem. Pharmacol.*, *38*:1 (1989).
57. J. Oravcova, V. Mlynarik, S. Bystricky, L. Soltes, P. Szalay, L. Bohacik, and T. Trnovec, Interaction of pirprofen enantiomers with human serum albumin, *Chirality*, *3* (1991).
58. I. Fitos, Z. Tegyey, M. Simonyi, J. Sjöholm, T. Larsson, and C. Lagercrantz, Stereoselective binding of 3-acetoxy and 3-hydroxy-1,4-benzodiazepin-2-ones to human serum albumin, *Biochem. Pharmacol.*, *35*:263 (1986).
59. U. Kragh-Hansen, Octanoate binding to indole- and benzodiazepine-binding region of human serum albumin, *Biochem. J.*, *273*:641 (1991).
60. W. E. Müller and U. Wollert, Characterisation of the binding of benzodiazepines to human serum albumin, *Naunyn-Schmiedeberg's Arch. Pharmacol.*, *280*:229 (1973).
61. W. E. Müller and U. Wollert, Binding of *d*- and *l*-oxazepam hemisuccinate to bovine serum albumin, *Res. Commun. Chem. Pathol. Pharmacol.*, *9*:413 (1974).
62. W. E. Müller and U. Wollert, Benzodiazepines, specific competitors for the binding of L-tryptophan to human serum albumin, *Naunyn-Schmiedeberg's Arch. Pharmacol.*, *288*:17 (1975).
63. I. Fitos, M. Simonyi, Z. Tegyey, L. Otvos, J. Kajtar, and M. Kajtar, Resolution by affinity chromatography: Stereoselective binding of racemic oxazepam esters to human serum albumin, *J. Chromatogr.*, *259*:494 (1983).
64. M. Simonyi, I. Fitos, J. Kajtar, and M. Kajtar, Application of ultrafiltration and CD spectroscopy for studying stereoselective binding of racemic ligands, *Biochem. Biophys. Res. Commun.*, *109*:851 (1982).
65. M. Simonyi, I. Fitos, Z. Tegyey, and L. Otvos, Stationary ultrafiltration; a radio-tracer technique for studying stereoselective binding of racemic ligands, *Biochem. Biophys. Res. Commun.*, *97*:1 (1980).
66. A. Konowal, G. Snatzke, T. Alebic-Kolbah, F. Kajfez, S. Rendic, and V. Sunjic, General approach to chiroptical characterisation of binding of prochiral and chiral 1,4-benzodiazepin-2-ones to human serum albumin, *Biochem. Pharmacol.*, *28*:3109 (1979).
67. I. Fitos, M. Simonyi, Z. Tegyey, M. Kajtar, and L. Otvos, Binding of

3-alkyl-1,4-benzodiazepin-2-ones to human serum albumin, *Arch. Pharm. (Weinheim)*, *319*:744 (1986).
68. T. Alebic-Kolbah, F. Kajfez, S. Rendic, V. Sunjic, A. Konowal, and G. Snatzke, Circular dichroism and gel filtration study of binding of prochiral and chiral 1,4-benzodiazepin-2-ones to human serum albumin, *Biochem. Pharmacol.*, *28*:2457 (1979).
69. C. Bertucci, E. Domenici, and P. Salvadori, Stereochemical features of 1,4-benzodiazepin-2-ones bound to human serum albumin: Difference UV and CD studies, *Chirality*, *2*:167 (1990).
70. T. Alebic-Kolbah, S. Rendic, Z. Fuks, V. Sunjic, and F. Kajfez, Enantioselectivity in the binding of drugs to serum proteins, *Acta Pharm. Jugoslav.*, *29*:53 (1979).
71. M. G. Wong, J. A. Defina, and P. R. Andrews, Conformational analysis of clinically active anticonvulsant drugs, *J. Med. Chem.*, *29*:562 (1986).
72. I. Fitos, M. Simonyi, and Z. Tegyey, Resolution of warfarin via enhanced stereoselective binding to human serum albumin induced by lorazepam acetate, *Chromatography '87* (H. Kalasz and L. S. Ettre, eds.), Akademiai Kiado, Budapest, 1988, p. 205.
73. T. A. G. Noctor, C. D. Pham, and I. W. Wainer, Exploration of the 'benzodiazepine binding site' of human serum albumin by high performance liquid affinity chromatography, *Mol. Pharmacol.*
74. D. C. Carter, X.-M. He, S. H. Munson, P. D. Twigg, K. M. Gernert, B. Broom, and T. Y. Miller, Three-dimensional structure of human serum albumin, *Science*, *244*:1195 (see also amendment, *Science*, *249*:302) (1989).
75. J. Jacobsen and R. Broderson, Albumin-bilirubin binding mechanism: Kinetic and spectroscopic studies of binding of bilirubin and xanthobilirubic acid to human serum albumin, *J. Biol. Chem.*, *258*:6319 (1983).
76. I. Fitos, C. Lagercrantz, T. Larsson, M. Simonyi, I. Sjöholm, and Z. Tegyey, Stereoselective binding of 3-acetoxy-1,4-benzodiazepin-2-ones and 3-hydroxy-1,4-benzodiazepin-2-ones to human serum albumin: Selective allosteric interaction with warfarin enantiomers, *Biochem. Pharmacol.*, *35*:263 (1986).
77. I. Fitos, Z. Tegyey, M. Simonyi, and M. Kajtar, Stereoselective allosteric interaction in the binding of lorazepam methylether and warfarin to human serum albumin, *Bio-Organic Heterocycles 1986—Synthesis, Mechanisms and Bioactivity* (H. C. van der Plas, M. Simonyi, F. C. Alderweireldt, and J. A. Lepoivre, eds.), Elsevier, Amsterdam, 1986, p. 275.
78. I. Fitos and M. Simonyi, Selective effect of clonazepam and (*S*)-uxepam on the binding of warfarin enantiomers to human serum albumin, *J. Chromatogr.*, *450*:217 (1988).
79. I. Fitos, J. Visy, A. Magyar, J. Kajtar, and M. Simonyi, Stereoselective effect of warfarin and bilirubin on the binding of 5-(o-chlorophenyl)-1,3-dihydro-3-methyl-7-nitro-2H-1,4-benzodiazepin-2-one enantiomers to human serum albumin, *Chirality*, *2*:161 (1990).
80. E. Domenici, C. Bertucci, P. Salvadori, G. Felix, I. Cahagne, S. Motellier, and

I. W. Wainer, Synthesis and chromatographic properties of an HPLC chiral stationary phase based upon immobilized human serum albumin, *Chromatographia, 29*:171 (1990).

81. P. A. Routledge, The plasma protein binding of basic drugs, *Br. J. Clin. Pharmacol., 22*:499 (1986).
82. P. van der Sluijs and D. K. F. Meijer, Binding of drugs with a quaternary ammonium group to alpha-1-acid glycoprotein, *J. Pharmacol. Exp. Ther., 234*: 703 (1985).
83. W. E. Müller and A. E. Stillbauer, Characterisation of a common binding site of basic drugs on human α_1-acid glycoprotein (orosmucoid), *Arch. Pharmacol., 322*:170 (1983).
84. F. Brunner and W. E. Müller, The stereoselectivity of the single drug binding site of human α_1-acid glycoprotein, *J. Pharm. Pharmacol., 39*:986 (1987).
85. F. Albani, R. Riva, M. Contin, and A. Baruzzi, Stereoselective binding of propranolol enantiomers to human α_1-acid glycoprotein and human plasma, *Br. J. Clin. Pharmacol., 18*:244 (1984).
86. J. Oravcova, S. Bystricky, and T. Trnovec, Different binding of propranolol enantiomers to human alpha-1-acid glycoprotein, *Biochem. Pharmacol., 38*: 2575 (1989).
87. H. Takahashi, H. Ogata, S. Kanno, and H. Takeuchi, Plasma protein binding of propranolol enantiomers as a major determinant of their stereoselective tissue distribution in rats, *J. Pharmacol. Exp. Ther., 252*:272 (1990).
88. T. Walle, J. G. Webb, E. E. Bagwell, U. K. Walle, H. B. Daniell, and T. E. Gaffney, Stereoselective delivery and actions of beta receptor antagonists, *Biochem. Pharmacol., 37*:115 (1988).
89. J. Oravcova, W. Lindner, P. Szalay, L. Bohacik, and T. Trnovic, Interaction of propafenone enantiomers with human α_1-acid glycoprotein, *Chirality, 3*:30, (1991).
90. J. J. Lima, G. L. Jungbluth, T. Devine, and L. W. Robertson, Stereoselective binding of disopyramide to human plasma protein, *Life Sci., 35*:835 (1984).
91. L. Valdivieso, K. M. Giacomini, W. L. Nelson, R. Pershe, and T. F. Blaschke, Stereoselective binding of disopyramide to plasma proteins, *Pharm. Res., 5*: 316 (1988).
92. J. J. Lima, H. Boudoulas, and M. Blanford, Concentration dependence of disopyramide binding to plasma protein and its influence on kinetics and dynamics, *J. Pharmacol. Exp. Ther., 219*:741 (1981).
93. J. J. Lima, Interaction of disopyramide enantiomers for sites on plasma protein, *Life Sci., 41*:2807 (1987).
94. H. Takahashi, H. Ogata, and Y. Seki, Binding interaction between enantiomers of disopyramide and mono-*N*-dealkyldisopyramide on plasma protein, *Drug Metab. Dispos., 19*:554 (1991).
95. P. Le Corre, D. Gibassier, P. Sado, and R. Le Verge, Stereoselective metabolism and pharmacokinetics of disopyramide enantiomers in humans, *Drug Metab. Dispos., 16*:858 (1988).
96. B. Vogelgesang and H. Echizen, Stereoselective protein binding of vera-

pamil isomers, *Naunyn-Schmeideberg's Arch. Pharmacol.*, *253*:98 (1985).
97. C. B. Eap, C. Cuendet, and P. Baumann, Binding of *d*-methadone, *l*-methadone and *dl*-methadone to proteins in plasma of healthy volunteers; Role of the variants of alpha-1-acid glycoprotein, *Clin. Pharmacol. Ther.*, *47*:338 (1990).
98. G. Sager, D. Sandes, A. Besseseb, and S. Jacobsen, Adrenergic ligand binding in human serum, *Biochem. Pharmacol.*, *34*:2812 (1985).
99. I. Fitos, J. Visy, and M. Simonyi, Binding of vinca alkaloid analogues to human serum albumin and to alpha-1-acid glycoprotein, *Biochem. Pharmacol.*, *41*:377 (1991).
100. E. Pike and B. Skuterud, Plasma binding variations of amitryptyline and nortryptyline, *Clin. Pharmacol. Ther.*, *32*:228 (1982).
101. M. Lemaire, S. Urien, E. Albengres, P. Riant, R. Zini, and J. P. Tillement, Lipoprotein binding of drugs, *Drug-Protein Binding* (M. M. Reidenberg and S. Erill, eds.), Praeger, New York, 1986, p. 93.
102. C. G. Kristensen, Imipramine serum protein binding in healthy subjects, *Clin. Pharmacol. Ther.*, *34*:689 (1983).
103. J. Godbillon, J. Richard, A. Gerardin, T. Meinertz, W. Kasper, and E. Janchen, Pharmacokinetics of the enantiomers of acenocoumarol in man, *Br. J. Clin. Pharmacol.*, *12*:621 (1981).
104. D. Ofori-Adjei, O. Ericsson, B. Linstrom, J. Hermansson, K. Adjepon-Yamoah, and F. Sjöquist, Enantioselective analysis of chloroquine and desethylchloroquine after oral administration of racemic chloroquine, *Ther. Drug Monit.*, *8*:457 (1986).
105. C. S. Cook, A. Karim, and P. Sollman, Stereoselectivity in the metabolism of disopyramide in rat and dog, *Drug Metab. Dispos.*, *10*:116 (1982).
106. J. J. Lima, Species dependent binding of disopyramide enantiomers, *Drug Metab. Dispos.*, *16*:563 (1988).
107. S. Caccia, M. Ballabio, and P. DePonte, Pharmacokinetics of fenfluramine enantiomers in man, *Eur. J. Drug Metab. Pharmacokin.*, *4*:129 (1979).
108. E. Schillinger, I. Ehrenberg, and K. Lubke, Stereospecific plasma binding of glifumide, a new antidiabetic drug, *Biochem. Pharmacol.*, *27*:651 (1978).
109. A. M. Evans, R. L. Nation, Sansom, F. Bochner, and A. A. Somogyi, Stereoselective plasma protein binding of ibuprofen enantiomers, *Eur. J. Clin. Pharmacol.*, *36*:283 (1989).
110. D. E. Drayer, Pharmacodynamic and pharmacokinetic differences between drug enantiomers in humans; an overview, *Clin. Pharmacol. Ther.*, *40*:125 (1986).
111. M. Romach, K. M. Piafsky, J. G. Abel, V. Khouw, and E. M. Sellers, Methadone binding to orosomucoid (α_1-acid glycoprotein): Determinant of the free fraction in plasma, *Clin. Pharmacol. Ther.*, *29*:211 (1981).
112. K. M. McErlane, L. Igwemezie, and C. R. Kerr, Stereoselective serum protein binding of mexiletine enantiomers in man, *Res. Commun. Chem. Pathol. Pharmacol.*, *56*:141 (1987).
113. H. Yamada, T. Ichihashi, K. Hirano, and H. Kinoshita, Plasma protein binding and urinary excretion of *R*- and *S*-epimers of an arylmalonylamino-1-oxacephem 1. In humans, *J. Pharm. Sci.*, *70*:112 (1981).

114. T. Niwa, Y. Tokuma, K. Nakagawa, H. Noguchi, Y. Yamazoe, and R. Kato, Stereoselective oxidation and plasma binding of nilvadipine, a new dihydropyridine calcium antagonist, in man, *Res. Commun. Chem. Pathol. Pharmacol.*, *60*:161 (1988).
115. C. E. Cook (cited in D. E. Drayer, Ref. 110).
116. P.-H. Hsyu and K. M. Giacomini, Stereoselective renal clearance of pindolol in humans, *J. Clin. Invest.*, *76*:1720 (1985).
117. H. Sullivan (cited in D. E. Drayer, Ref. 110).
118. D. Notterman, D. E. Drayer, and M. M. Reidenberg, Stereoselective renal tubular secretion of quinine and quinidine, *Pharmacologist*, *26*:185 (1984).
119. A. J. Sedman, D. C. Bloedow, and J. Gal, Serum binding of tocainide and its enantiomers in human subjects, *Res. Commun. Chem. Pathol. Pharmacol.*, *38*: 165 (1982).
120. M. Eichelbaum, G. Mikus, and B. Vogelgesang, Pharmacokinetics of (+)- and (−)- and (±)-verapamil after intravenous administration, *Br. J. Clin. Pharmacol.*, *17*:453 (1984).
121. S. B. Earle and J. J. Mackichan, Separation and protein binding of verapamil enantiomers, *Clin. Pharmacol. Ther.*, *41*:233 (1987).
122. A. S. Gross, B. Heuer, and M. Eichelbaum, Stereoselective protein binding of verapamil enantiomers, *Biochem. Pharmacol.*, *37*:4623 (1988).
123. M. E. Jones, B. C. Sallustio, Y. Purdie, and P. J. Meffin, Enantioselective disposition of 2-arylpropionic acid non-steroidal anti-inflammatory drugs. II. 2-Phenylpropionic acid binding, *J. Pharmacol. Exp. Ther.*, *238*:288 (1986).
124. R. A. O'Reilly, W. F. Trager, C. H. Motley, and W. Howald, Stereoselective interaction of phenylbutazone with (12C/13C) warfarin pseudoracemates in man, *J. Clin. Invest.*, *65*:746 (1980).
125. R. A. O'Reilly, W. F. Trager, C. H. Motley, and W. Howald, Interaction of secobarbital with warfarin pseudoracemates, *Clin. Pharmacol. Ther.*, *28*:187 (1980).
126. R. A. O'Reilly, W. F. Trager, A. E. Rettie, and D. A. Goulart, Interaction of amiodarone with warfarin and its separated enantiomorphs in humans, *Clin. Pharmacol. Ther.*, *42*:290 (1987).
127. D. W. Clark, Genetically determined variability in acetylation and oxidation: Therapeutic implications, *Drugs*, *29*:342 (1985).
128. J. Feely and T. Grimm, A comparison of drug protein binding and α_1-acid glycoprotein concentration in Chinese and Caucasians, *Br. J. Clin. Pharmacol.*, *31*:551 (1991).
129. H. J. Zhou, R. P. Roshakji, D. Silberstein, G. R. Wilkinson, and A. J. J. Wood, Racial differences in drug response, *New Eng. J. Med.*, *320*:565 (1989).
130. H. J. Zhou, S. D. Shay, and A. J. J. Wood, Ethnic differences in stereoselective binding of propranolol, *Clin. Pharmacol. Ther.*, *49*:177 (1991).
131. Z. Tegyey, G. Maksay, J. Kardos, and L. Otvos, Comparison of dihydrodiazepam enantiomers: Metabolism, serum binding and brain receptor binding, *Experientia*, *36*:1031 (1980).

13

FDA Perspective on the Development of New Stereoisomeric Drugs: Chemistry, Manufacturing, and Control Issues

Wilson H. De Camp *Center for Drug Evaluation and Research, U.S. Food and Drug Administration, Rockville, Maryland*

One of the most technically challenging aspects of the manufacturing and control of chiral drug substances and drug products is the detection and quantitation of one enantiomer of a chiral drug substance in the presence of the other. This is true for both racemic and enantiopure drug substances, as well as the drug products manufactured from them. The fact that diastereomers have differing physical and chemical properties renders the regulatory considerations related to their analytical controls, manufacturing processes, and stability fundamentally similar to those for nonchiral drugs.

The relevance of this analytical methodology to the development of new stereoisomeric drugs has been recently addressed by the U.S. Food and Drug Administration (FDA) in all areas of development, ranging from manufacturing control through pharmacology to clinical evaluation (1). It is stated therein "[t]his policy statement is intended to address these issues thereby assisting manufacturers in the development of new stereoisomeric drugs." The relevance of these same concerns to the generic duplicate versions of marketed drug products is reflected in a statement reflecting awareness that the standards for identity of the active ingredient in such drugs must take stereochemistry into consideration (2 at 17958–9).

I. DEFINITIONS

A few of the finer points of technical language are worthy of special emphasis, if only as a reminder. The term *chiral* is used to refer to both racemates and pure enantiomers. From a regulatory viewpoint, the standards for identity, strength, quality, and purity of both the drug substance and drug product are the same for both chiral and nonchiral drugs. It is only in specific technical details, such as the choice of a regulatory method or the establishment of regulatory specifications, that our approaches may diverge.

The term *racemate* refers to any mixture of equal or nearly equal amounts of enantiomeric molecules. In reference to solid state properties, the most common type of racemate is a racemic compound, in which each crystal, no matter how small, is racemic. The racemic conglomerate or mixture, such as Pasteur resolved, is much less commonly encountered. In this case, each crystal is composed entirely of the same enantiomer, but the bulk sample approaches a 1:1 ratio.

To describe a bulk material that is composed of one enantiomer with no significant amount of the other, the term *enantiopure* has been recently proposed by Eliel and Wilen (3). For reasons of clarity, this term is preferable to older usages, such as "optically pure".

The terms *eutomer* and *distomer* have been coined to refer to the members of an enantiomeric pair that have, respectively, the greater and lesser physiological activities. It is generally understood, but often unstated, that this reference is necessarily related to the single activity being studied. The eutomer for one effect may well be the distomer when another is studied. For purposes of clarity, these terms should be avoided in regulatory writing, though they may be quite technically appropriate in other circumstances.

II. GENERAL POLICY

The central regulatory objective for a chiral drug is knowledge of its composition. Whether racemic, enantiopure, or intermediate, the quantitative isomeric composition must be known for the drug supplies used in studies of its pharmacology, toxicology, and clinical effectiveness. Since these findings lead to the development of specifications for a marketed product, they establish the stereochemically sensitive analytical method as the nexus of chiral drug development.

The justification for the development of analytical methodology for quantitation of one enantiomer in the presence of the other is not derived solely from its importance as a control on the manufacturing process. It is equally essential for those investigations of absorption, distribution, biotransformation, and excretion that constitute the pharmacokinetic profile of the chiral drug (1,4).

Because diastereomeric impurities and degradation products, as well as geometric isomers, present an analytical challenge that is substantially similar to that of other impurities, the FDA policy statement specifically excludes these stereochemical problems. The considerations here are limited to those that are the direct consequence of the analytical problem of simultaneously quantitating molecules that are identical in their physical (except for the rotation of the plane of polarization of light) and most chemical properties.

Although the policy statement primarily addresses the development of new stereoisomeric drugs, its significance is not limited to this phase. The manufacturing controls developed during this time form the basis for demonstrating the batch-to-batch consistency of the manufacturing process. They are, therefore, a key part of compliance with good manufacturing practices (GMPs).

III. ANALYTICAL CONTROL METHODS AND SPECIFICATIONS FOR CHIRAL DRUGS

A. Drug Substances

The specifications for both enantiopure and racemic chiral drug substances should be sufficient to establish both chemical and stereochemical aspects of identity, strength, quality and purity. This implies both that the identity test use a stereochemically specific method and that the assay method be stereochemically selective.

B. Specifications for Stereochemical Identity or Purity

Regulatory specifications for the identity of a bulk drug substance have two major applications. First, they are used as controls on the manufacture of the bulk drug substance. Second, they document compliance with GMPs by verifying the identity of the drug component before its release for further manufacturing. Such identity tests as the infrared spectrum or chromatographic retention time are unlikely to be capable of discriminating between stereochemical forms of a drug in most cases. An IR spectrum may be suitable for establishing stereochemical identity if there are obvious differences between the racemic and enantiopure material, but this may be difficult to demonstrate. Melting ranges may also be suitable, but there appears to be a trend toward categorically eliminating these specifications from compendial monographs in favor of instrumental methods. In any event, a conclusion that an identity test is stereochemically sensitive should be supported by test results on the other stereochemical form(s).

Many chiral compounds are known by the chemist to be racemic because of the lack of stereoselective influences on the synthesis or to be enantiopure because of natural origin. Such knowledge, while based on a sound technical foundation, may not be suitable for regulatory purposes. For example, the commercial availability of a racemate or the "opposite" enantiomer may make its substitution for the approved component conceivable. Therefore, it may be necessary to bring other factors (e.g., synthetic feasibility or commercial sources) into consideration to establish whether a stereochemically specific identity test is necessary.

Certain identity tests intended to establish the stereochemical identity of a drug are known. Their applicability to a specific drug substance may vary depending upon the magnitude of differences between the values to be determined. These are discussed below in the following order (which does not suggest the suitability of the technique for regulatory purposes):

1. Optical rotation
2. Melting range
3. Infrared spectrum
4. X-ray powder diffraction
5. Optical rotatory dispersion
6. Chiral chromatography (including liquid, gas, and thin-layer, as well as chiral capillary electrophoresis)

Optical rotation has the dual advantages of historical use and widespread recognition in the compendia. For an enantiopure material, it defines its configuration when used in conjunction with other valid chemical tests. However, optical rotation has been used ineffectively when the primary analytical goal is the determination of stereochemical purity. The limits selected for the specification seem to be unrelated to the purity required by other methods. For example, the compendial monograph for naproxen requires that the drug substance meet a specification of "between +63.0 and +68.5°" in a chloroform solution. Based on the published specific rotation, this corresponds to a stereochemical purity of 95.5 to 103.7%, compared to the assay limits of 98.5 to 100.5%, determined by titration with sodium hydroxide (5).

The traditional sodium D wavelength was (historically) chosen for practical reasons associated with the use of visual polarimeters rather than for any technical advantage. Current technology makes the photoelectric polarimeter the instrument of choice, permitting the use of several lines in the mercury spectrum, in addition to the classical choice of the sodium D wavelength. Chafetz (6) has provided a variety of examples of the higher specific rotations (and resulting increased sensitivity) obtained at the shorter wavelengths. Thus, specific rotation should be determined at several wavelengths, and the choice of which to use should be justified.

A major disadvantage of optical rotation as a measure of stereochemical purity is that interpretation of the single value lacks significance if any impurities are chiral. To insure the validity of an optical rotation specification, it should be supported by specific rotation measurements on any known chiral impurity or degradation product. This is, of course, in addition to the usual studies of the effects of solvent, concentration, and temperature on the specific rotation of the principal analyte.

A relatively common use of optical rotation is as an identity test for a racemate with specification limits that are symmetrical around zero. Such a specification has little if any regulatory significance. Its validation necessarily depends upon knowledge of the specific rotation and thus requires the resolution of the racemate on a laboratory scale. Furthermore, even with such supporting data, the method is dependent upon the accuracy of the sample preparation, since a solvent blank would also show a rotation of zero. Other analytical methods are far more appropriate for the stereochemically specific identification of racemates.

As a stereochemically specific identity test, a measurement of melting range is complementary to optical rotation. It can suffice to identify the racemic form but cannot distinguish enantiopure materials of opposite handedness. The melting range of the unwanted stereochemical form must also be known if such a test is to be valid. Chafetz (6) has cited examples of melting ranges of official drugs, some of which have received compendial recognition, even though they are of questionable validity. However, it should not be assumed that nothing is learned from these measurements.

For example, racemic pseudoephedrine melts at 118°C, while the pure enantiomer melts at 119°C (7). This near identity of melting ranges, while of little regulatory value, suggests that the racemate may exist as a crystalline conglomerate, rather than as a racemic compound. Because of such situations, prudence in the development of analytical specifications suggests that the minimum difference in the melting ranges of racemic and enantiopure material should be at least twenty degrees Celsius for use as a regulatory identity specification for a chiral drug. Such a difference will, in general, be sufficient to distinguish between a crystalline racemic compound and racemic conglomerate (8).

Infrared spectra, where they are used as regulatory specifications, generally make use of a reference standard. The specification is usually written to require that the sample spectrum "exhibits maxima only at the same wavelengths" as the reference spectrum. This type of specification

discussed on the basis of individual drugs, since one cannot provide a general guideline for the entire class. Nevertheless, the concept offers new opportunities in the field of drug development that must be explored and wisely exploited.

The decision to develop a stereochemically pure compound must be based on pharmacodynamic, safety, pharmacokinetic and pharmaceutic considerations. Therefore, attempts will be made to discuss the issue of racemate vs. enantiomer from these viewpoints, as well as those of regulatory affairs. This chapter is not intended to be exhaustive, hence, many more examples may exist than those to which we refer.

I. PHARMACODYNAMIC AND TOXICITY CONSIDERATIONS

From the therapeutic point of view, racemic drugs are divided into three groups (a detailed listing can be found in Ref. 4): (1) racemates with equipotent enantiomers, (2) those with one enantiomer possessing the majority or all of the beneficial properties, (3) racemates with properties superior to those of their stereoisomers, and (4) racemates with enantiomers possessing stereospecific effects (i.e., enantiomers having entirely different pharmacodynamic properties). For the latter group of racemic drugs, there appears to be little controversy as to the need for separation of the enantiomers. Propoxyphene is an example of a drug with stereospecific effects [(−)-enantiomer analgesic while (+)-enantiomer antitussive].

It must be pointed out that many drugs may belong to more than one of the above categories, and with the ever-growing knowledge of stereochemistry of drugs' action and disposition, they may be more appropriately classified into different categories. For example, the β-adrenoceptor blocker, propranolol, may be classified under the second category with respect to its cardiovascular effects, whereas as a spermicidal agent, it belongs in the first category (5).

A. Equipotent Enantiomers

There are not many known racemic drugs with equipotent enantiomers (4). Examples include the antiasthmatic drug proxyphylline, the antineoplastic cyclophosphamide, the antimalarial agent primaquine, and the antiarrhythmic agents propafenone, flecainide, and disopyramide. For these drugs there seems to be little rationale for separating the two enantiomers, as the known target receptors do not distinguish between the two chemical entities. However, although the enantiomers may be qualitatively and quantitatively similar with respect to the main therapeutic properties for which they are indicated, they may have subtle and sometimes unknown differences. For example, although enantiomers of propranolol have equal potencies in reducing sperm motility, the β-adrenoceptor property of the drug is attributed mainly to its *S* enantiomer (4). Hence, as a spermicide, *R*-propranolol may be a safer drug as it reduces sperm motility without much of an effect on the cardiovascular system. A similar scenario has been observed for another β-blocker, timolol: Both enantiomers appear to be effective in the treatment of open-angle glaucoma (6). However, topical use of the available *S*-timolol has been associated with some systemic side-effects such as asthmatic attacks (7). On the other hand, the *R* enantiomer that only has weak β-blocking activity increases retinal/choroidal blood flow that can improve the treatment (8). Therefore, the availability of an optical *R*-timolol seems to be timely.

Enantiomers of propafenone, a potent antiarrhythmic drug, are equipotent with respect to their sodium channel-blocking activity. However, *R*(−)-propafenone has significantly less β-blocking potency. Hence, in patients who do not need β-blocking activities, the stereochemical pure *R*(−)-propafenone may be a safer alternative to the available racemate as it is expected to produce less nonspecific peripheral vasodilation (9). However, this remains to be proven. Nevertheless, the pharmacokinetics of propafenone are complicated due to its extensive stereoselectivity, first-pass metabolism, and genetically controlled metabolism (9). The time course of the drug is also altered by the presence of many other drugs (e.g., 10) and by food (11). Although the stereoselective nature of these alterations is not fully explored, a less complicated formulation, that is, a stereochemically pure product, may demonstrate a more predictable pharmacokinetic/dynamic pattern than the racemate.

Interestingly, the secondary effects of enantiomers may be masked by the presence of their antipodes and only surface after the therapeutic and toxicity properties of the enantiomers are compared with their respective racemate. For example, enantiomers of disopyramide illicit equal antiarrhythmic effects, but the *S* enantiomer possesses a greater extent of anticholinergic side-effects (12). Intuitively, a formulation composed of the *R* enantiomer would seem to be a safer alternative. However, it appears that the single enantiomer of disopyramide possesses additional side-effects that are minimal after administration of the racemic drug.

B. One Enantiomer with All or Most Activity

Many racemic drugs belong to this group for which one enantiomer is known to possess the majority or all the beneficial properties; for example, important drugs in this group are many of the chiral nonsteroidal anti-

nontoxic isomer. These hypothetical situations have not been demonstrated in clinical practice, although enantiomer-selective absorption clearly occurs with dihydroxyphenylalanine (dopa) (12) and methotrexate (13) and enantiomer-selective renal drug elimination has been demonstrated for the β-adrenoceptor antagonist, pindolol (14).

C. Mixtures of Isomers with Opposing Pharmacodynamic Activities

Dihydropyridine calcium antagonist drugs, with the exception of nifedipine that lacks an assymetric center, have been uniformly developed as racemic mixtures. Blockade of the potential operated calcium channel has been consistently associated with one enantiomer, with the other generally considered inert. Hof et al. demonstrated rather conclusively for one dihydropyridine, isopropyl 4-(2,1,3,-benzoxadiazol-4-yl)-1,4-dihydro-2,6-dimethyl-5-nitro-3-pyridinecarboxylate, that the (−)-isomer inhibited calcium flux through the potential operated calcium channel, while the (+)-isomer enhanced calcium flux and these effects resulted in the inhibition and enhancement, respectively, of rabbit aortic ring contraction (15). In this case, when the racemic mixture was studied at low concentrations, aortic ring contraction was enhanced, whereas at higher concentrations, contraction was inhibited. Interpretation of the pharmacodynamic activity of the compound was only possible by study of the enantiomers separately, and obviously development of such a compound as a therapeutically useful drug would require development of the enantiomer with the desired pharmacodynamic effect.

An interesting variation on this theme is the rather conclusive demonstration that each of the enantiomers of a number of opiates including methadone, codeine, and morphine bind selectively to different opiate receptors in mouse brain, with high binding stereoselectivity by each enantiomer for the specific receptor (16). The possible clinical correlate is unique pharmacodynamic activity attributable to occupation of the respective receptors with *l*-pentazocine, a more potent analgesic drug, and *d*-pentazocine associated with dysphoric anxiety (17). Again, administration of the racemic mixture of the synthetic opiates, the drugs available for therapeutic use, leads to a complex summation of the pharmacodynamic effects of each of the enantiomers (18). Among the opiate and synthetic narcotics, characterization of the various central and peripheral receptors supports the likelihood that differing pharmacodynamics of each isomer are related to occupation of specific and different receptors by the enantiomers. Development of racemic mixtures of these drugs may provide an example in which therapeutic usefulness was impaired, and development of specific isomers that occupy specific opiate receptors would provide drugs with a different pharmacodynamic effect or an improved therapeutic index.

D. Mixtures of Isomers with Similar Pharmacodynamic Effects, Although Often with Different Potency and/or Pharmacokinetic Characteristics

Coumarin-derivative anticoagulants provide an instructive example of isomers with similar pharmacodynamic effects. However, it is an example in which one isomer is much more potent [in the case of *S*- and *R*-warfarin an 8:1 potency ratio (19)] and the isomers have different pharmacokinetic characteristics (20). In clinical practice, when therapy is administered to a pharmacodynamic endpoint such as prolongation of plasma prothrombin time, the use of a pure isomer has no advantage over the racemic mixture. However, to understand drug–drug interactions involving warfarin, the use of pure isomers is critical. In the case of a recently reported diltiazem-warfarin pharmacokinetic interaction, although the clearance of racemic warfarin was markedly impaired, no pharmacodynamic effect was noted due to diltiazem-induced inhibition of the clearance of the relatively inactive *R*-warfarin (20). In contrast, in the pharmacokinetic interaction between phenylbutazone and warfarin, in which the clearance of racemic warfarin is similarly impaired, a striking pharmacodynamic effect was noted as a result of phenylbutazone-induced inhibition of *S*-warfarin (21).

Although warfarin and many other drugs that are extensively bound to plasma proteins exhibit stereoselectivity in this binding (19,22), the transient nature of stereoselective protein binding interactions in the absence of changes in drug clearance leads to few demonstrable clinical consequences when a racemic drug is administered in contrast to a pure isomer. However, the stereoselectivity of plasma protein binding may dictate the extent of tissue distribution of each isomer, as in the case of propranolol (23). Here, if the isomers displace each other from binding, one could envision the potency or tissue-selective toxicity of an active isomer in a racemic drug mixture to be different when the active isomer is administered alone. However, no examples of such a situation have been reported. A related example is the case of disopyramide, in which plasma protein binding of the isomers exhibits stereoselectivity, and binding of each isomer is concentration-dependent (nonlinear) (24). When administered separately, the *d* and *l* isomers of disopyramide had similar clearance, distribution, and elimination half-life. However, when administered as the racemic mixture, they had lower total clearance and a smaller apparent volume of distribution was noted for *l*-disopyramide. Due to the nonlinear protein binding of each isomer, with the increased free fraction of each at higher concentrations, one would anticipate that the difference in clearance (and steady-state concentration during chronic dosing) would diminish at higher doses and concentrations of racemic disopyramide. This example is explained in some detail to outline potential pharmacokinetic complexity. However, clinically if a pharmacodynamic measure (such as arrhythmia suppression) is used to determine dose, the usefulness of a racemic drug is not significantly impaired. In contrast, during drug development a clear understanding of such pharmacokinetic complexity is important to establish dose interval and dose-concentration-effect relationships.

III. AREAS REQUIRING INVESTIGATION

A. Stereoselective Pharmacokinetics in Disease States

During drug development, pharmacokinetics in patients with liver disease, renal disease, and when appropriate cardiac disease are routinely studied. At the present time, little information is available for drugs administered as racemic mixtures to determine if the stereoselective disposition observed in healthy volunteers remains or changes in the presence of these states. In the absence of data, one can only speculate. However, these diseases do have marked effects on a number of drug disposition processes that exhibit marked stereoselectivity. With severe liver disease, it seems unlikely that each of the stereoselective oxidative processes will decline in parallel. At present no data are available to determine if racemic plasma drug concentrations after single or multiple drug doses are composed of the same ratios of drug isomers.

Similarly, in the case of renal disease, Giacomini and others have clearly demonstrated the stereoselectivity of renal drug excretory processes (14,27). Again with decreased renal function, including glomerular filtration, proximal reabsorption, and secretion, no information is available to determine whether the stereoselective processes decline in parallel with nonstereoselective mechanisms, such that after the administration of a racemic drug, the isomeric ratio in plasma remains the same.

Finally, absorption and first-pass hepatic extraction of racemic mixtures of high-clearance drugs have been demonstrated to have marked stereoselectivity when studied in healthy volunteers. Examples are verapamil (28) and tertbutaline (29). In the case of verapamil, the pharmacologically more active isomer, *l*-verapamil, undergoes much more extensive first-pass hepatic extraction after oral administration. Thus, when racemic-mixture plasma verapamil concentrations are measured, a markedly less pharmacodynamic effect at a given plasma drug concentration is observed after oral drug administration. For tertbutaline, in which (−)-tertbutaline is the pharmacologically active isomer, the oral bioavailability of (−)-tertbutaline was reported to be 14.8% and that of (+)-tertbutaline 7.5%. Since tertbutaline is predominantly excreted unchanged (55%) in the urine, the authors interpreted this stereoselectivity in bioavailability to be predominantly at the level of drug absorption. In the case of patients with liver disease, in which first-pass hepatic extraction is diminished (30), there are no data available to predict whether or not the stereoselectivity of first-pass extraction of drugs remains constant as the extraction ratio is diminished due to disease. Congestive heart failure in patients has been demonstrated to alter both first-pass hepatic drug extraction and the apparent volume of distribution of selected drugs such as lidocaine (31). However, again there are no data to allow the prediction of stereoselectivity of either hepatic extraction or drug distribution when racemic mixtures of drugs are administered to patients with congestive heart failure.

B. Stereoselective Pharmacokinetics in Aging

It is now generally accepted that drug pharmacokinetics are altered in aging humans (32). Preliminary reports indicate that otherwise healthy aging individuals experience altered stereoselective disposition of the barbiturate derivatives hexobarbital (33) and mephobarbital (34). For mephobarbital, *R*-mephobarbital had much higher clearance than *S*-mephobarbital in young subjects, whereas in the elderly, the clearance of *R*-mephobarbital was decreased and *S*-mephobarbital clearance was not age-related. Such findings support the likelihood that the stereoselective drug disposition for other racemic drugs may change with age as well.

C. Stereoselective Pharmacodynamics in Disease States and Aging

Levy and others (35,36) have developed an extensive literature which demonstrates in animal models of disease that the pharmacodynamics of a number of drugs are altered, even after controlling for pharmacokinetic changes. Studies with drugs administered as racemic mixtures in humans with renal dysfunction, hepatic dysfunction, and other disease states that address the issue of stereoselective pharmacodynamics are lacking. However, the cited animal studies suggest that such a line of investigation would be fruitful.

IV. RACEMIC DRUGS AND THERAPEUTIC DRUG MONITORING

The development of consistent and reproducible relationships between plasma drug concentration and therapeutic and/or toxic drug effect across patient populations is an important requirement before considering the use of plasma drug concentrations to guide pharmacotherapy. This approach is considered less desirable than the use of clinical and/or pharmacodynamic endpoints to establish drug dose and the presence of therapeutic effects or toxicity if these parameters are clinically obvious, predictable, or measurable by a simple laboratory test (e.g. plasma prothrombin time). However, to properly use drugs with a low therapeutic index, for example, digoxin, theophylline, phenytoin, phenobarbital, aminoglycosides, and a number of antiarrhythmic drugs, plasma drug concentration monitoring is essential to properly treat patients. Fortunately for the clinician, the antiarrhythmic agents are the only major group of drugs administered in most instances as racemic mixtures and for which plasma drug concentration monitoring is routinely used to guide therapy. As has been suggested by Reidenberg (37) and others for these selected few occasions, the availability of either concentrations of the active isomer, or preferably administration of the active isomer only, is necessary for optimal therapeutic drug monitoring. In contrast, for most drugs administered as racemic mixtures, such as warfarin, calcium antagonists, β-adrenoceptor blockers, and tricyclic antidepressants, therapy is most appropriately guided by the use of pharmacodynamic and clinical endpoints. Therefore, to suggest that the use of racemic drugs markedly impairs the usefulness of therapeutic drug monitoring is not justified.

V. CLINICAL RESEARCH IN DRUG DEVELOPMENT

Much has been written regarding a proper approach to the development of racemic mixtures of drugs, and whether such drugs are even candidates for development (1,8,22,37–42). As clinical pharmacologists and medical practitioners, we shall advance a point of view that represents a compromise among the various camps, but one that we believe can effectively serve medical therapeutics now and in the future. The previous material in this paper represents a summary, not a comprehensive review, of the database from which attitudes, opinions, and therapeutic approaches can be formulated.

As has been observed on repeated occasions, many drugs in therapeutic use are administered as racemic mixtures. The therapeutic efficacy of these drugs has generally been demonstrated and the clinical practitioner is familiar with their use. Particularly for drugs in this group with a wide therapeutic index such as the β-adrenoceptor blockers, effective therapeutics would only be hampered by taking issue simply because they are administered as racemic mixtures. Therefore, we adopt the view that such drugs should be "grandfathered" and fully accepted as racemic mixtures.

That is not to say continued research with pure isomers of such drugs should be hampered. Instead, many important pharmacological and physiological questions can be explored using such pure isomers. In addition, such investigations may enhance the therapeutic usefulness of the drug racemic mixture.

How best to serve therapeutics for racemic mixtures of drugs in development or being considered for clinical development? It seems a desirable objective to characterize the pharmacokinetics and pharmacodynamics of pure isomers of drug candidates and to select a pure isomer on the basis of these studies for further development. This tactic would prevent the various difficulties as summarized earlier in this chapter when attempts are made to characterize the disposition, efficacy, and toxicity of drug candidates. The sponsor of a drug candidate should have the opportunity to request an exception from such a policy, and a number of situations can be envisioned in which such a request should be viewed favorably. They are the following:

1. Development of a pure isomer (in contrast to a racemic mixture) would markedly delay the development of a drug that may represent a breakthrough in the treatment of a life-threatening disease. The drug sponsor would then be expected to develop pharmacokinetic and pharmacodynamic data on both the racemic mixture and the individual isomers concurrent with the regulatory evaluation and approval process.
2. Synthesis and/or resolution of isomers in commercial quantities for a drug candidate viewed as a significant advance for a therapeutic indication is prohibitively expensive, thereby in fact preventing development of the putative therapeutic advance.

In contrast, the more common situation is the proposed development of the racemic mixture of a drug that is an extension of a chemical or therapeutic class of drugs. Examples are β-adrenoceptor blockers, calcium antagonists, and nonsteroidal antiinflammatory drugs. The development of such drugs has been recognized as important, as often changes in pharmacokinetic properties may be therapeutically advantageous, the therapeutic index may be improved, and on occasion individual patients will have an improved therapeutic response to one member of a drug class. Such drug candidates, however, do not generally lead to therapeutic breakthroughs in the treatment of disease. Therefore, we believe in such cases drug sponsors should be required to develop an isomerically pure drug.

The future is bright for the development of single stereoisomers of drugs for two reasons. First, the present situation in which large numbers of drug racemic mixtures are in therapeutic use simply represents a lack of synchrony in scientific progress. The synthesis of complex organic molecules with both asymmetry and potent pharmacological activity moved ahead of chiral synthesis. At the present time, important advances in chiral synthesis (43) suggest that the synthesis of pure isomers of drug candidates is rapidly coming abreast of the identification of drug targets for which a candidate may be effective. Second, the rapid development of drugs that are prepared by biological systems synthesizing only one isomer of a complex molecule, and in commercial quantities, obviates the racemic mixture issue for an increasingly large number of drug candidates.

VI. CONCLUSION

At the present time, the clinical pharmacologist and practitioner utilize a number of racemic mixtures of drugs to optimally provide pharmacotherapy to patients. Unfortunately, this is in effect administering drug mixtures, each component of which may have unique pharmacokinetics and pharmacodynamics. The efficacy of these drug mixtures has generally been established, and acceptance of their continued use seems the appropriate stance. In contrast, chiral drug candidates either in development or being considered for such should generally be developed only as pure isomers.

Acknowledgment. The authors gratefully recognize support from the National Institutes of Health (Grant No. R01-AG08226) and the Pharmaceutical Manufacturers Association Foundation. Sincere appreciation to Grazina Kulawas for administrative support and manuscript preparation is also expressed.

REFERENCES

1. D. J. Birkett, "Racemates or enantiomers: Regulatory approaches," *Clin. Exp. Pharmacol. Physiol.*, *16*:479–483 (1989).
2. E. J. Ariëns, E. W. Wuis, and E. J. Veringa, "Stereoselectivity of bioactive xenobiotics, a pre-Pasteur attitude in medicinal chemistry, pharmacokinetics and clinical pharmacology," *Biochem. Pharmacol.*, *37*:9–18 (1988).
3. A. M. Barrett and V. A. Cullum, "The biological properties of the optical isomers of propranolol and their effects on cardiac arrhythmias," *Br. J. Pharmacol.*, *34*:43–55 (1968).
4. S. S. Adams, P. Bresloff, and G. G. Mason, "Pharmacological difference between the optical isomers of ibuprofen: Evidence for the metabolic inversion of the (−) isomer," *J. Pharm. Pharmacol.*, *28*:256–257 (1976).

5. R. I. Brittain, G. M. Drew, and G. P. Levy, "The alpha- and beta-adrenoceptor blocking potencies of labetalol and its individual stereoisomers in anaesthetized dogs and in isolated tissues," *Br. J. Pharmacol.*, *77*:105–114 (1982).
6. D. R. Abernethy, J. B. Schwartz, J. R. Plachetka, E. L. Todd, and J. M. Egan, "Comparison in young and elderly patients of pharmacodynamics and disposition of labetalol in systemic hypertension," *Am. J. Cardiol.*, *60*:697–702 (1987).
7. D. M. Tenero, M. B. Bofforff, B. D. Given, W. G. Kramer, M. B. Affrime, J. E. Patrick, and R. L. Lalonde, "Pharmacokinetics and pharmacodynamics of dilevalol," *Clin. Pharmacol. Ther.*, *46*:646–656 (1989).
8. M. S. Lennard, "Clinical pharmacology through the looking glass: Reflections on the racemate vs enantiomer debate," *Brit. J. Clin. Pharmacol.*, *31*:623–625 (1991).
9. J. A. Clark, H. J. Zimmerman, and L. A. Tanner, "Labetalol hepatoxicity," *Ann. Int. Med.*, *113*:210–213 (1990).
10. S. Fabro, R. L. Smith, and R. T. Williams, "Toxicity and teratogenecity of optical isomers of thalidomide," *Nature*, *215*:296 (1967).
11. P. F. White, J. Ham, W. L. Way, and A. J. Trevor, "Pharmacology of ketamine isomers in surgical patients," *Anesthesiology*, *52*:231–239 (1980).
12. D. N. Wade, P. T. Mearrick, and J. L. Morris, "Active transport of L-dopa in the intestine," *Nature*, *242*:463–465 (1973).
13. J. Hendel and H. Brodthagen, "Enterohepatic cycling of methotrexate estimated by use of the D-isomer as a reference marker," *Eur. J. Clin. Pharmacol.*, *26*:103–107 (1984).
14. P.-H. Hsyu and K. M. Giacomini, "Stereoselective renal clearance of pindolol in humans," *J. Clin. Invest.*, *76*:1720–1726 (1985).
15. R. P. Hof, U. T. Rüegg, A. Hof, and A. Vogel, "Stereoselectivity at the calcium channel: Opposite action of the enantiomers of a 1,4-dihydropyridine," *J. Cardiovasc. Pharmacol.*, *7*:689–693 (1985).
16. B. R. Martin, J. S. Katzen, J. A. Woods, H. L. Tripathi, L. S. Harris, and E. L. May, "Stereoisomers of [^{3}H]-*N*-allylnormetazocine bind to different sites in mouse brain," *J. Pharmacol. Exp. Ther.*, *231*:539–544 (1984).
17. W. H. Forrest, J. Beer, and J. W. Bellville, "Analgesic and other effects of the *d* and *l*-isomers of pentazocine," *Clin. Pharmacol. Ther.*, *10*:468–476 (1969).
18. G. D. Olsen, H. A. Wendel, J. D. Livermore, R. M. Leger, R. K. Lynn, and N. Gerber, "Clinical effects and pharmacokinetics of racemic methadone and its optical isomers," *Clin. Pharmacol. Ther.* *21*:147–157 (1977).
19. S. Toon and W. F. Trager, "Pharmacokinetic implications of stereoselective changes in plasma-protein binding: Warfarin-sulfinpyrazone," *J. Pharm. Sci.*, *73*:1671–1673 (1984).
20. D. R. Abernethy, L. S. Kaminsky, and T. H. Dickinson, "Selective inhibition of warfarin metabolism by diltiazem in humans," *J. Pharmacol. Exp. Ther.*, *257*: 411–415 (1991).
21. R. J. Lewis, W. F. Trager, K. K. Chan, A. Breckenridge, M. Orme, M. Rowland, and W. Schary, "Warfarin: Stereochemical aspects of its metabolism and interaction with warfarin," *J. Clin. Invest.*, *53*:1607–1617 (1974).

22. D. E. Drayer, "Pharmacodynamic and pharmacokinetic differences between drug enantiomers in humans: An overview," *Clin. Pharmacol. Ther.*, *40*:125–133 (1986).
23. H. Takahashi, H. Ogata, S. Kanno, and H. Takeuchi, "Plasma protein binding of propranolol enantiomers as a major determinant of their stereoselective tissue distribution in rats," *J. Pharmacol. Exp. Ther.*, *252*:272–278 (1990).
24. K. M. Giacomini, W. L. Nelson, R. A. Pershe, L. Valdivieso, K. Turner-Tamiyasu, and T. F. Blaschke, "In vivo interaction of the enantiomers of disopyramide in human subjects," *J. Pharmacokinet. Biopharm.*, *14*:335–356 (1986).
25. D. D. Breimer and J. M. van Rossum, "Pharmacokinetics of (+)-, (−)- and (plus or minus)-hexobarbitone in man after oral administration," *J. Pharm. Pharmacol.*, *25*:762–764 (1973).
26. M. Eichelbaum, G. Miksu, and B. Vogelgesang, "Pharmacokinetics of dextro, levo, and racemic verapamil after intravenous administration," *Br. J. Clin. Pharmacol.*, *17*:453–458 (1984).
27. D. A. Notterman, D. E. Drayer, L. Metakis, and M. M. Reidenberg, "Stereoselective renal tubular secretion of quinidine and quinine," *Clin. Pharmacol. Ther.*, *40*:511–517 (1986).
28. H. Echizen, B. Vogelgesang, and M. Eichelbaum, "Effects of *d*, *l* verapamil on atrioventricular conduction in relation to its stereoselective first-pass metabolism," *Clin. Pharmacol. Ther.*, *38*:71–76 (1985).
29. L. Borgstrom, L. Nyberg, S. Jonsson, C. Lindberg, and J. Paulson, "Pharmacokinetic evaluation in man of tertbutaline given as separate enantiomers and as the racemate," *Br. J. Clin. Pharmacol.*, *27*:49–56 (1989).
30. A. Somogyi, M. Albrecht, G. Kliems, K. Schafter, and M. Eichelbaum, "Pharmacokinetics, bioavailability, and ECG response of verapamil in patients with liver cirrhosis," *Br. J. Clin. Pharmacol.*, *12*:51–60 (1981).
31. P. D. Thomson, K. L. Melmon, J. A. Richardson, K. Cohn, W. Steinbrum, R. Cudihee, and M. Rowland, "Lidocaine pharmacokinetics in advanced heart failure, liver disease, and renal failure in humans," *Ann. Int. Med.*, *78*:499–508 (1973).
32. D. J. Greenblatt, E. M. Sellers, and R. I. Shader, "Drug disposition in old age," *N. Engl. J. Med.*, *306*:1081–1088 (1982).
33. M. H. H. Chandler, S. R. Scott, and R. A. Blouin, "Age-associated stereoselective alterations in hexobarbital metabolism," *Clin. Pharmacol. Ther.*, *43*:436–441 (1988).
34. W. D. Hooper and M. S. Qing, "The influence of age and gender on the stereoselective metabolism and pharmacokinetics of mephobarbital in humans," *Clin. Pharmacol. Ther.*, *48*:633–640 (1990).
35. M. Danhof, M. Hisaoka, and G. Levy, "Kinetics of drug action in disease states. II. Effect of experimental renal dysfunction on phenobarbital concentrations in rats at onset of loss of righting reflex," *J. Pharmacol. Exp. Ther.*, *230*: 627–631 (1984).
36. M. Danhof, M. Hisaoka, and G. Levy, "Kinetics of drug action in disease states. XII: Effect of experimental liver diseases on the pharmacodynamics of phenobarbital and ethanol in rats," *J. Pharm. Sci.*, *74*:321–324 (1985).

37. M. M. Reidenberg, "The therapeutic use of stereochemically pure drugs—a pragmatic point of view," *Drug Stereochemistry* (I. W. Wainer and D. E. Drayer, eds.), Marcel Dekker, New York, 1988, pp. 365–369.
38. F. Jamali, R. Mehvar, and F. M. Pasutto, "Enantioselective aspects of drug action and disposition: Therapeutic pitfalls," *J. Pharm. Sci.*, *78*:695–715 (1989).
39. E. J. D. Lee and K. M. Williams, "Chirality—clinical pharmacokinetic and pharmacodynamic considerations," *Clin. Pharmacokinet.*, *18*:339–345 (1990).
40. G. T. Tucker and M. S. Lennard, "Enantiomer specific pharmacokinetics," *Pharmacol. Ther.*, *45*:309–329 (1990).
41. D. B. Campbell, "Stereoselectivity in clinical pharmacokinetics and drug development," *Eur. J. Drug Metab. Pharmacokinet.*, *15*:109–125 (1990).
42. E. J. Ariëns, "Racemic therapeutics—ethical and regulatory aspects," *Eur. J. Clin. Pharmacol.*, *41*:89–93 (1991).
43. J. D. Morrison, ed., *Asymmetric Synthesis*, Academic Press, New York, Vols. 1–5 (1983–1985).

16

Development of Stereoisomeric Drugs

An Industrial Perspective

Mitchell N. Cayen *Schering-Plough Research Institute, Kenilworth, New Jersey*

I. INTRODUCTION

In previous chapters, analytical, biological, and other scientific aspects of drug stereochemistry were described. The purpose of this chapter is to discuss pharmaceutical drug development for those drugs containing a stereogenic center. It is currently estimated that it takes approximately 12 yr to bring a new chemical entity (NCE) to regulatory clearance in the United States, at a cost (including discovery and development) averaging $231 million (1). With such a major resource commitment by the pharmaceutical industry to research and development, it is critical that NCEs be subjected to a rational development program to appropriately evaluate their safety and efficacy. The development of drugs with chiral centers presents specific challenges that must be addressed at various stages from discovery through clinical evaluation and finally to market.

II. DRUG DEVELOPMENT PROGRAM

A typical drug development program is outlined in Fig. 1. NCEs with a chiral center are typically synthesized and initially tested for desired pharmacological activity in animal models and/or in vitro screens as a racemic mixture. Should no pharmacological activity be demonstrated, the NCE is no longer considered a viable candidate drug. However, should some activity emerge, the first decision related to the NCE's chirality is required. Subsequent biological tests are usually more specific regarding drug activity and often require more material for evaluation. What should

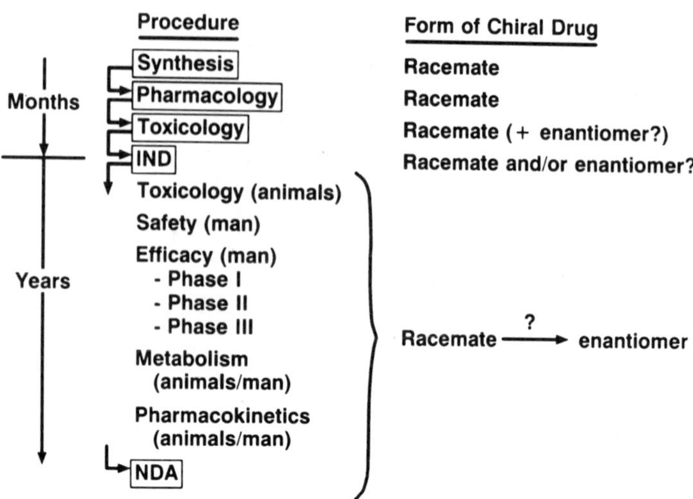

FIGURE 1 Typical drug development scheme (simplified).

be tested in these followup measurements: the racemic mixture or each of the component enantiomers? Depending on the chemistry of what is now a drug candidate, the synthesis/resolution of the individual enantiomers may be somewhat difficult and thus time-consuming, thereby delaying further testing. Since time is a critical resource in drug development and, at this stage, sufficient data are required for a decision on whether or not toxicity studies should be initiated, delays in synthesis of sufficient amounts of the pure enantiomers may be contraindicated. After all, should the racemate not be sufficiently active in the subsequent biological tests, the candidate drug would not be further evaluated. The earlier such information is available, the more efficient the overall process. However, should the drug product still demonstrate desirable activity, the next step would be initiation of the safety evaluation studies. Again, a critical decision is required. What should be tested? It would certainly be desirable to know whether the desired pharmacological activity resides in a single enantiomer (which is more common), or whether both enantiomers are required (rare). Such information may indeed be available as a result of synthesis of sufficient amounts of enantiomers on a laboratory scale. However, should activity be due to a single enantiomer, such small-scale synthesis may not be adequate for the large amounts required to initiate toxicological evaluations. Numerous techniques are emerging for the large-scale preparation of single enantiomers (2) and the fact that such entities need to be rapidly available for appropriate testing has provided

(2) new mixture of stereoisomers (standard biological differences and/or potential interactions between component stereoisomers); (3) enantiomer of marketed mixture of stereoisomers (if distomer active, document eutomer as NCE; if distomer inactive, eutomer submitted as generic application and all claimed advantages should be verified); and (4) generic application of marketed mixture (appropriate bioequivalence studies for *both* enantiomers).

It can be seen that the current positions of the different regulatory agencies are not comprehensive and, in some instances, not particularly clear. This can create problems for both sponsoring companies (who generally register drugs internationally and must, therefore, develop programs to satisfy all major regulatory agencies), as well as reviewers within agencies who must evaluate submitted data on chiral drugs. It would appear that the Japanese are not working on formal guidelines, and that the FDA has decided to issue a policy statement rather than an actual guideline. The CPMP is the only major regulatory body working on formal guidelines for chiral drugs. Current work on international harmonization is focusing on drug safety evaluation studies, and there has been no initiative to harmonize chiral drug submissions among the major regulatory bodies. The challenge rests with the drug sponsors to be cognizant of the regulatory atmosphere and expectations in the absence of formal guidelines.

V. INDUSTRIAL ISSUES IN DRUG DEVELOPMENT

Although this chapter focuses on chirality in drug development, the racemate vs. enantiomer "debate" also impacts on racemic drugs already in therapeutic use. Lennard (25) has recently argued that, for many marketed racemic drugs, such as the β-adrenoceptor antagonists, available evidence indicates that the therapeutically inactive enantiomer is not harmful, and that the patient would not benefit from receiving the pure active enantiomer. On the other hand, there is no doubt about the compelling reasons for chiral drugs that are used as single enantiomers, such as dopa [the (−)-form is less toxic], methotrexate [the (−)-form is more active], propoxyphene [the (+)-form is the analgesic, and the (−)-form an antitussive], and pentazocine [the (−)-form is more analgesic, the (+)-form causes a greater degree of anxiety].

From the drug development perspective, to state dogmatically that all chiral drugs should be developed as single enantiomers is simplistic and unrealistic. There is no doubt that there are strong compelling reasons to develop single enantiomers (5,14,26), as listed in Table 2. The principal reason is that, in most instances, the desired therapeutic activity resides

TABLE 2 Reasons to Develop a Single Enantiomer

1. Therapeutic activity resides principally in one enantiomer
2. Low therapeutic index
3. Toxicity associated with distomer
4. No chiral inversion
5. Economic feasibility

principally in a single isomer and, therefore, there is no clinical advantage to proceed with the racemate. This becomes even more acute if the active enantiomer exhibits a low therapeutic index or there is clinically significant toxicity associated with the distomer. The absence of chiral inversion assures enantiomeric purity within the dosage form and of the administered drug within the body. Finally, the bulk production of the single enantiomer both for testing and marketing should be sufficiently economically feasible so as to support reasonable cost structure to the patient.

Although the list of reasons to justify development of a racemate is numerically longer than that for the enantiomer (Table 3), the justifications for such a decision need to be convincing. In those relatively rare instances when there is additive or synergistic activities with both isomers, then the racemate is potentially the better drug, provided there is no enantioselective toxicity. If there is a high therapeutic ratio, this could support racemate development. The principal safety concern is the liability of the distomer insofar that the less active enantiomer should not impart significant toxicity to the racemate. Should chiral inversion occur to give rise to a fixed ratio of the two enantiomers and depending on the route/reversibility of the inversion, this could support a decision to proceed with a racemate.

TABLE 3 Justification for the Development of a Racemate

1. Additive or synergistic therapeutic activities of both enantiomers
2. High therapeutic index
3. Negligible/low toxicity of the distomer
4. Chiral inversion yields fixed ratio of enantiomers
5. Physicochemical characteristics
6. Novelty of drug
7. Indication for life-threatening disease
8. Cost-ineffective to manufacture active enantiomer on a large scale

Occasionally, the physicochemical characteristics (which determine dissolution and solubility characteristics) of the racemate may impart higher and/or less variable bioavailability than after administration of the single enantiomer; solubility is also a determining factor for intravenous or other routes of administration. If racemate development is perceived to be more rapid than that of the eutomer, then the novelty of the drug product or indication for life-threatening disease may be attempted as a justification for marketing the racemate. Cost-ineffective bulk synthesis has also been listed; however, although this could be proposed as an "excuse" a decade ago because the technology to manufacture bulk drug economically had not been extensively developed, such is rarely the case today. Process chemists have made significant advances in recent years to improve the technology such that cost-ineffective large-scale production would now need to be convincingly justified.

VI. RACEMATE OR ENANTIOMER: FACTORS TO CONSIDER DURING DRUG DEVELOPMENT

When a chiral compound synthesized as a racemate is found to have desirable pharmacological activity in initial tests, the decision-making process is begun as to whether the development should continue as a racemate or single enantiomer. Some issues to be addressed in this process are listed in Table 4. Since it is the rule rather than the exception that enantiomers exhibit different responses in the chiral environment of biological systems, such pharmacokinetic and pharmacodynamic differences should be assessed as early as possible. Often, delays can occur in such evaluations due to the unavailability of sufficient quantities for testing. Pharmacokinetic differences are assessed using a stereospecific assay, the development and validation of which may be time-consuming.

Should an early decision be made to develop the eutomer, then the drug development program would be the same as for conventional NCEs, with the possible exception that assessment of in vitro and/or in vivo chiral inversion may be desirable. However, if development continues with the racemate, time, cost, and staff resource commitments become magnified. For example, a very important variable to consider is species differences in enantiomer exposure. Appropriate toxicokinetic studies are advisable in order to assure that, at toxicological doses, the animal species tested have attained sufficient plasma concentrations of *each* enantiomer to support clinical evaluation at therapeutic doses in humans. The enantiomeric ratio (based on maximum drug concentrations [Cmax] and/or area under the plasma drug concentration-time curve [AUC]) should be evaluated, and

Table 4 Racemate or Enantiomer: Factors to Consider During Drug Development

1. Similarity/differences between enantiomers
   Pharmacodynamics (pharmacology and toxicity)
   Metabolic disposition
   Pharmacokinetics
2. Biological factors
   Chiral inversion
   Species differences in exposure to enantiomers
   Therapeutic index of the eutomer
   Liability of the distomer
3. Technical factors
   Development of stereospecific assay
   Timing: how much information available for IND?
   Physicochemical properties
   Difficulty/expense of bulk production
4. Therapeutic need/novelty
5. How is the patient best served?

any differences observed between animals and humans should be addressed in the overall safety evaluation of the drug.

The therapeutic index of the eutomer will also impact on the decision. Because the distomer provides no therapeutic advantage to the patient, the toxicity/liability/side-effects of this stereoisomer need to be evaluated. It is, therefore, probably wise to conduct at least some safety evaluation study(ies) with the distomer if the racemate is being considered for development.

Technical and other practical factors also merit consideration. Are there indeed technical difficulties in the manufacture of a bulk drug that would render marketing of the eutomer cost-prohibitive? Accumulation of relevant data for each stereoisomer is a time-consuming process. Are these delays justifiable therapeutically? When an IND is submitted for a racemate, there may be limited or negligible safety and/or pharmacokinetic data on the individual enantiomers. Should this delay initiation of phase I clinical trials?

The issue centers around how the patient is best served. The goal of both drug sponsors and drug regulators is to develop safe and effective drugs. The practical consideration with the high costs involved is the optimum utilization of time and cost resources. Each drug is an individual entity, and the decision must be based on the respective characteristics of the drug in question.

VII. CONCLUDING COMMENTS: WHERE DO WE STAND?

The enantiomer vs. racemate issue is still being termed a "debate." It is the author's view that a debate no longer exists. The science is well recognized. The chemical and analytical technology has seen dramatic advances in the past decade. Basic scientists, pharmaceutical developers, and regulatory authorities are well aware of the issues. Development of a single enantiomer is wise and prudent. Patients would not be exposed to a drug mixture. Pharmaceutical companies are very much aware that, in the 1990s, racemate development will be cost-ineffective and time-consuming. Regulatory agencies are asking the right questions and taking a very hard look at racemate NCE's; the justification must be convincing. Therefore, why ask for trouble?

It is relatively simple dogma to state that only single enantiomers should be candidates for future drugs. As discussed, there are and will be scenarios whereby the marketing of a racemic mixture will better serve the patient. The right questions need to be asked, the appropriate well-designed studies should be conducted and interpreted, and a balanced view needs to be taken. It is apparent that few new drugs will be marketed as racemates. However, our responsibility is to use good science and common sense in the decision-making process, and to keep in mind that dogma and prudence may not result in the best attainable therapeutic agent.

REFERENCES

1. P. R. Vagelos, "Are prescription drug prices high?", *Science*, *252*:1080–1084 (1991).
2. R. A. Sheldon, "The industrial synthesis of pure enantiomers," *Drug Inform. J.*, *24*:129–139 (1990).
3. E. J. Ariens, "Stereochemistry, a basis for sophisticated nonsense in pharmacokinetics and clinical pharmacology," *Eur. J. Clin. Pharmacol.*, *26*:663–668 (1984).
4. M. Gross, "Significance of drug stereochemistry in modern pharmaceutical research and development," *Ann. Reps. Med. Chem.*, *25*:323–331 (1990).
5. M. N. Cayen, "Racemic mixtures and single stereoisomers: Industrial concerns and issues in drug development," *Chirality*, *3*:94–98 (1991).
6. D. H. Deutsch, "Chiral drugs: The coming revolution," *Chemtech.*, *21*:157–159 (1991).
7. B. Testa and J. M. Mayer, "Stereoselective drug metabolism and its significance in drug research," *Progress. Drug. Res.*, *32*:249–298 (1988).
8. G. T. Tucker and M. S. Lennard, "Enantiomer specific pharmacokinetics," *Pharmacol. Ther.*, *45*:309–329 (1990).
9. J. Caldwell, S. M. Winter, and A. J. Hutt, "The pharmacological and toxico-

About the Editor

Irving W. Wainer is Professor and Director of the Pharmacokinetics Division in the Department of Oncology at McGill University, Montreal, Quebec, Canada. The author or coauthor of more than 150 professional papers, and 4 books, his research interests include the chromatographic resolution of enantiomeric compounds, pharmacological and therapeutic consequences of stereoisomerism, and the development and application of HPLC chiral stationary phases. He lectures widely at seminars and conferences in the United States and Europe and is a member of the International Society for the Study of Xenobiotics, the Association for Research in Vision and Ophthalmology, the American Chemical Society, and the American Association for the Advancement of Science. He is founding Editor-in-Chief of the journal *Chirality* and North American Editor of the *Journal of Chromatography–Biomedical Applications*. In 1992 he was awarded the Chromatographic Society's A. J. P. Martin Medal for contributions to the development of chromatographic science. Dr. Wainer received the B.S. degree (1965) in chemistry from Wayne State University, Detroit, Michigan, and the Ph.D. degree (1970) in organic chemistry from Cornell University, Ithaca, New York.